Australian Signpost MATHS

6

Alan McSeveny
Alan Parker
Diane McSeveny-Foster
Rachel McSeveny
Bob Collard

Short Contents

Pearson Australia
(a division of Pearson Australia Group Pty Ltd)
459–471 Church Street, Level 1, Building B, Richmond, Victoria 3121
PO Box 23360, Melbourne, Victoria 8012
www.pearson.com.au

First published 2018 by Pearson Australia
2025 2024 2023 2022
10 9 8 7 6 5 4 3 2 1

Content Learning Specialist: Julian Lumb
Project Manager: Shelly Wang
Production Manager: Elizabeth Gosman & Aptara
Editor: Aptara
Designer: Anne Donald & Jennifer Johnston
Cover Designer: Jennifer Johnston
Desktop operator: Aptara
Rights & Permissions Editor: Samantha Russell-Tulip
Senior Publishing Services Analyst: Rob Curulli
Cover art: Michael Barter
Cover background: rf_vector/shutterstock.com
Illustrators: Michael Barter, Wendy Gorton, Nives Porcellato, Andrew Craig
Printed in Malaysia (CTP-VVP)

ISBN 978 1 4886 2193 2
Pearson Australia Group Pty Ltd ABN 40 004 245 943

Acknowledgements
All material identified by ACARA AC is material subject to copyright under the *Copyright Act 1968* and is owned by the Australian Curriculum, Assessment and Reporting Authority 2018. ACARA neither endorses nor verifies the accuracy of the information provided and accepts no responsibility for incomplete or inaccurate information. In particular, ACARA does not endorse or verify that:
- The content descriptions are solely for a particular year and subject;
- All the content descriptions for that year and subject have been used; and
- The author's material aligns with the Australian Curriculum content descriptions for the relevant year and subject.

You can find the unaltered and most up to date version of this material at http://www.australiancurriculum.edu.au/ This material is reproduced with the permission of ACARA.

We thank the following for their contributions to our text book:
123RF: texelart, p. 29 (game console).
Shutterstock: Valentyna Chukhlyebova, p. 29 (coffee machine); cobalt88, p. 29 (TV).

Every effort has been made to trace and acknowledge copyright. However, if any infringement has occurred, the publishers tender their apologies and invite the copyright holders to contact them.

What is Australian Signpost Maths?

Australian Signpost Maths is a mathematics activity book series for students from Foundation to Year 6. The series has been written to meet the requirements of the Australian Curriculum.

The components of the series include Student Books, Teacher's Books, Mentals Books and an interactive Website. Teachers can select an appropriate program for every student from the rich and varied material provided.

The content has been carefully sequenced within each year level and across the series to take into account students' likely mathematical development.

Student Books

Teacher's Books

Mentals Books

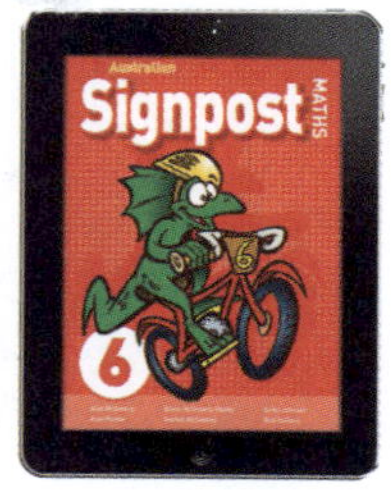

Website

Structure of Australian Signpost Maths

Australian Signpost Maths emphasises the curriculum's syllabus content as well as problem-solving strategies, language development and the use of technology.

The syllabus is organised into three content strands and four proficiency strands:

Content Strands

- Number and Algebra
- Measurement and Geometry
- Statistics and Probability

Proficiency Strands (see page v)

- Understanding
- Fluency
- Problem Solving
- Reasoning

The curriculum's **general capabilities** are developed throughout the Australian Signpost Maths program. These are:

- literacy
- numeracy
- information and communication technologies (ICT)
- critical and creative thinking.

Australian Signpost Maths also provides opportunities to develop other general capabilities, such as personal and social competence and intercultural understanding.

The cross-curriculum dimensions of the syllabus – 'Aboriginal and Torres Strait Islander histories and cultures', 'Asia and Australia's engagement with Asia' and 'Sustainability' – are embedded in the program.

To maximise the benefits of the program, the Student Book, Teacher's Book, Mentals Book and Website should be used together.

The structure of the **Student Book** allows teachers to determine both the order and the extent of content covered. Strands are organised separately so that the teacher, not the Student Book, decides the content of the next lesson. However, a suggested term program (see page xi of this book) and a detailed program (see the Teacher's Book and Website) are also provided.

The **Teacher's Book** also provides lesson plans for each page of the Student Book and blackline masters to assist teachers in implementing the program.

The **Mentals Book** mixes examples from all strands. It revises the content covered in the Student Book. Each content strand is thoroughly covered, with the proficiency strands incorporated within each section. A special feature woven throughout the Mentals Book is the tables program in the four operations.

The innovative **Website** help teachers to bring mathematics alive with technology. The website provides interactive maths tools, games and practice opportunities as well as relevant resource masters and worksheets for all year levels. These can be used for whole-class, small-group and individual learning. The website also includes **Concept Check-In**, a new diagnostic screener.

Student Book pages are colour-coded by section.

Number and Algebra A

Measurement and Geometry A

Statistics and Probability

Number and Algebra B

Measurement and Geometry B

Answers

Australian Curriculum Proficiency Strands

The proficiency strands of the Australian Curriculum describe how content is explored or developed – that is, the 'thinking and doing' of mathematics.

Understanding

Learning the concepts

*Students build a robust knowledge of adaptable and transferable mathematical concepts. They make connections between related concepts and progressively apply the familiar to develop new ideas. They develop an understanding of the relationship between the 'why' and the 'how' of mathematics.**

Conceptual understanding of maths ideas includes the explanation of a concept using text and diagrams. This occurs throughout Australian Signpost Maths at the top of many pages and is indicated by the Concepts icon.

Fluency

Using the concepts

*Students develop skills in choosing appropriate procedures, carrying out procedures flexibly, accurately, efficiently and appropriately, and recalling factual knowledge and concepts readily.**

The practice of maths skills to build fluency occurs on every page of Australian Signpost Maths.

Problem Solving

Applying concepts and strategies to develop solutions to problems

*Students develop the ability to make choices, interpret, formulate, model and investigate problem situations, and communicate solutions effectively.**

Problem solving provides opportunities for students to use strategies and skills such as investigating and questioning, to collaborate with others and to communicate their findings to different audiences. Such activities are often indicated throughout Australian Signpost Maths by the Activity and Investigation icons.

Reasoning

Coherent and logical thought

*Students develop an increasingly sophisticated capacity for logical thought and actions, such as analysing, proving, evaluating, explaining, inferring, justifying and generalising.**

Students require opportunities to explain their mathematical thinking and can do so through both diagrams and written explanations. Reasoning questions are located throughout Australian Signpost Maths.

* The Australian Curriculum: Mathematics, v1.2 – Content structure

Special Features of Australian Signpost Maths

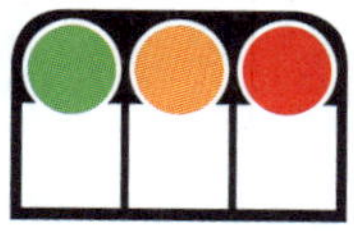

- **Traffic Light** system allows students to reflect on their work and highlight any units that they are having trouble understanding. They tick the red for units they feel they still don't understand, and green for those they feel they understand fully.
- Exercises are **well graded**. New work is reinforced in the Mentals Book.
- **Answers** are supplied in the back of this book as well as in the Teacher's Book.
- **Concept Check-In** diagnostic screener (on the Website) provides a snapshot of the class' conceptual understandings to aid in classroom management. It also allows teachers to measure progress over time.
- The eight **Diagnostic Tests** (now also in the back of this book) allow the teacher to discover each student's strengths and weaknesses, and the cross-references direct students to the pages where that work is introduced. Answers are supplied in the Teacher's Book.
- The **Dictionary** at the beginning of this Student Book will help students to learn the language of mathematics.
- **ID Cards** (in the Mentals Book, Teacher's Book and Website) review the language of mathematics by asking students to identify common terms, shapes and symbols.
- Important **rules and concepts** are clearly highlighted.
- **Worked examples** and explanations are given throughout the Student Book where new ideas are introduced.
- The use of **colour** makes emphasis clear and is highly motivating.
- **Cartoons** give instruction and friendly advice.
- **Interactive activities** are provided on the website for whole-class, small-group and individual learning.

Australian Signpost Icons

Signpost icons are used throughout the book as cues to the essential nature of exercises and activities, and as a guide to ways of engaging with them. These icons often indicate alternative or more concrete approaches to dealing with concepts.

This icon highlights **important rules and concepts** occurring throughout the book. It often appears with worked examples.

Activities provide **applications and enrichment**. These activities usually involve the use of concrete materials and partner or group work.

These enjoyable activities are used to **motivate and involve** students in mathematical pursuits. They usually involve games and puzzles.

Investigations allow students to **explore and discover** maths concepts.

This icon indicates the use of computers, calculators or other **information and communications technology**.

6 Contents and Syllabus Overview

KEY

- Number and Algebra (blue)
- Measurement and Geometry (green)
- Statistics and Probability (orange)

Number and Algebra A

Page	Unit	Title	Sub-strand	Number and place value	Fractions and decimals	Patterns and algebra	Content	Counting and numeration	Place value	Fractions	Decimals	Percentages	Number patterns	Suggested progress
1	1:01	Numbers Above One Million		■				●	●					Term 1
2	1:02	Square Numbers		■				●						
3	1:03	Square Numbers		■		■		●					●	
4	1:04	Percentages			■					●	●	●		
5	1:05	Percentages			■					●	●	●		
6	1:06	Percentages			■					●	●	●		T1, T2*
7	1:07	Powers of Ten		■				●	●					
8	1:08	Negative Numbers		■				●						Term 2
9	1:09	Positive and Negative Numbers		■		■		●	●					
10	1:10	Improper Fractions and Mixed Numbers			■					●				
11	1:11	Patterns with Fractions and Decimals			■	■				●	●		●	
12	1:12	Using the Number Line		■		■		●					●	T3, T4*
13	1:13	Positive and Negative Numbers		■				●						
14	1:14	Using Negative Numbers		■				●						
15	1:15	Geometric Patterns		■		■		●					●	Term 3
16	1:16	Fractions of a Group			■					●				
17	1:17	Fractions of a Group			■					●				
18	1:18	Addition and Subtraction of Fractions			■					●				
19	1:19	Subtracting Fractions			■					●				
20	1:20	Equivalent Fractions			■					●				T5, T6*
21	1:21	Equivalent Fractions			■	■				●			●	
22	1:22	Equivalent Fractions			■					●				
23	1:23	Addition and Subtraction of Fractions			■					●				
24	1:24	Addition and Subtraction of Fractions			■					●				
25	1:25	Problems Using Fractions			■					●				Term 4
26	1:26	Finding Percentages			■					●	●	●		T7, T8*
27	1:27	Finding Percentages			■					●	●	●		
28	1:28	Prime and Composite Numbers		■		■		●						
29	1:29	Prime and Composite Numbers		■				●						

* Suggested progress for Diagnostic Tests 1 to 8 is found in the Teacher's Book.
The first of each pair of tests covers the first half of the period.

Number and Algebra B

Page	Unit	Title	Sub-strand: Number and place value	Money and financial mathematics	Patterns and algebra	Content: Addition	Subtraction	Multiplication	Division	Place value	Number patterns	Operations with decimals	Suggested progress
30	2:01	Multiplication Review	■		■			●			●		Term 1
31	2:02	Division	■					●	●				
32	2:03	Addition Review	■			●							
33	2:04	Subtraction Review	■				●						
34	2:05	Order of Operations	■			●	●	●	●				
35	2:06	Order of Operations	■		■	●	●	●	●		●		
36	2:07	Multiplying 10s, 100s and 1 000s	■					●					
37	2:08	Multiplication of Larger Numbers	■	■				●					
38	2:09	Mental Strategies for Subtraction	■				●						
39	2:10	Problem Solving	■			●	●	●	●				T1, T2*
40	2:11	Division Review	■						●				
41	2:12	Division	■						●				
42	2:13	Division Involving Fractions	■						●				Term 2
43	2:14	Averages	■			●			●				
44	2:15	Addition to 999 999	■			●							
45	2:16	Subtraction to 999 999	■				●						
46	2:17	5-Digit Subtraction from Thousands	■	■			●						
47	2:18	Money	■	■		●	●						
48	2:19	Travel Maths	■			●	●	●	●				
49	2:20	Order of Operations	■			●	●	●	●				
50	2:21	Strategies with Decimals	■	■		●						●	
51	2:22	Addition of Decimals	■			●						●	T3, T4*
52	2:23	Addition of Decimals	■			●						●	
53	2:24	Subtraction of Decimals	■				●					●	
54	2:25	Multiplying Thousands	■					●					
55	2:26	Multiplication of Decimals	■					●				●	
56	2:27	Multiplication of Decimals	■	■				●				●	
57	2:28	Multiplication of Decimals	■	■				●				●	
58	2:29	Division of Numbers Larger than 999	■	■					●				Term 3
59	2:30	Division with Zeros in the Answer	■						●				
60	2:31	Division of Large Numbers by 10	■	■					●				
61	2:32	Multiplying and Dividing by Powers of 10	■					●	●	●		●	
62	2:33	Dividing by a Multiple of 10	■	■					●				
63	2:34	Constructing Number Sentences	■		■	●	●	●	●			●	

Number and Algebra B			Sub-strand	Number and place value	Money and financial mathematics	Patterns and algebra	Content	Addition	Subtraction	Multiplication	Division	Place value	Number patterns	Operations with decimals	Suggested progress
Page	**Unit**	**Title**													
64	2:35	Division of Decimals		■	■						●			●	
65	2:36	Division of Decimals		■							●			●	
66	2:37	Extending Multiplication Facts		■		■				●			●		T5, T6*
67	2:38	Multiplying Numbers Ending in Zeros		■						●					
68	2:39	Multiplication by 2-Digit Numbers		■				●		●					
69	2:40	Multiplication by 2-Digit Numbers		■						●					
70	2:41	Multiplication by 2-Digit Numbers		■	■				●	●					
71	2:42	Multiplication by 2-Digit Numbers		■						●					
72	2:43	Number Sentences		■		■		●	●	●					Term 4
73	2:44	Divisibility and Factors		■						●	●				
74	2:45	Estimation with Decimals		■				●	●	●				●	
75	2:46	Using Rounding to Check Answers		■				●	●	●	●			●	
76	2:47	A Strategy for Multiplying Decimals		■						●		●		●	
77	2:48	Division Strategies		■							●			●	T7, T8*
78	2:49	Addition of Large Numbers		■	■			●							
79	2:50	Subtraction of Large Numbers		■					●						
80	2:51	Problem Solving		■				●	●	●	●			●	
81	2:52	Using Strategies to Solve Problems		■		■		●					●		

* Suggested progress for Diagnostic Tests 1 to 8 is found in the Teacher's Book.
The first of each pair of tests covers the first half of the period.

Measurement and Geometry A			Sub-strand	Using units of measurement	Content	Length	Area	Volume	Capacity	Mass	Time	Suggested progress
Page	**Unit**	**Title**										
82	3:01	Centimetres and Millimetres				●						Term 1
83	3:02	Kilometres				●						
84	3:03	Converting Measurements				●						
85	3:04	Converting Measurements				●						
86	3:05	Units of Length				●						
87	3:06	Variation in Measurement				●						T1, T2*
88	3:07	Perimeter				●						Term 2
89	3:08	Perimeter				●						
90	3:09	Elapsed Time									●	
91	3:10	Timetables									●	
92	3:11	Square Centimetres					●					
93	3:12	Square Metres					●					
94	3:13	Comparing Perimeter and Area				●	●					T3, T4*
95	3:14	Packing and Stacking						●	●			
96	3:15	Volume of Prisms						●				Term 3
97	3:16	Volume of Rectangular Prisms						●				
98	3:17	Cubic Centimetres						●				
99	3:18	Cubic Metres						●				
100	3:19	Millilitres and Litres							●			
101	3:20	Kilolitres and Megalitres							●			
102	3:21	Cubic Centimetres and Millilitres						●	●			
103	3:22	Millilitres and Litres						●	●			T5, T6*
104	3:23	Cubic Centimetres and Litres						●	●			
105	3:24	Time Lines									●	
106	3:25	Tonnes								●		Term 4
107	3:26	Units of Mass								●		
108	3:27	Units of Mass								●		
109	3:28	Introducing the Milligram								●		T7, T8*
110	3:29	Hectares					●					
111	3:30	Square Kilometres					●					
112	3:31	Time Zones									●	

Measurement and Geometry B			Sub-strand			Content					Suggested progress
Page	Unit	Title	Shape	Location and transformation	Geometric reasoning	2D Space	3D Space	Location	Transformations	Angles	
113	4:01	Space Review	■	■		●	●				Term 1
114	4:02	Angles and Informal Units			■					●	
115	4:03	The Protractor			■					●	
116	4:04	Angle Types			■					●	
117	4:05	Angles			■					●	T1, T2*
118	4:06	Angles on a Straight Line			■					●	
119	4:07	Angles at a Point			■					●	
120	4:08	Vertically Opposite Angles			■					●	Term 2
121	4:09	2D Shapes	■			●					
122	4:10	Finding Your Way		■				●			
123	4:11	Following Directions – The Compass		■				●			
124	4:12	Compass Directions		■				●			T3, T4*
125	4:13	Compass Directions		■				●			
126	4:14	Using Maps		■				●			
127	4:15	Using Coordinates		■				●			
128	4:16	Sections and Elevations of 3D Objects	■				●				Term 3
129	4:17	Naming 3D Objects	■				●				
130	4:18	Nets of Prisms	■				●				
131	4:19	Nets of Pyramids	■				●				
132	4:20	The Number Plane		■				●			T5, T6*
133	4:21	The Number Plane – Four Quadrants		■				●			
134	4:22	Using Coordinates		■				●			
135	4:23	Using Maps		■				●			Term 4
136	4:24	Transformations		■					●		
137	4:25	Transformations		■					●		
138	4:26	Making Patterns on a Computer		■					●		T7, T8*
139	4:27	Constructing Regular Shapes	■		■	●				●	
140	4:28	Drawing and Recognising 3D Objects	■				●				
141	4:29	Cones, Cylinders and Spheres	■				●				

* Suggested progress for Diagnostic Tests 1 to 8 is found in the Teacher's Book. The first of each pair of tests covers the first half of the period.

Page	Unit	Title	Sub-strand: Chance	Sub-strand: Data representation and interpretation	Content: Chance	Content: Investigation	Content: Data representation	Suggested progress
142	5:01	Reading Graphs		■			●	Term 1
143	5:02	Line Graphs		■			●	
144	5:03	Tables and Graphs		■			●	
145	5:04	Reading Graphs and Tallies		■			●	T1, T2*
146	5:05	The Probability of Events	■		●			Term 2
147	5:06	Ordering the Probability of Events	■		●			
148	5:07	Divided Bar and Sector Graphs		■			●	
149	5:08	Drawing Sector Graphs (Pie Charts)		■			●	
150	5:09	Using Samples	■	■	●		●	T3, T4*
151	5:10	Dot Plots		■			●	
152	5:11	Misleading Displays		■			●	
153	5:12	Misleading Displays		■			●	
154	5:13	Possible Outcomes to Experiments	■	■	●	●		Term 3
155	5:14	Chance Using Two Dice	■	■	●	●	●	
156	5:15	Using Samples	■	■	●	●	●	
157	5:16	Collecting Information		■		●	●	T5, T6*
158	5:17	Repeating an Experiment	■	■	●	●		
159	5:18	Chance: Expected Results	■	■	●	●		Term 4
160	5:19	Tree Diagrams	■		●			
161	5:20	Likelihood of an Outcome	■		●			T7, T8*
162	5:21	Planning a Good Survey		■		●	●	
163	5:22	Unusual Graphs		■			●	

Suggested Program

	Weeks 1–10	Weeks 11–20	Weeks 21–30	Weeks 31–end
Number and Algebra A	1:01–1:07	1:08–1:14	1:15–1:24	1:25–1:29
Number and Algebra B	2:01–2:12	2:13–2:28	2:29–2:42	2:43–2:52
Measurement and Geometry A	3:01–3:06	3:07–3:14	3:15–3:24	3:25–3:31
Measurement and Geometry B	4:01–4:07	4:08–4:15	4:16–4:22	4:23–4:29
Statistics and Probability	5:01–5:04	5:05–5:12	5:13–5:17	5:18–5:22

The eight Diagnostic Tests are found in the Teacher's Book.
See the Contents and Syllabus Overview on pages vi–xi for suggested placement of each test.

Contents Cross-reference

Number and Algebra

1 Number and place value: Whole numbers	Pages	Australian Curriculum Reference
Large numbers and place value	1, 7	Recognise, represent and order numbers to at least tens of thousands (ACMNA072); Apply place value to partition, rearrange and regroup numbers to at least tens of thousands to assist calculations and solve problems (ACMNA073) *[Progression]*
Prime and composite numbers, factors and multiples	3, 28, 29, 30, 73	Identify and describe properties of prime, composite, square and triangular numbers (ACMNA122)
Powers of ten	7, 62, 84	Recognise, represent and order numbers to at least tens of thousands (ACMNA072) *[Progression]*; Select and apply efficient mental and written strategies and appropriate digital technologies to solve problems involving all four operations with whole numbers (ACMNA123)
Estimation and rounding	32, 38, 44, 46, 47, 48, 54, 71, 79, 126	Select and apply efficient mental and written strategies and appropriate digital technologies to solve problems involving all four operations with whole numbers (ACMNA123)
Square and triangular numbers	2, 3, 15, 30, 35	Identify and describe properties of prime, composite, square and triangular numbers (ACMNA122)
Positive and negative numbers	8, 9, 12, 13, 14	Investigate everyday situations that use integers. Locate and represent these numbers on a number line (ACMNA124)
2 Number and place value: Addition		
Mental strategies	63, 72	Select and apply efficient mental and written strategies and appropriate digital technologies to solve problems involving all four operations with whole numbers (ACMNA123)
Written strategies	32, 44, 47, 78	Select and apply efficient mental and written strategies and appropriate digital technologies to solve problems involving all four operations with whole numbers (ACMNA123)
Problem solving	32, 39, 43, 48, 78, 104	Select and apply efficient mental and written strategies and appropriate digital technologies to solve problems involving all four operations with whole numbers (ACMNA123)
3 Number and place value: Subtraction		
Mental strategies	33, 38, 63, 72	Select and apply efficient mental and written strategies and appropriate digital technologies to solve problems involving all four operations with whole numbers (ACMNA123)
Written strategies	33, 45, 46, 47, 79	Select and apply efficient mental and written strategies and appropriate digital technologies to solve problems involving all four operations with whole numbers (ACMNA123)
Problem solving	33, 38, 39, 45, 46, 47, 48, 79, 80, 104	Select and apply efficient mental and written strategies and appropriate digital technologies to solve problems involving all four operations with whole numbers (ACMNA123)
4 Number and place value: Multiplication		
Multiplication facts	30, 66	Select and apply efficient mental and written strategies and appropriate digital technologies to solve problems involving all four operations with whole numbers (ACMNA123)
Multiplication of large numbers by one-digit numbers	36, 37, 54, 66	Select and apply efficient mental and written strategies and appropriate digital technologies to solve problems involving all four operations with whole numbers (ACMNA123)
Multiplication of large numbers by two-digit numbers	36, 66, 67, 68, 69, 70, 71	Select and apply efficient mental and written strategies and appropriate digital technologies to solve problems involving all four operations with whole numbers (ACMNA123)
Mental strategies	63, 66, 67, 68, 72, 76	Select and apply efficient mental and written strategies and appropriate digital technologies to solve problems involving all four operations with whole numbers (ACMNA123)
Multiples, factors and square numbers	28, 29, 30	Identify and describe properties of prime, composite, square and triangular numbers (ACMNA122)
Problem solving	30, 36, 39, 48, 49, 54, 68, 69, 70, 71, 80, 109	Select and apply efficient mental and written strategies and appropriate digital technologies to solve problems involving all four operations with whole numbers (ACMNA123)

5	Number and place value: Division	Pages	Australian Curriculum Reference
	Division facts linked with multiplication	31, 63	Select and apply efficient mental and written strategies and appropriate digital technologies to solve problems involving all four operations with whole numbers (ACMNA123)
	Dividing by one-digit numbers	31, 40, 41, 42, 43, 58, 59, 73	Select and apply efficient mental and written strategies and appropriate digital technologies to solve problems involving all four operations with whole numbers (ACMNA123)
	Mental strategies	29, 31, 60, 62, 63, 72, 73, 77	Select and apply efficient mental and written strategies and appropriate digital technologies to solve problems involving all four operations with whole numbers (ACMNA123)
	Problem solving	31, 39, 40, 41, 42, 43, 48, 49, 58, 59, 60, 80	Select and apply efficient mental and written strategies and appropriate digital technologies to solve problems involving all four operations with whole numbers (ACMNA123)
6	**Fractions and decimals**		
	Compare and order fractions	10, 11, 20	Compare fractions with related denominators and locate and represent them on a number line (ACMNA125)
	Equivalent fractions	19, 20, 21, 22, 23, 24	Compare fractions with related denominators and locate and represent them on a number line (ACMNA125)
	Fractions on the number line	10, 18, 20	Compare fractions with related denominators and locate and represent them on a number line (ACMNA125); Continue and create sequences involving whole numbers, fractions and decimals. Describe the rule used to create the sequence (ACMNA133)
	Finding a fraction of a quantity	16, 17, 20, 65, 83	Find a simple fraction of a quantity where the result is a whole number, with and without digital technologies (ACMNA127)
	Addition and subtraction of fractions	18, 19, 23, 24	Solve problems involving addition and subtraction of fractions with the same or related denominators (ACMNA126)
	Solve problems involving fractions	17, 25	Find a simple fraction of a quantity where the result is a whole number, with and without digital technologies (ACMNA127)
	Add and subtract decimals	11, 25, 50, 51, 52, 53, 74, 75, 82, 89	Add and subtract decimals, with and without digital technologies, and use estimation and rounding to check the reasonableness of answers (ACMNA128); Continue and create sequences involving whole numbers, fractions and decimals. Describe the rule used to create the sequence (ACMNA133)
	Using estimation and rounding	47, 51, 52, 74, 75	Add and subtract decimals, with and without digital technologies, and use estimation and rounding to check the reasonableness of answers (ACMNA128)
	Multiply and divide decimals	55, 56, 57, 61, 64, 65, 74, 75, 76, 77, 82, 83, 84, 85, 86, 89, 100, 101, 102, 103, 107, 108	Multiply decimals by whole numbers and perform divisions by non-zero whole numbers where the results are terminating decimals, with and without digital technologies (ACMNA129); Multiply and divide decimals by powers of 10 (ACMNA130)
	Equivalence of fractions, decimals and percentages	4, 5, 6, 11, 157	Make connections between equivalent fractions, decimals and percentages (ACMNA131)
7	**Money and financial mathematics**		
	Operations with money	14, 26 27, 36, 37, 39, 47, 50, 56, 57, 70	Select and apply efficient mental and written strategies and appropriate digital technologies to solve problems involving all four operations with whole numbers (ACMNA123); Add and subtract decimals, with and without digital technologies, and use estimation and rounding to check the reasonableness of answers (ACMNA128); Multiply decimals by whole numbers and perform divisions by non-zero whole numbers where the results are terminating decimals, with and without digital technologies (ACMNA129); Investigate and calculate percentage discounts of 10%, 25% and 50% on sale items, with and without digital technologies (ACMNA132)
	Calculate percentage discounts of 10%, 25% and 50% on sale items	26, 27	Investigate and calculate percentage discounts of 10%, 25% and 50% on sale items, with and without digital technologies (ACMNA132)
8	**Patterns and algebra**	**Pages**	**Australian Curriculum Reference**
	Patterns and pattern rules	3, 10, 11, 15, 30, 35, 66, 72	Continue and create sequences involving whole numbers, fractions and decimals. Describe the rule used to create the sequence (ACMNA133)
	Number sentences with missing elements	15, 63, 72	Select and apply efficient mental and written strategies and appropriate digital technologies to solve problems involving all four operations with whole numbers (ACMNA123)
	Solving unusual problems	37, 54, 71, 78, 80, 81, 144	Select and apply efficient mental and written strategies and appropriate digital technologies to solve problems involving all four operations with whole numbers (ACMNA123)

Order of operations	34, 35, 49	Explore the use of brackets and order of operations to write number sentences (ACMNA134)
Number lines	8, 9, 10, 12, 14, 18, 21	Select and apply efficient mental and written strategies and appropriate digital technologies to solve problems involving all four operations with whole numbers (ACMNA123); Investigate everyday situations that use integers and locate and represent these numbers on a number line (ACMNA124); Compare fractions with related denominators and locate and represent them on a number line (ACMNA125)
9 Technology		
Using a calculator	8, 14, 15, 20, 26, 27, 37, 44, 50, 51, 55, 64, 66, 74, 78, 80, 84, 124	Select and apply efficient mental and written strategies and appropriate digital technologies to solve problems involving all four operations with whole numbers (ACMNA123)
Using a computer	106, 109, 118, 119, 130, 131, 136, 137, 138, 139	Select and apply efficient mental and written strategies and appropriate digital technologies to solve problems involving all four operations with whole numbers (ACMNA123)

Measurement and Geometry

1 Using units of measurement	Pages	Australian Curriculum Reference
Reading scales	8, 13, 14, 82, 84, 103, 104	Connect decimal representations to the metric system (ACMMG135); Convert between common metric units of length, mass and capacity (ACMMG136); Connect volume and capacity and their units of measurement (ACMMG138)
Length	48, 82, 83, 84, 85, 86, 87, 88, 89, 92, 94, 121	Connect decimal representations to the metric system (ACMMG135); Convert between common metric units of length, mass and capacity (ACMMG136); Solve problems involving the comparison of lengths and areas using appropriate units (ACMMG137)
Perimeter	82, 88, 89, 94	Connect decimal representations to the metric system (ACMMG135); Convert between common metric units of length, mass and capacity (ACMMG136); Solve problems involving the comparison of lengths and areas using appropriate units (ACMMG137)
Area	92, 93, 94, 110, 111	Solve problems involving the comparison of lengths and areas using appropriate units (ACMMG137)
Volume	95, 96, 97, 98, 99, 102, 103, 104	Connect volume and capacity and their units of measurement (ACMMG138)
Capacity	95, 100, 101, 102, 103, 104	Connect decimal representations to the metric system (ACMMG135); Convert between common metric units of length, mass and capacity (ACMMG136); Connect volume and capacity and their units of measurement (ACMMG138)
Mass	106, 107, 108, 109	Connect decimal representations to the metric system (ACMMG135); Convert between common metric units of length, mass and capacity (ACMMG136)
Temperature	8, 12, 13, 14	Investigate everyday situations that use integers. Locate and represent these numbers on a number line (ACMNA124)
Time	54, 90, 91, 105, 112	Interpret and use timetables (ACMMG139)
12- and 24-hour time	90, 91, 112	Interpret and use timetables (ACMMG139)
Converting between units	60, 82, 83, 84, 85, 86, 100, 101, 102, 103, 106, 107, 108, 109, 110, 111	Convert between common metric units of length, mass and capacity (ACMMG136)
2 Shape		
2D shapes	113, 121, 139	Compare and describe two dimensional shapes that result from combining and splitting common shapes, with and without the use of digital technologies (ACMMG088) *[Progression]*
3D objects	113, 128, 129, 130, 131, 140, 141	Construct simple prisms and pyramids (ACMMG140)
Prisms and pyramids	113, 128, 129, 130, 131, 140, 141	Construct simple prisms and pyramids (ACMMG140)

3 Location and transformation	Pages	Australian Curriculum Reference
Maps, scales and legends	85, 86, 112, 122, 124, 125, 126, 127, 134, 135	Use simple scales, legends and directions to interpret information contained in basic maps (ACMMG090); Use a grid reference system to describe locations, and describe routes using landmarks and directional language (ACMMG113) *[Progression]*
Using compass directions	122, 123, 124, 125, 126, 127, 134, 135	Use simple scales, legends and directions to interpret information contained in basic maps (ACMMG090) *[Progression]*; Introduce the Cartesian coordinate system using all four quadrants (ACMMG143)
The number plane: coordinates and grid references	122, 125, 126, 127, 132, 133, 135	Introduce the Cartesian coordinate system using all four quadrants (ACMMG143)
Describe routes	125, 126, 135	Use a grid reference system to describe locations, and describe routes using landmarks and directional language (ACMMG113) *[Progression]*; Introduce the Cartesian coordinate system using all four quadrants (ACMMG143)
Translations, reflections and rotations	113, 136, 137, 138	Investigate combinations of translations, reflections and rotations, with and without the use of digital technologies (ACMMG142)
4 Geometric reasoning		
Classifying angles	114, 116	Investigate, with and without digital technologies, angles on a straight line, angles at a point and vertically opposite angles. Use results to find unknown angles (ACMMG141)
Measuring and drawing angles	114, 115, 116, 117, 121, 139	Investigate, with and without digital technologies, angles on a straight line, angles at a point and vertically opposite angles. Use results to find unknown angles (ACMMG141)
Estimating angles	114, 117	Investigate, with and without digital technologies, angles on a straight line, angles at a point and vertically opposite angles. Use results to find unknown angles (ACMMG141)
Angles on a straight line	118	Investigate, with and without digital technologies, angles on a straight line, angles at a point and vertically opposite angles. Use results to find unknown angles (ACMMG141)
Angles at a point	119, 139, 149	Investigate, with and without digital technologies, angles on a straight line, angles at a point and vertically opposite angles. Use results to find unknown angles (ACMMG141)
Vertically opposite angles	120	Investigate, with and without digital technologies, angles on a straight line, angles at a point and vertically opposite angles. Use results to find unknown angles (ACMMG141)

Statistics and Probability

1 Chance	Pages	Australian Curriculum Reference
Chance experiments	142, 150, 154, 155, 158, 159	Conduct chance experiments with both small and large numbers of trials using appropriate digital technologies (ACMSP145)
Describing likelihood	146, 147, 150, 154, 155, 158, 159, 160, 161	Describe probabilities using fractions, decimals and percentages (ACMSP144)
Experimental and expected frequencies	147, 150, 154, 155, 156, 158, 159	Compare observed frequencies across experiments with expected frequencies (ACMSP146); Interpret secondary data presented in digital media and elsewhere (ACMSP148)
2 Data representation and interpretation		
Reading graphs and tables	44, 90, 91, 111, 142, 143, 144, 145, 148, 151, 152, 163	Interpret and compare a range of data displays, including side-by-side column graphs for two categorical variables (ACMSP147); Interpret secondary data presented in digital media and elsewhere (ACMSP148); Interpret and use timetables (ACMMG139)
Drawing graphs and tables	142, 143, 144, 145, 148, 149, 150, 151, 157	Interpret and compare a range of data displays, including side-by-side column graphs for two categorical variables (ACMSP147)
Comparing displays of data	142, 144, 152, 153	Interpret and compare a range of data displays, including side-by-side column graphs for two categorical variables (ACMSP147); Interpret secondary data presented in digital media and elsewhere (ACMSP148)
Interpreting data	44, 91, 111, 112, 142, 143, 144, 150	Interpret and compare a range of data displays, including side-by-side column graphs for two categorical variables (ACMSP147); Interpret secondary data presented in digital media and elsewhere (ACMSP148)
Bias in presenting data	152, 153	Interpret secondary data presented in digital media and elsewhere (ACMSP148)
Surveys, sample and census	150, 153, 156, 157, 162	Conduct chance experiments with both small and large numbers of trials using appropriate digital technologies (ACMSP145)

6 Dictionary

abacus

An instrument used for counting and calculating.

546 = 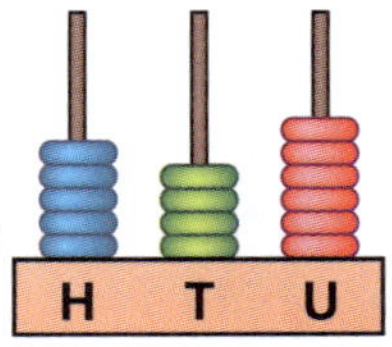

am (ante meridiem)

Any time between midnight and midday.

- The time is 25 past 7 in the morning. It is 7:25 am.

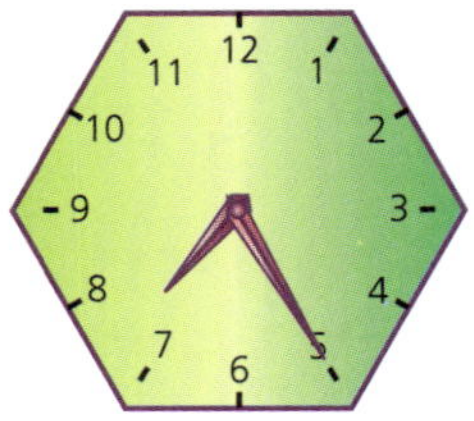

Morning

analogue time

The time shown on a clock face.

- 13 minutes to 6 is the time shown on this analogue clock.

angle

The amount of turning between two arms about a common point.

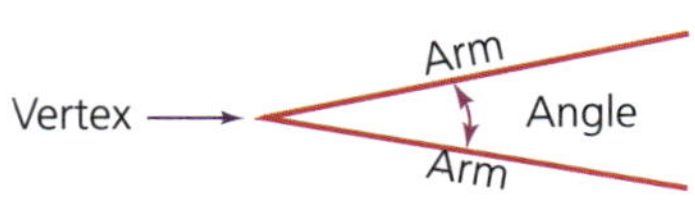

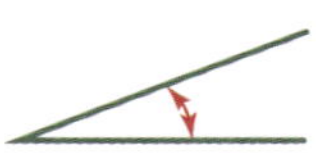

Acute Angle < 90°

Right Angle = 90°

90° < Obtuse Angle < 180°

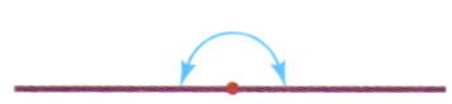

Straight Angle = 180°
(Half Turn)

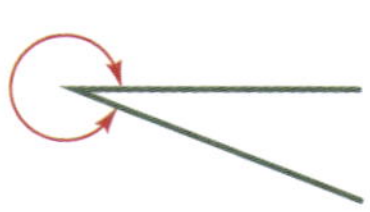

180° < Reflex Angle < 360°

Revolution = 360°
(Full Turn)

area

The size of a surface.

Area is measured in square units.

- **square centimetres: cm^2**
- **square metres: m^2**
- 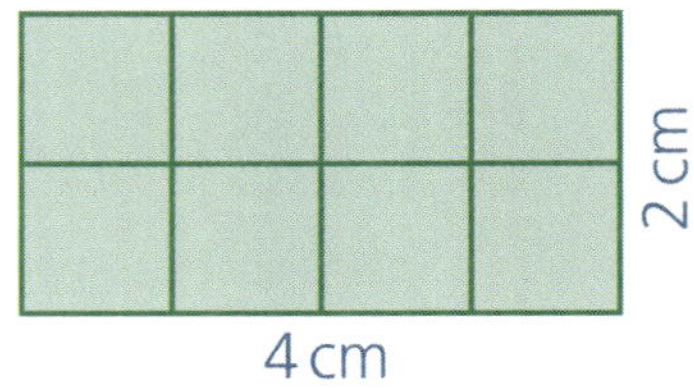

Area = 8 cm^2

ascending order

Arranged in order from least value to greatest value.

- $1.65, $2.30, $4.75, $5.96
  (least) (most)

average

A fair share.

- Average = (sum of scores) ÷ (number of scores)

axis of symmetry

A line that divides a picture in half so that each half is the mirror image of the other part.

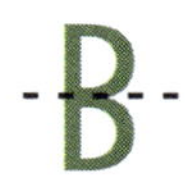

- One Axis of Symmetry

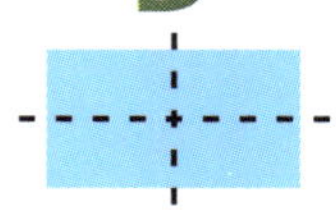

- Two Axes of Symmetry

The plural of *axis* is *axes*.

capacity

The amount that a container can hold.

- The capacity of this juice bottle is 250 mL.

census (used in surveys)

A collection of data (information) involving everybody in the population being studied.

centimetre (cm)

A unit of length equal to one hundredth of a metre.

- 100 cm = 1 m

chance

The chance (or probability) of something happening is its likelihood of happening.

- If you toss a coin, there is an even chance of tossing a head.

See *probability*.

compass directions

The needle of a compass points north (N).

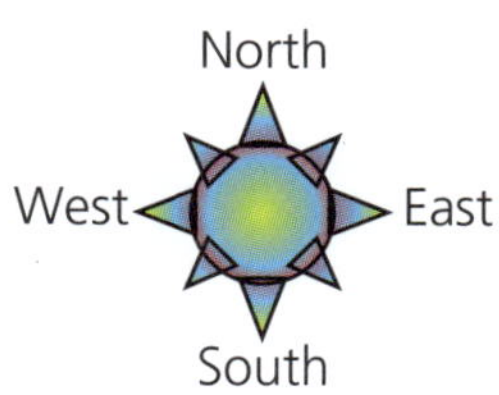

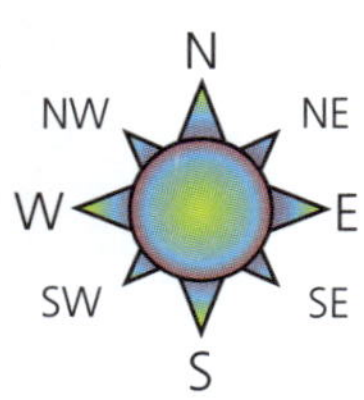

composite number

A number that has more than two factors.

- 9 is composite as it has three factors, 1, 3 and 9.

cone

A three-dimensional object with a circular base that tapers to a point.

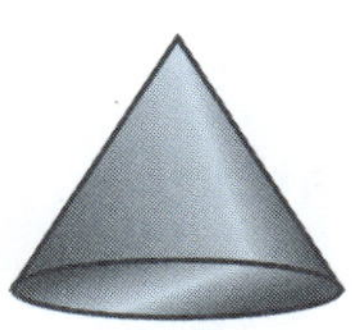

coordinates

Pairs of letters or numbers used to show position on a grid.

Two numbers are used on a number plane, e.g. (3, 3).

- This position is D3 or (D, 3).

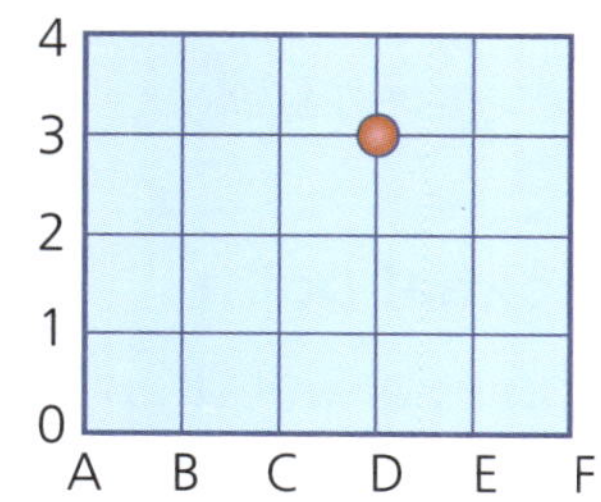

cross-section

A face that is exposed when a solid object is cut through.

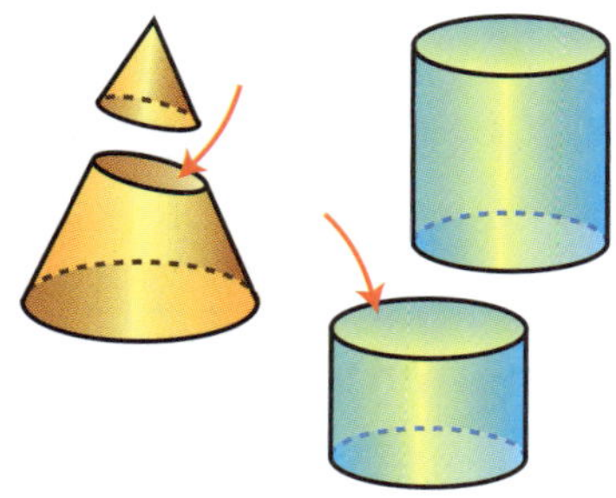

cube

A three-dimensional object that has six equal square faces, eight vertices and twelve equal edges.

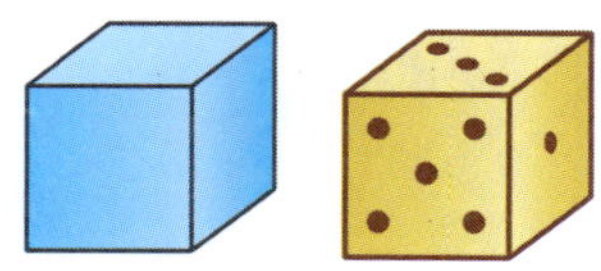

cube number

- 2 cubed = 2^3 ← Index

  = 2 × 2 × 2

  = 4 × 2

  = 8

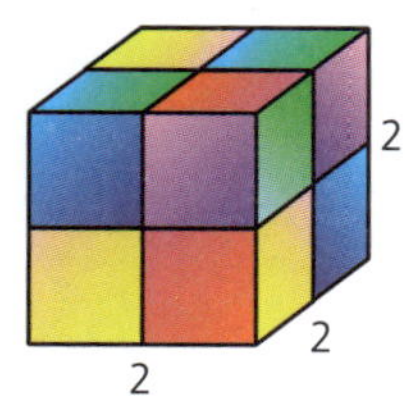

cubic centimetre (cm^3)

A unit of volume equal to the volume of a cube of side length 1 cm.

cubic metre (m^3)

A unit of volume equal to the volume of a cube of side length 1 m.

cylinder

A three-dimensional object with two equal circular faces and one curved surface.

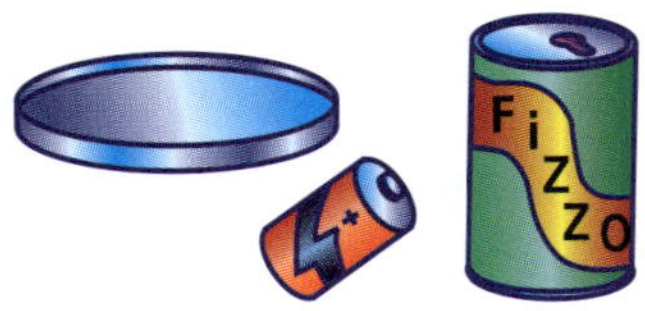

decimal point

A point or dot used to separate a decimal from a whole number part.

36.9

↑

Decimal Point

decrease

To make smaller or less.

denominator

The bottom number of a fraction.
It tells the number of equal parts there are in the whole.

Numerator ⟶ 7

Denominator ⟶ 8

$\frac{7}{8}$

descending order

Arranged in order from greatest value to least value.

- $5.96, (most) $4.75, $2.30, $1.65 (least)

diagonal

A line that joins any two non-adjacent corners of a polygon.

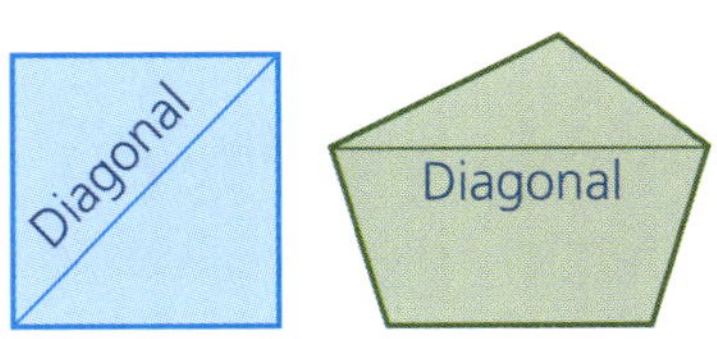

digital time

ime expressed using digits.

- This digital clock shows 24 minutes past 10.

division (÷)

Breaking up groups into equal parts.

- 6 ÷ 3

  a Sharing 2 each

  b Grouping 2 groups

divisible

To have no remainder when divided.

- 30 is divisible by 3.

elevations

Views of a 3D object from the front, top and side.

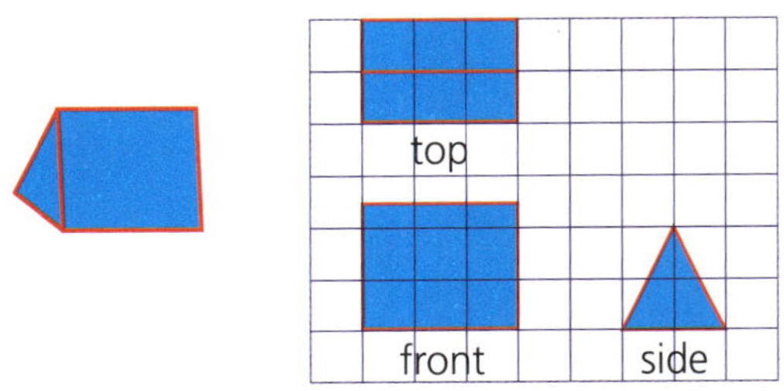

enlargement

To make bigger or smaller (reduction) using a scale factor.

- A scale factor of 2 doubles lengths.

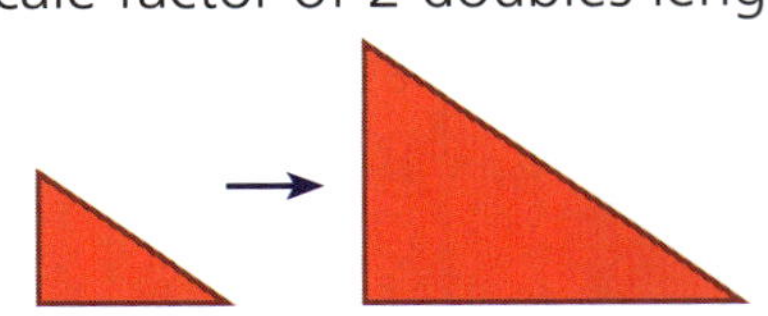

- A scale factor of $\frac{1}{2}$ halves lengths.

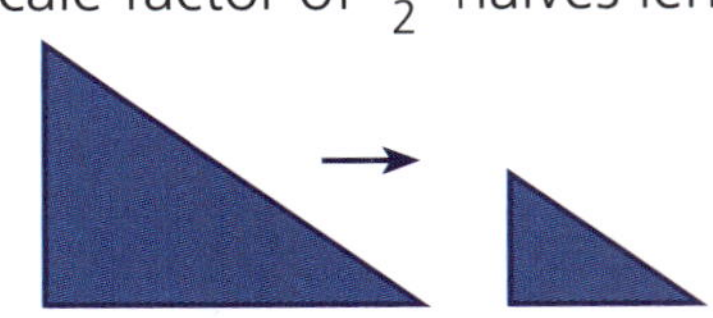

estimate (estimation)

A good guess.

expanded notation

A way of writing numerals to show the place value of each digit.

- $137 = (1 \times 100) + (3 \times 10) + 7$

face

A flat surface of a three-dimensional object.

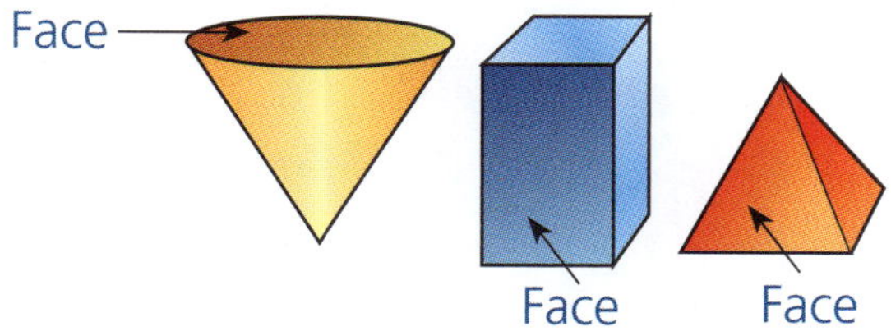

factor

A factor of a number divides the number exactly, leaving no remainder.

- The factors of 12 are 1, 12, 2, 6, 3 and 4.

flip (reflection)

To turn over.

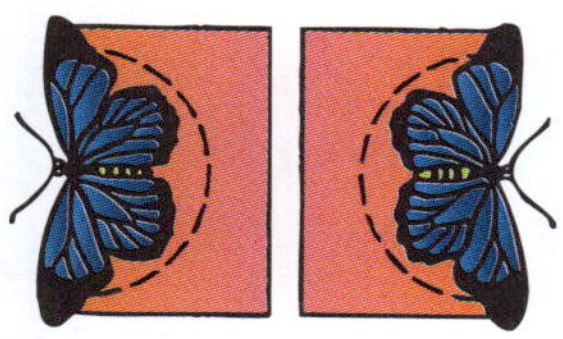

fraction

A part of a whole or group.

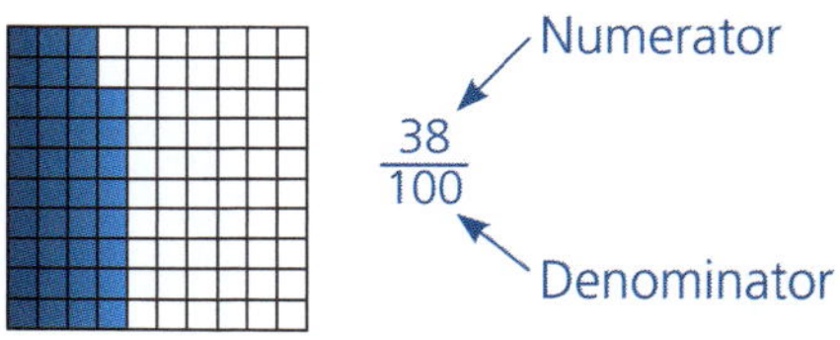

Equivalent fractions

Fractions that have the same value.

- $\frac{1}{2} = \frac{5}{10} = \frac{15}{30} = \ldots$

Improper fraction

A fraction which has a numerator that is bigger than the denominator.

- $\frac{9}{8}$

Mixed number

A number that has a whole number part and a fraction part.

- $1\frac{2}{3}$

gram (g)

A unit of mass.

- 1 kilogram = 1 000 grams

graphs

- Bar Graph

  A graph which uses horizontal bars to compare the size of groups.

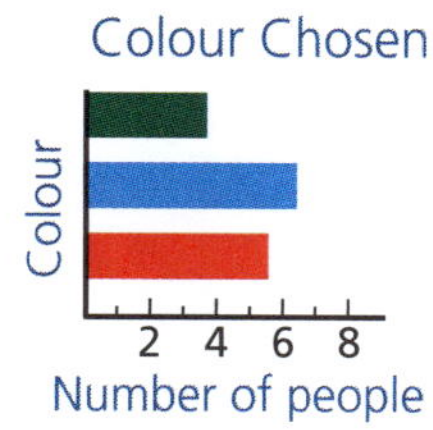

- Column Graph

  Groups are compared using the heights of columns.

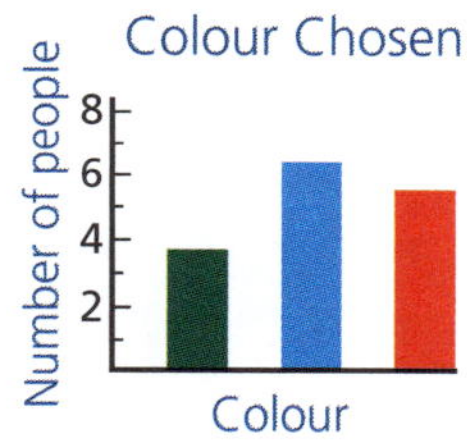

- Divided Bar Graph

  A bar is divided to show the make-up of the data.

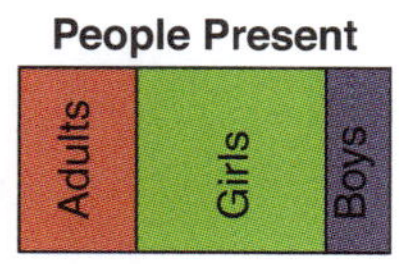

- Dot Plot

  A graph which uses dots to compare the size of groups.

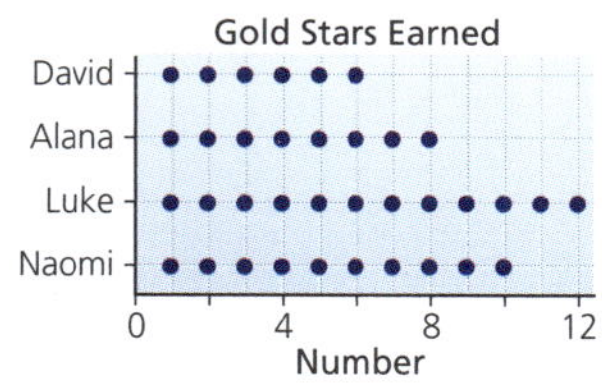

- Line Graph

  A continuous line shows the connection between variables.

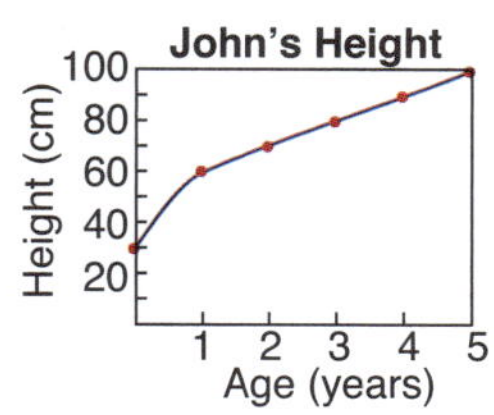

- Picture Graph

  A picture is used as a unit to show how many.

- Sector Graph

  A circle is cut into sectors to show the parts of a whole.

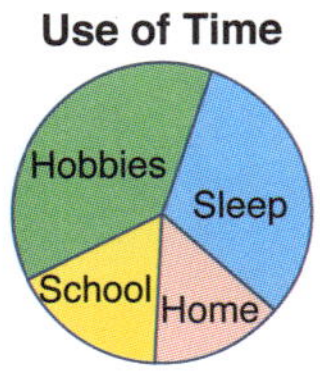

greater than (>)

A way of showing that a number is larger than another number.

- $7 > 3$

See also *less than (<)*.

hectare (ha)

A unit of area equal to a square with sides 100 m.

- 1 ha = 10 000 m^2

horizontal

- Parallel to the horizon.
- Level or flat.
- Any direction at right angles to the vertical.

increase

To make larger or more.

inverse operations

Adding 8 is the opposite of subtracting 8.

- $100 + 8 - 8 = 100$

Multiplying by 2 is the opposite of dividing by 2.

- $4 \times 2 \div 2 = 4$

jump strategy

Adding or subtracting numbers, jumping by tens and ones.

- $52 - 14 = 38$

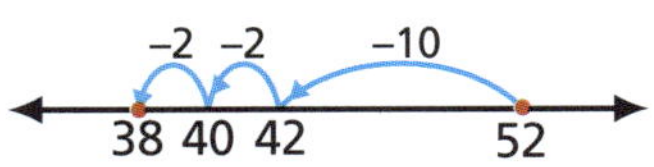

kilo (k)

Kilo means 1000.

kilogram (kg)

The basic unit of mass, equal to 1 000 grams.

- 1 kg = 1 000 g

kilolitre (kL)

A unit of capacity equal to 1 000 litres.

- 1 kL = 1 000 L

kilometre (km)

A unit of length equal to one thousand metres.

- 1 km = 1 000 m

less than (<)

A way of showing that a number is smaller than another number.

- $3 < 7$

See also *greater than (>)*.

line of symmetry

A line that divides something in half so that each half is a mirror image of the other.

Line of Symmetry

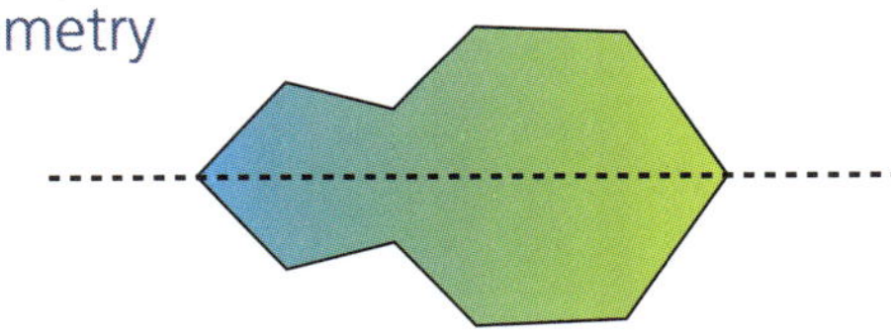

litre (L)

A unit of capacity (or volume) used for the measurement of liquids.

- 1 L = 1 000 mL
- 1 L = 1 000 cm^3

map or plan

A picture of an area viewed from above.

mass

The amount of matter in an object, a measure of how heavy something is.

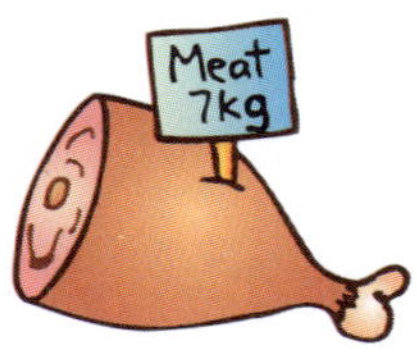

mean

The arithmetic average.

- mean = $\dfrac{\text{sum of scores}}{\text{number of scores}}$

See also *average*.

median

The middle number of a set of numbers that have been written in order from smallest to largest.

megalitre (ML)

A unit of capacity equal to one million litres.

- 1 ML = 1 000 000 L
- 1 ML = 1 000 kL

metre (m)

The basic unit of length, equal to 100 centimetres.

- 1 m = 100 cm

milligram (mg)

A unit of mass equal to one thousandth of a gram.

- 1 000 mg = 1 g

millilitre (mL)

A unit of capacity equal to one thousandth of a litre.

- 1 000 mL = 1 L
- 1 mL = 1 cm^3

millimetre (mm)

A unit of length equal to one thousandth of a metre.

- 10 mm = 1 cm
- 1 000 mm = 1 m

mixed numeral

A numeral that has a whole number part and a fraction part.

- $4\frac{1}{8}$

mode

The number that occurs the most often in a set of numbers.

- 2, 3, 3, 3, 4, 4, 5, 7

  The mode is 3.

multiple

The result of multiplying a counting number by another counting number.

- The multiples of 5 are 5, 10, 15, 20, …

negative number

A number below zero.

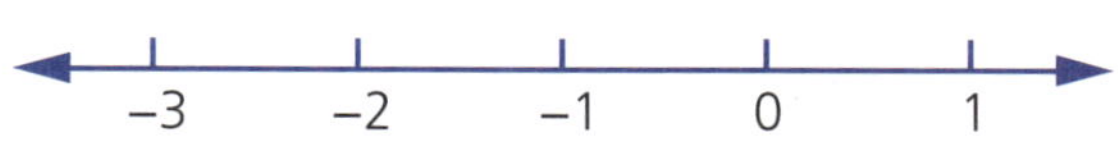

–3 (spoken 'minus 3') is a negative number.

net

A flat shape that can be folded to make a three-dimensional object.

Net of a Cube

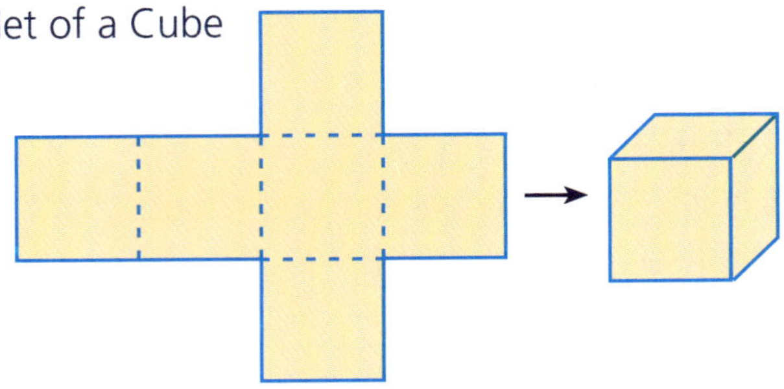

number plane

A grid using numbers on both axes.

The coordinates (3, 7) are 3 units to the right of the origin (0, 0) and 7 up.

object

The term used to describe a three-dimensional shape.

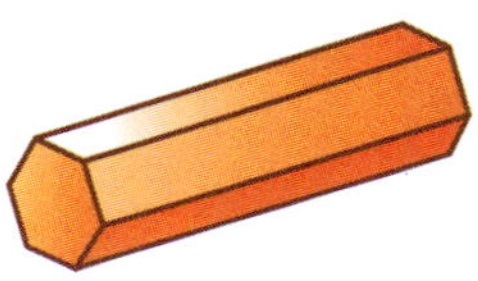

Hexagonal Prism

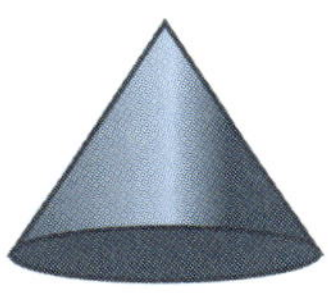

Cone

octagon

A polygon with eight sides.

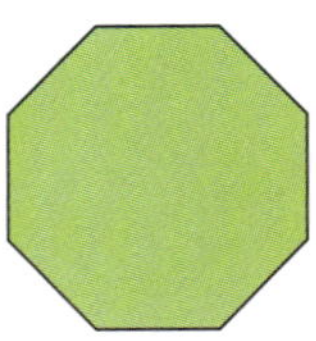

Regular Octagon

Irregular Octagon

See also *polygon*.

order of operations

1 Do operations inside brackets.

2 Do × and ÷, going from left to right.

3 Do + and −, going from left to right.

parallel lines

Straight lines on the same flat surface that do not meet.

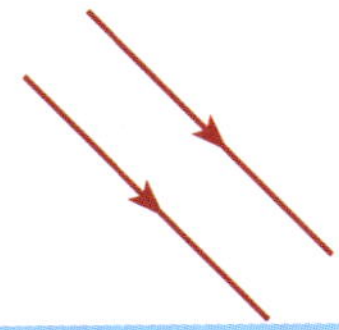

parallelogram

A shape with four sides, opposite sides parallel and opposite sides equal.

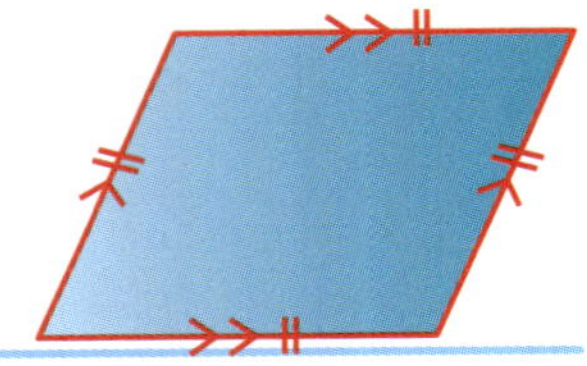

pattern (number)

Set of numbers in order, made using a rule.

- The rule for 5, 9, 13, 17, … is **Add 4**.

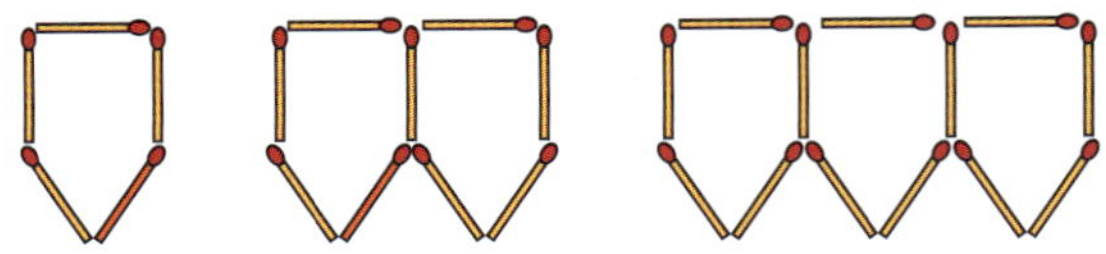

pentagon

A polygon with five sides.

Regular Pentagon

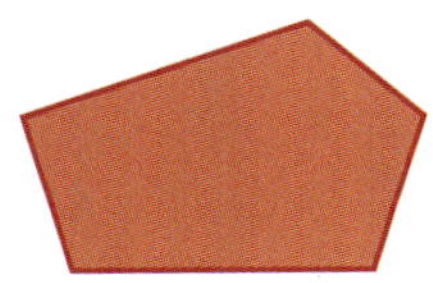

Irregular Pentagon

See also *polygon*.

per cent (%)

Out of one hundred.

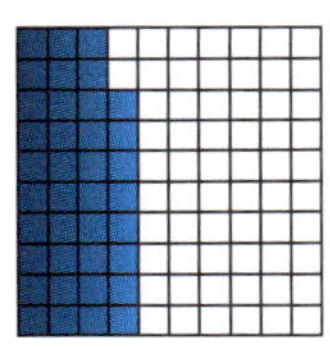

$\frac{38}{100} = 0{\cdot}38 = 38\%$

perimeter

The distance around the outside of a shape; the boundary.

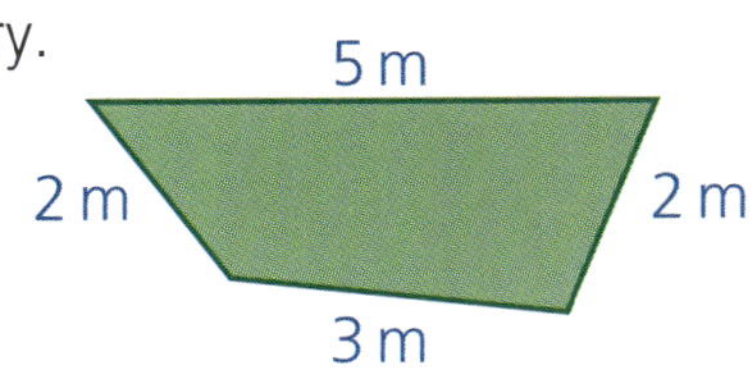

- Perimeter = 2 m + 3 m + 2 m + 5 m
  = 12 m

perpendicular lines

Lines that meet at right angles.

place value

The column value of a digit.

- 396 =

Hundreds	Tens	Ones
3	9	6

pm (post meridiem)

Any time between midday and midnight.

- The time is 20 past 1 in the afternoon. It is 1:20 pm.

Afternoon

polygon

A two-dimensional shape with three or more straight sides such as a triangle, quadrilateral, pentagon etc.

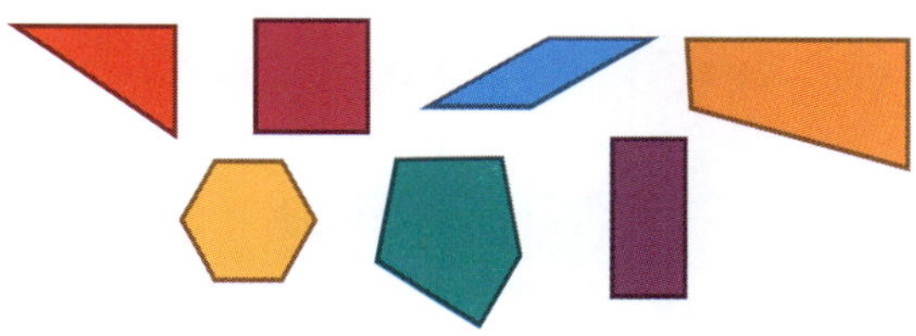

positive number

A number above zero.

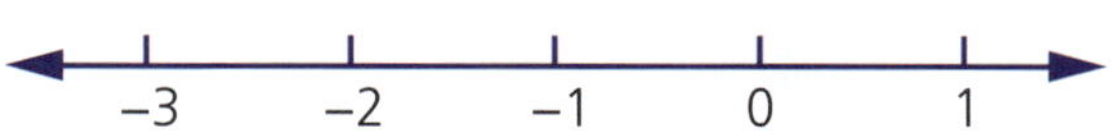

- 1 is a positive number.

power

A power (or index) is a way of showing that a number has been multiplied by itself a number of times.

- $3^4 = 3 \times 3 \times 3 \times 3$
- $10^3 = 10 \times 10 \times 10$

prime number

A number that has only two factors.

- 7 has only two factors, 1 and 7.

prism

A three-dimensional object with a uniform cross-section. The ends are identical shapes and all other faces are rectangles. Prisms are named by the shape of their ends.

Triangular Prism   Hexagonal Prism

probability

The probability (or chance) of something happening is its likelihood of happening.

- The probability of tossing an even number on a dice is 50%.

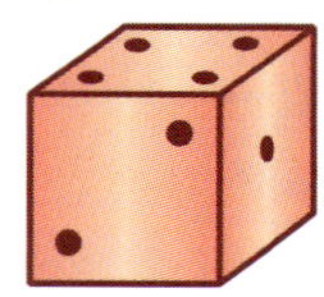

product

The answer to a multiplication question.

- The product of 8 and 9 is 72.

protractor

An instrument used for measuring and drawing angles.

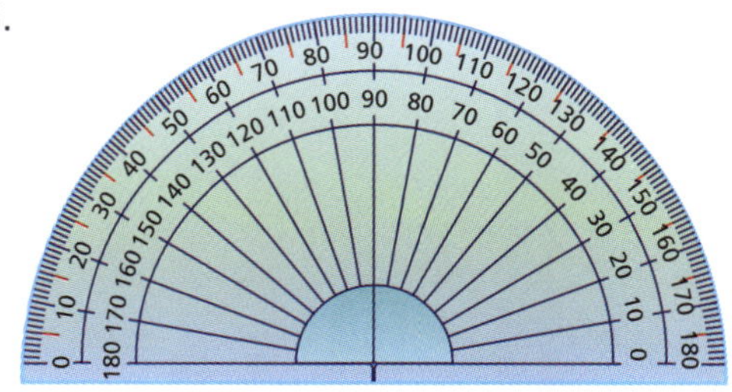

pyramid

A three-dimensional object that has a polygon for a base and triangles for all other faces. Pyramids are named by the shape of their base.

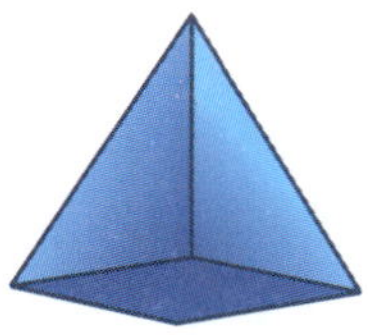
Square Pyramid

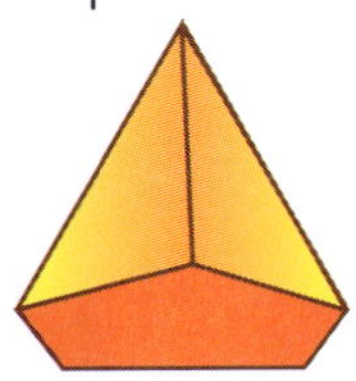
Pentagonal Pyramid

quadrilateral

A two-dimensional shape with four straight sides, such as rectangle, square or kite.

questionnaire or survey

A list of questions used to obtain statistical information or opinions from people.

quotient

The answer when one number is divided by another.

random selection

Choosing without looking.

Each item has an equal chance of being chosen.

range

The difference between the highest and lowest numbers in a set.

- 2, 3, 5, 8, 9, 10, 10

  Range = 10 – 2 = 8

reflection

See *flip*.

regular and irregular shapes

Regular shapes have all sides and all angles equal. Irregular shapes do not.

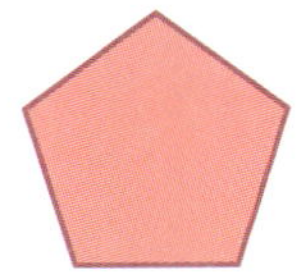
Regular Shape

Irregular Shape

remainder

The number that is left over after sharing or dividing.

- 22 cups shared among 5 people gives 4 cups each, remainder 2.

$$5\overline{)22}\ \text{gives}\ 4\ \text{r}\ 2$$

rhombus

A shape with four sides, opposite sides parallel and all sides equal.

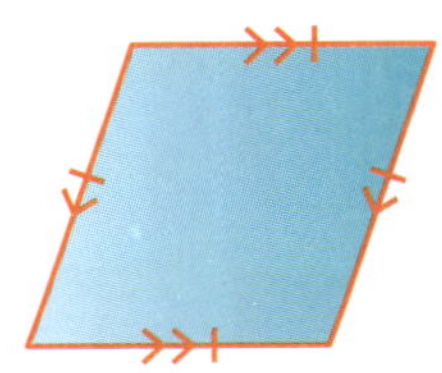

rigid shape

A model that cannot be pushed out of shape because triangles have been used in its construction.

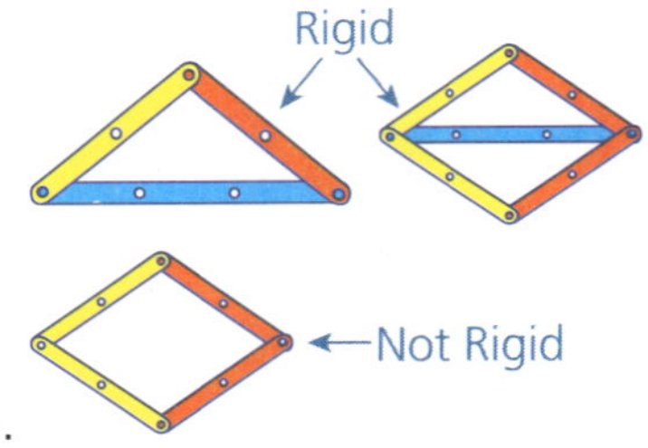

Roman numerals

A number system devised by the ancient Romans.

Roman numerals use letters for numbers:

I	V	X	L	C	D	M
1	5	10	50	100	500	1000

- XXVIII = 28

rotation

See *turn*.

rounding

Rewriting a number to the nearest 5, 10 or 100.

sample (used in surveys)

A part of a population that represents the whole.

skip counting

Counting on, adding the same number each time.

- 5, 10, 15, 20, 25, … is skip counting by 5.

slide (translation)

To move a shape in any direction without changing the original orientation of the shape.

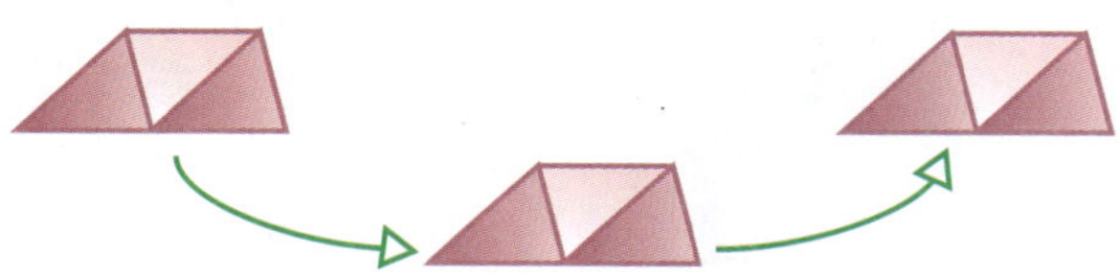

solid

A term used to describe a three-dimensional shape.

Hexagonal Prism Cone

sphere

A three-dimensional object that is ball-shaped and round. All points on the surface of a sphere are the same distance from its centre.

split strategy

Adding numbers by splitting them into their parts.

- $36 + 52 = 30 + 6 + 50 + 2$
  $= (30 + 50) + (6 + 2)$
  $= 80 + 8$
  $= 88$

square centimetre (cm^2)

A unit of area equal to a square with sides of 1 cm.

square kilometre (km^2)

A unit of area equal to a square with sides of 1 km.

- $1\ km^2 = 1\,000\,000\ m^2$
- $1\ km^2 = 100$ ha

square metre (m^2)

A unit of area equal to a square with sides of 1 m.

- $10\,000\ m^2 = 1$ ha

square number

A whole number times itself.

- $7 \times 7 = 49$
  49 is a square number.

sum

The answer when you add numbers.

surface

The outside layer of a three-dimensional object. A surface can be flat or curved.

See also *face*.

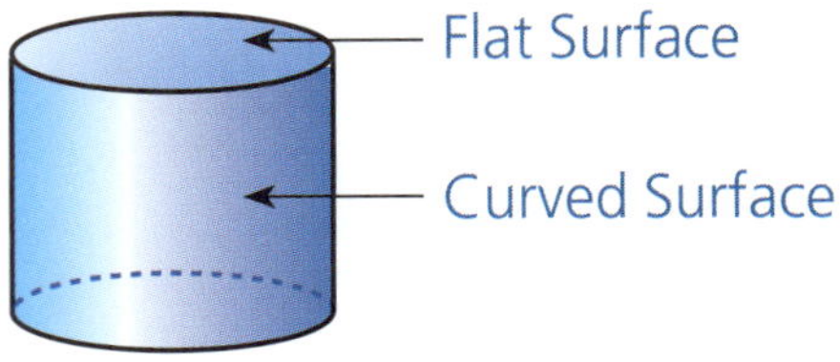

survey or **questionnaire**

A list of questions used to obtain statistical information or opinions from people.

symmetry

A balanced arrangement.

- Line Symmetry

  A property of a figure where one half is the mirror image of the other.
- Line (or Axis) of Symmetry

  A line that divides a figure into two parts that are mirror images of each other.
- Rotational Symmetry

  A property of a figure where it can be spun about a point so that it repeats its shape more than once in a rotation.

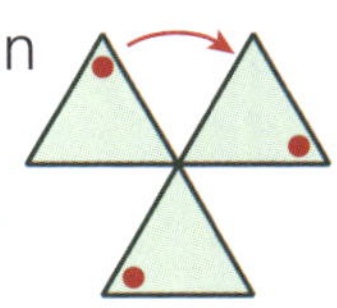

tally

To keep count by making a mark to represent each item.

To make counting easy, the marks are drawn in groups of five with each fifth mark crossing the preceding four marks.

- ~~IIII~~ ~~IIII~~ ~~IIII~~ III = 18

temperature

A measure of how hot or cold something is.

Temperature is usually measured in degrees Celsius (°C).

- Water freezes at 0°C.
- Water boils at 100°C.

tessellation

A pattern of identical shapes that fit together without gaps or overlaps.

thermometer

An instrument used for measuring temperature.

three-dimensional (3D) objects

Objects are three-dimensional. They have length, width and height.

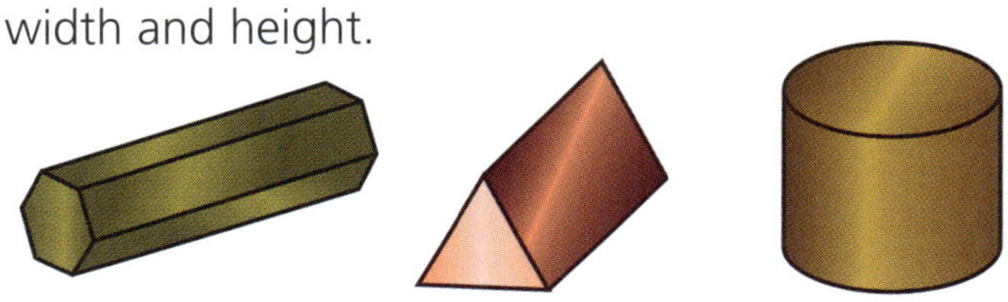

tonne (t)

A unit of mass equal to 1 000 kg.

transformation

Reflection (flip), translation (slide) and rotation (turn).

translation

See *slide*.

trapezium

A quadrilateral with one pair of parallel sides.

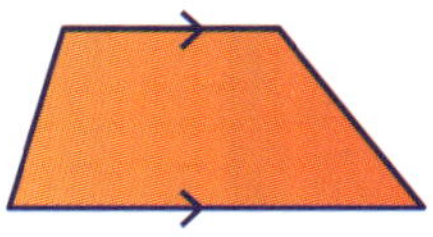

tree diagram

A diagram used to show all possible outcomes.

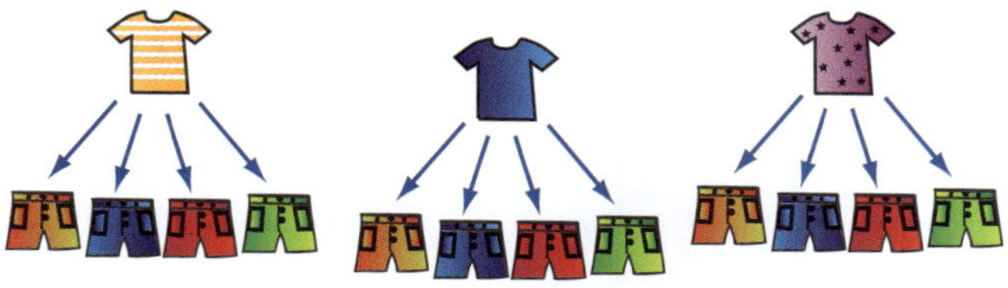

triangle

A two-dimensional shape with three straight sides and three angles.

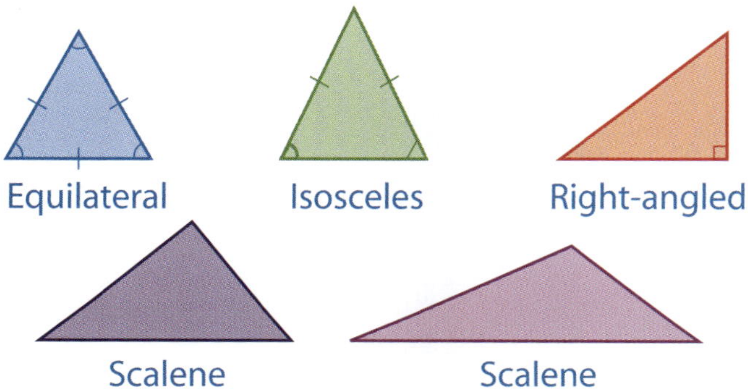

Scalene triangles have no sides equal.

See also *polygon*.

turn (rotation)

To rotate a shape about a given point.

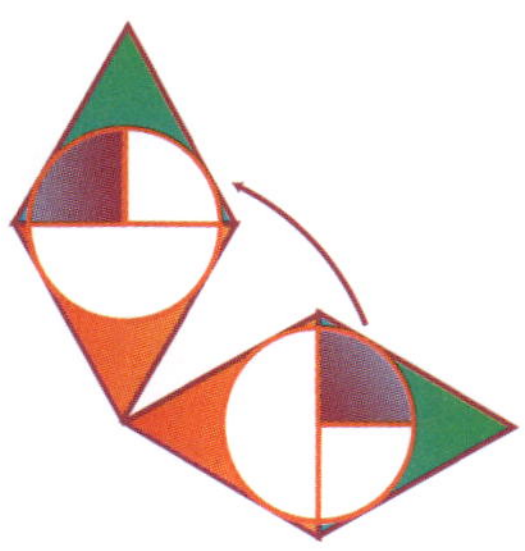

twenty-four hour time

Time given as a 4-digit number, the first two digits indicating the hour and the second two digits indicating minutes.

- 13:20 is 20 past 1 in the afternoon, or 1:20 pm.

two-dimensional (2D) shapes

Plane shapes are two-dimensional. They have length and width.

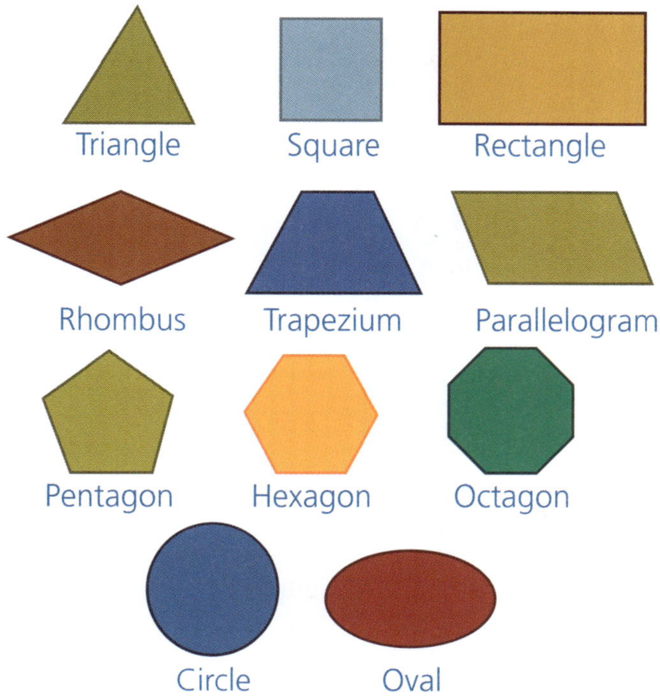

two-way tables

Tables of data that show categories down the side as well as across the top.

	Blue	Red	White
Bus	2	5	0
Car	8	10	21
Truck	1	6	2

vertex

A point at which two or more lines meet to form a corner on a 2D shape or 3D object.

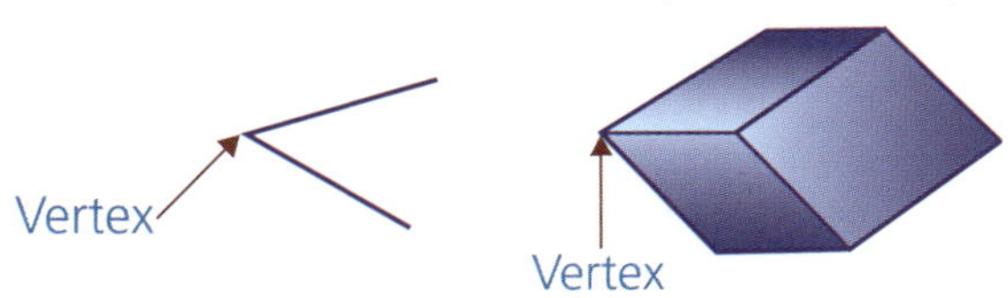

The plural of *vertex* is *vertices*.

vertical

- At right angles to the horizontal.
- Straight up and down.
- The direction in which an object falls under gravity.

vertically opposite angles

Opposite angles formed when two straight lines cross.

They are always equal.

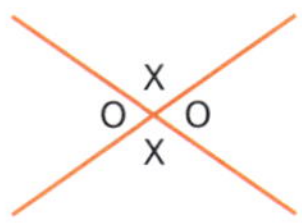

volume

The amount of space occupied by a 3D object.

- 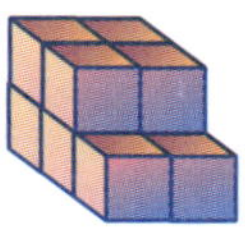 Volume = 10 cubic units

width or breadth

The distance from side to side.

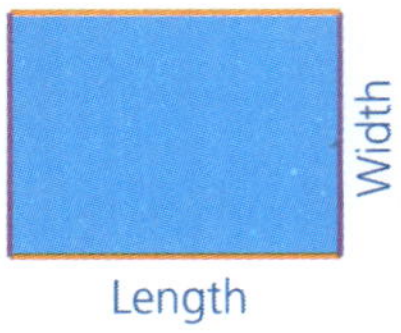

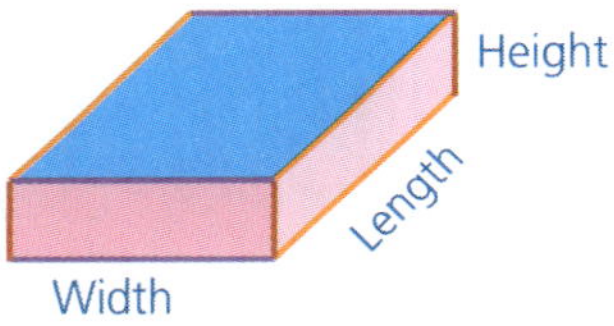

1:01 Numbers Above One Million

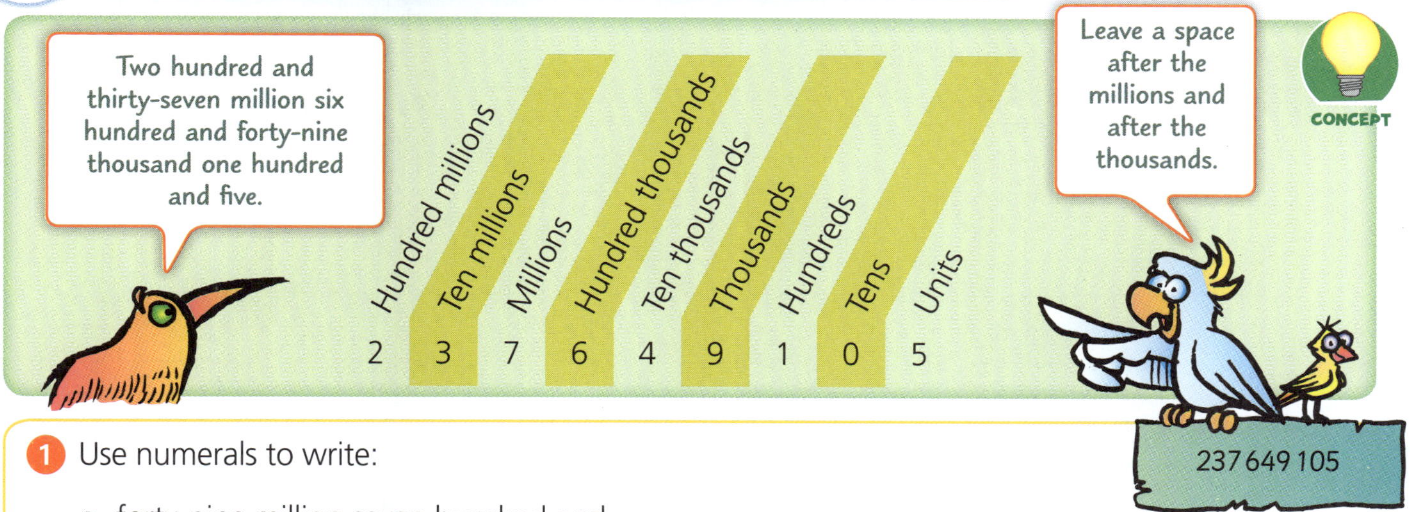

CONCEPT

1 Use numerals to write:

a forty-nine million seven hundred and sixty thousand six hundred and twenty-one

b eighty-three million one hundred and thirty-two thousand five hundred and forty-nine

2 Write the value for each coloured digit.

a 37 **4**68 901 **b** 2**3** 674 768 **c** 43 1**6**9 235

d **9**6 347 607 **e** 6**7** 911 213 **f** 16**5** 273 406

3 Arrange each group of numbers in ascending order.

a 26 349 721 62 419 637 43 296 714

b 65 375 670 63 497 624 56 811 769

c 32 693 475 41 623 912 17 634 658

4 Is each number below closer to 30 000 000 or 40 000 000?

a 32 645 762 **b** 34 177 624 **c** 36 396 408

INVESTIGATION

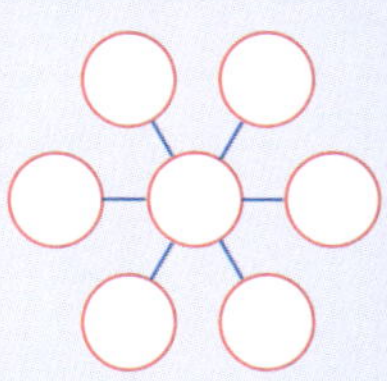

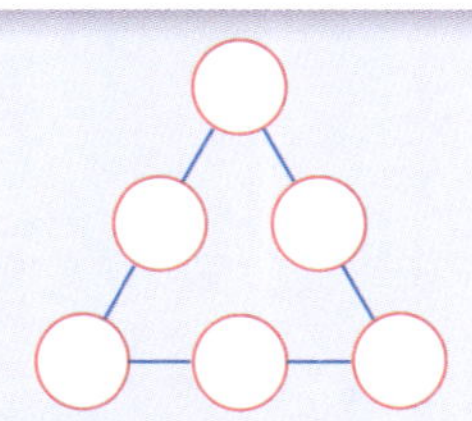

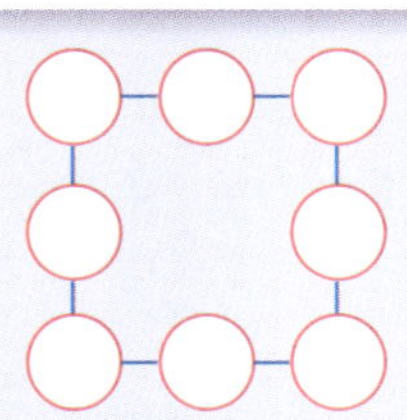

5 **a** Use the digits 1, 2, 3, 4, 5, 6, 7. Write one digit in each space so that all the lines add up to the same sum.

b Use the digits 1, 2, 3, 4, 5, 6. Write one digit in each space so that the sum of the numbers along each side is the same.

c Use the digits 1, 2, 3, 4, 6, 7, 8, 9. Write one digit in each space so that the sum of the numbers along each side is the same.

 • *AUSTRALIAN SIGNPOST MATHS 6* • ISBN 978 1 4886 2193 2

1:02 Square Numbers

1 Draw a square on each coloured side.

a

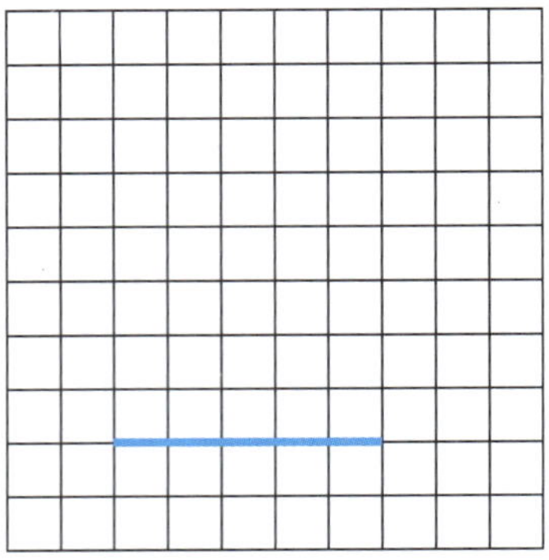

Area of blue square = ☐ small squares

b

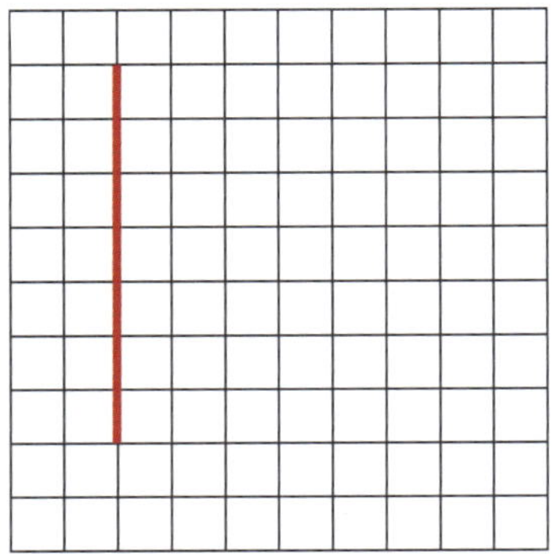

Area of red square = ☐ small squares

c

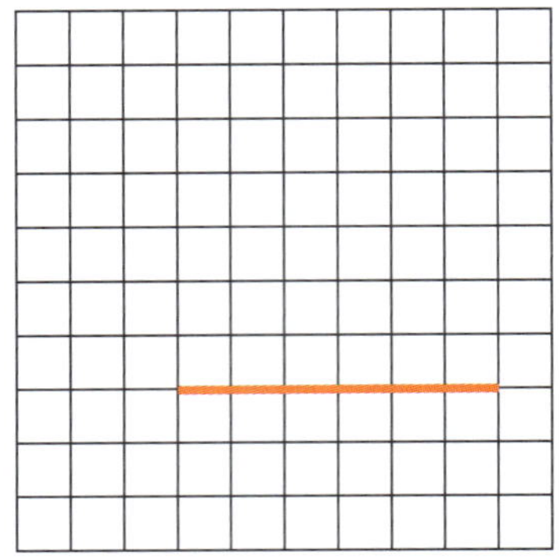

Area of orange square = ☐ small squares

d

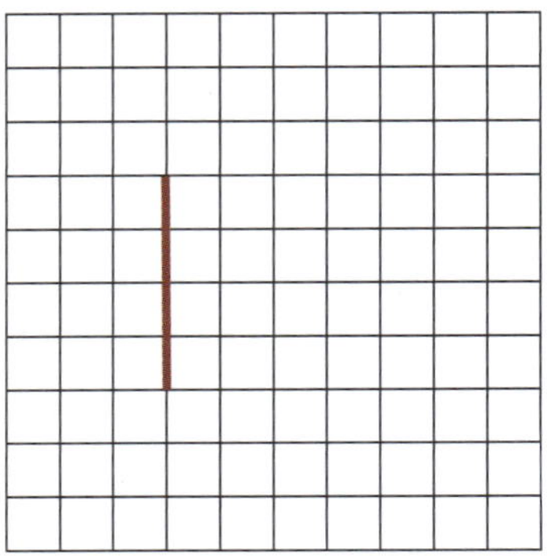

Area of brown square = ☐ small squares

e

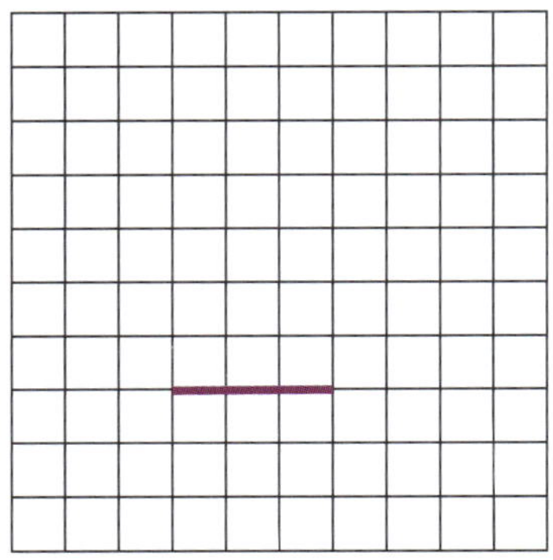

Area of purple square = ☐ small squares

f

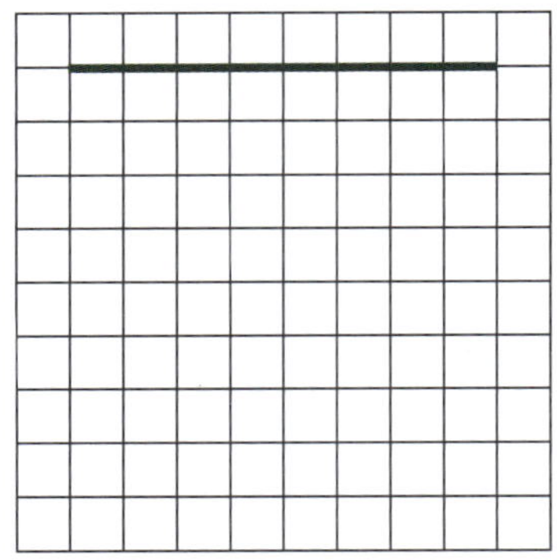

Area of green square = ☐ small squares

CONCEPT

The result of multiplying a number by itself is called a **square number**.

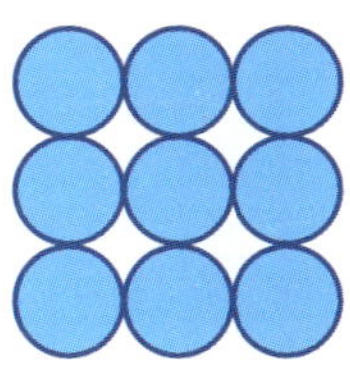

2 **a** Here we have 3 rows of 3 counters.

$3 \times 3 =$ ☐

b What shape does this array look like?

c Make square arrays using 4, 9, 16 and 25 counters.

3 List all of the square numbers up to 100.

4 Use place-value blocks or a calculator to find at least seven more square numbers.

ICT

5 Explain why the numbers 1, 4, 9, 16, … are called square numbers.

 • *AUSTRALIAN SIGNPOST MATHS 6* • ISBN 978 1 4886 2193 2

1:03 Square Numbers

CONCEPT

A whole number times itself gives a **square number**.

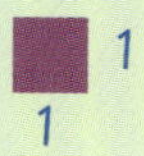

1 squared

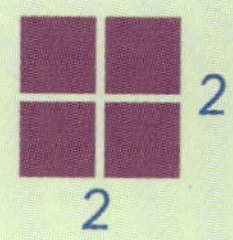

2 squared

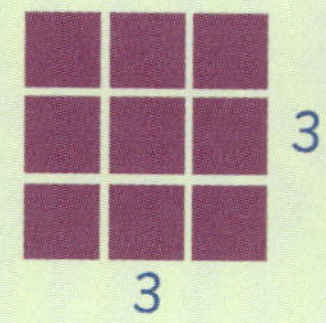

3 squared

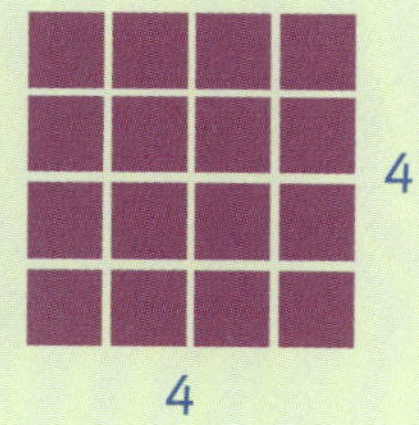

4 squared

The number of blocks in a square pattern is a square number. 3 squared is 3 × 3.

1 = 1 × 1

4 = 2 × 2 | 9 = 3 × 3

16 = 4 × 4 | ☐ = 5 × 5 | ☐ = 6 × 6

1 Use place-value blocks to find these square numbers.

a 5 squared = ☐ **b** 6 squared = ☐ **c** 7 squared = ☐

d 8 squared = ☐ **e** 9 squared = ☐ **f** 10 squared = ☐

g 2 × 2 = ☐ **h** 4 × 4 = ☐ **i** 7 × 7 = ☐ **j** 3 × 3 = ☐

k 6 × 6 = ☐ **l** 10 × 10 = ☐ **m** 1 × 1 = ☐ **n** 8 × 8 = ☐

2 Look carefully at the first ten square numbers below.

1 (+3) 4 (+5) 9 (+7) 16 (+9) 25 ☐ 36 ☐ 49 ☐ 64 ☐ 81 ☐ 100

a Complete the pattern shown here.

b Write down the next two square numbers after 100. ☐, ☐

c Write down the square numbers from those above that are also even.

☐, ☐, ☐, ☐, ☐

3 Complete:

a 9 = ☐ squared **b** 25 = ☐ squared

c 16 = ☐ squared **d** 36 = ☐ squared

4 Use blocks to find all factors of:

a 25: ☐, ☐ and ☐ **b** 9: ☐, ☐ and ☐ **c** 49: ☐, ☐ and ☐

d 4: ☐, ☐ and ☐ **e** 16: ☐, ☐, ☐, ☐ and ☐

1:04 Percentages

CONCEPT

25%

25 hundredths

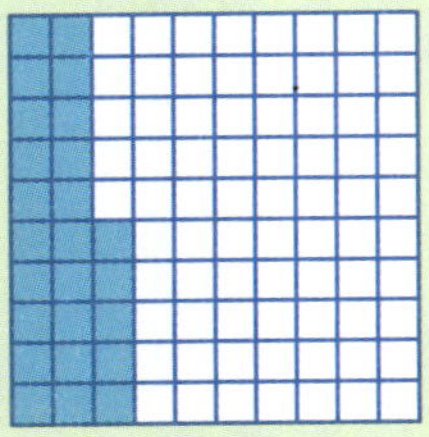

$\frac{25}{100}$

0·25

1 What percentage of each square is coloured?

a 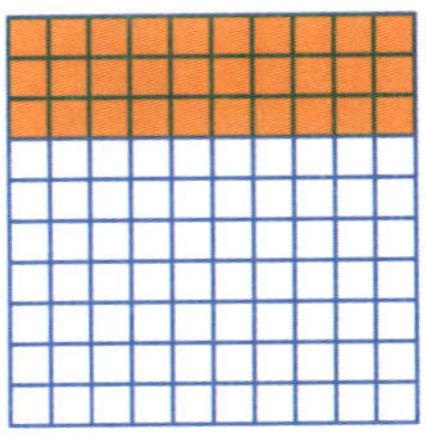

b 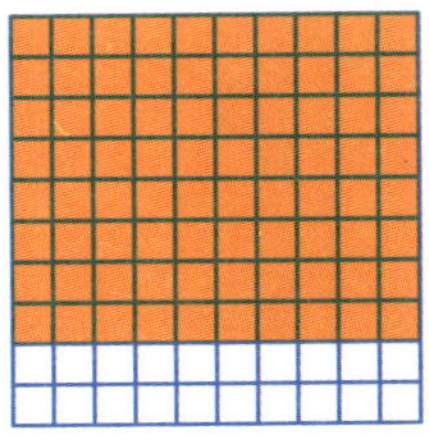

c 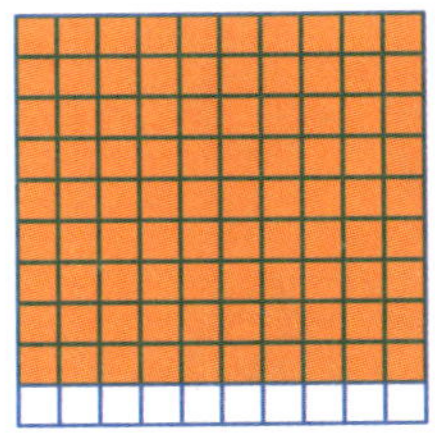

d 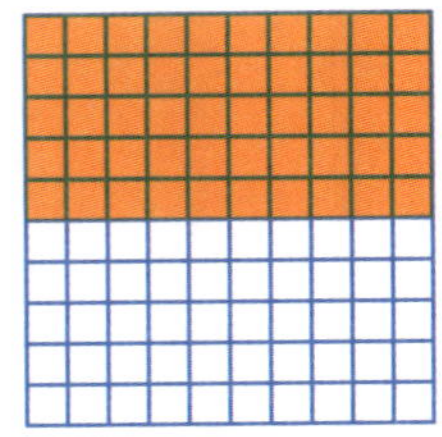

e 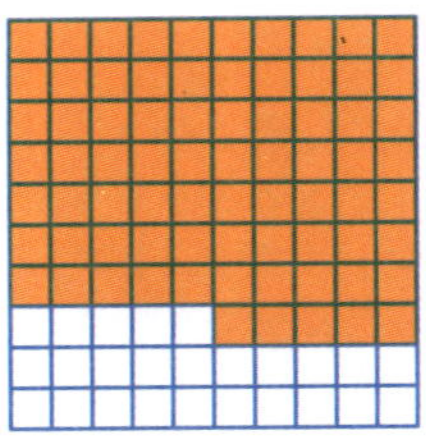

f 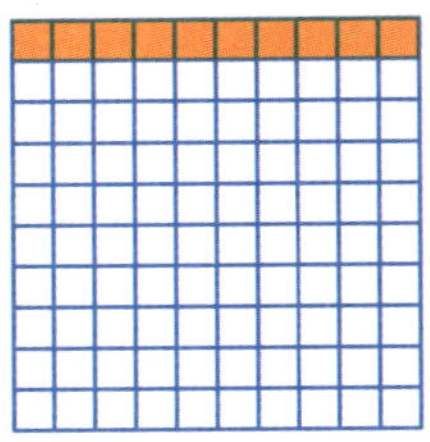

g 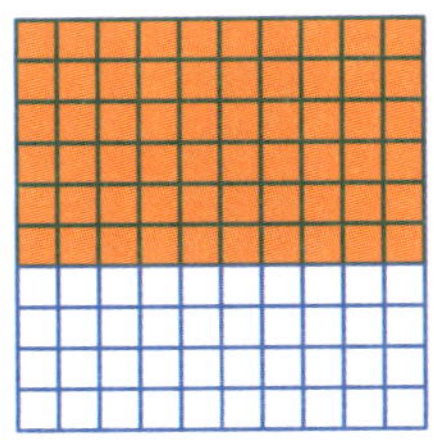

h 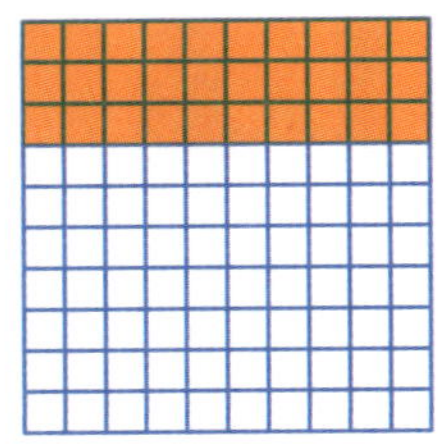

2 What percentage of each square is not coloured in Question 1?

a ____ b ____ c ____ d ____
e ____ f ____ g ____ h ____

3 Complete the following.

a	0·25	$\frac{\square}{100}$	☐ %
b	0·35	$\frac{\square}{100}$	☐ %
c	0·65	$\frac{\square}{100}$	☐ %
d	0·75	$\frac{\square}{100}$	☐ %
e	0·15	$\frac{\square}{100}$	☐ %
f	0·55	$\frac{\square}{100}$	☐ %
g	0·90	$\frac{\square}{100}$	☐ %
h	0·40	$\frac{\square}{100}$	☐ %
i	0·80	$\frac{\square}{100}$	☐ %

Percentages in the Environment

ACTIVITY

- Collect examples of percentages from newspapers and packets.
- Discuss the different ways in which percentages are used.

 • *AUSTRALIAN SIGNPOST MATHS 6* • ISBN 978 1 4886 2193 2

1:05 Percentages

1 What percentage of each square is coloured?

a 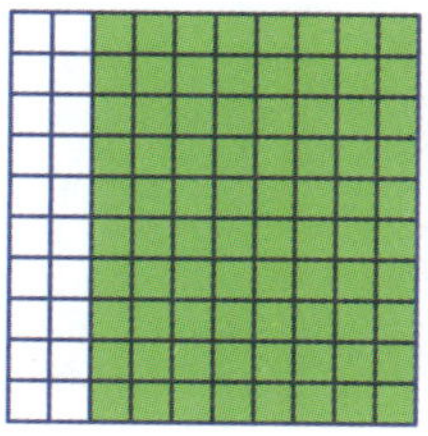 ☐

b 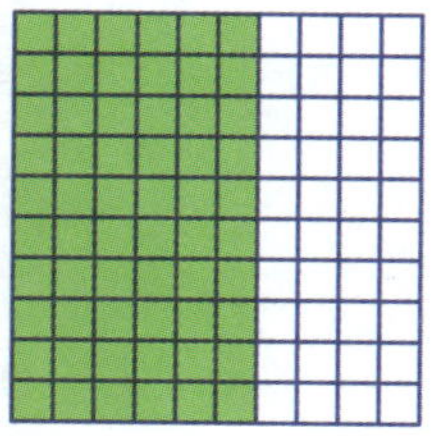 ☐

c 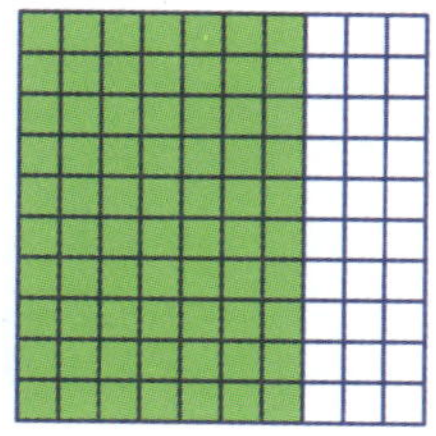 ☐

d 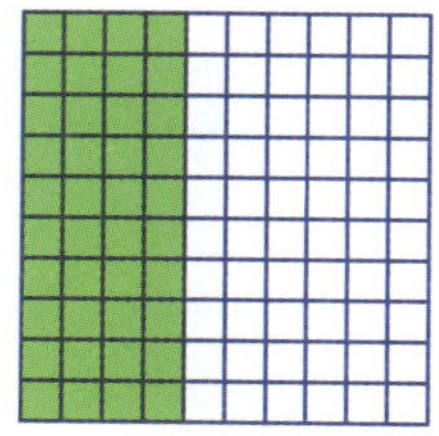 ☐

e 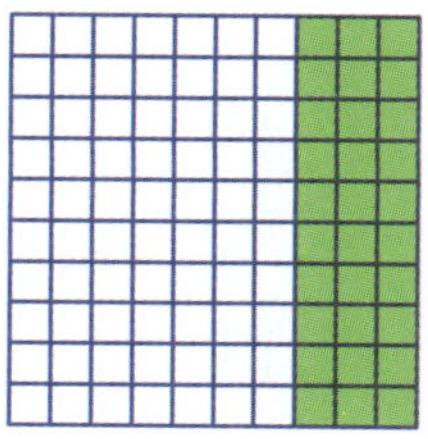 ☐

f 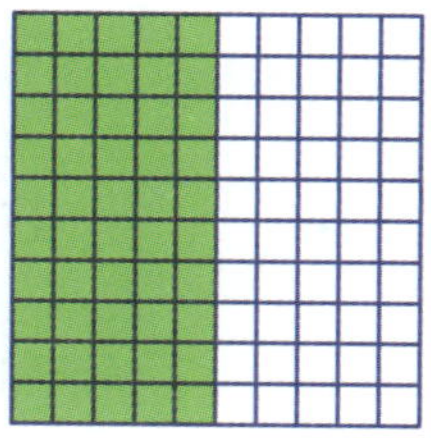 ☐

g  ☐

h 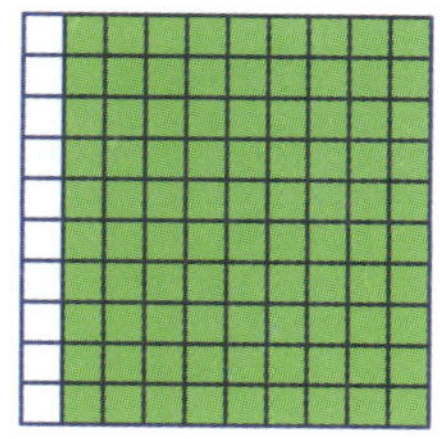 ☐

2 What percentage of each square is not coloured in Question 1?

a ☐ **b** ☐ **c** ☐ **d** ☐

e ☐ **f** ☐ **g** ☐ **h** ☐

3 Complete the following.

a	$\frac{25}{100}$	0·☐	☐%	**b**	$\frac{55}{100}$	0·☐	☐%	**c**	$\frac{75}{100}$	0·☐	☐%
d	$\frac{95}{100}$	0·☐	☐%	**e**	$\frac{65}{100}$	0·☐	☐%	**f**	$\frac{45}{100}$	0·☐	☐%
g	$\frac{9}{10}$	0·☐	☐%	**h**	$\frac{3}{10}$	0·☐	☐%	**i**	$\frac{7}{10}$	0·☐	☐%
j	$\frac{4}{10}$	0·☐	☐%	**k**	$\frac{5}{10}$	0·☐	☐%	**l**	1	☐·0	☐%

4 Draw lines to connect the equivalent numbers.

a	0·25	45%	**b**	0·7	55%	**c**	0·35	85%	**d**	0·3	65%
	0·5	60%		0·55	70%		0·1	90%		0·65	40%
	0·45	25%		0·8	95%		0·85	10%		0·4	30%
	0·6	50%		0·95	80%		0·9	35%		1	100%

1:06 Percentages

CONCEPT

40 out of 100

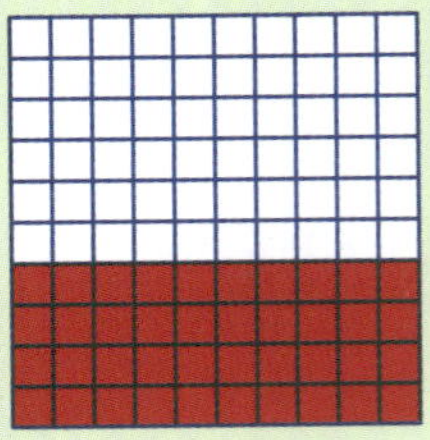

We can write a fraction as a decimal or a percentage.

1 Write the percentage equivalent for each common fraction.

a $\frac{3}{10}$ ______ **b** $\frac{9}{10}$ ______ **c** $\frac{75}{100}$ ______ **d** $\frac{6}{10}$ ______ **e** $\frac{25}{100}$ ______

f $\frac{4}{10}$ ______ **g** $\frac{65}{100}$ ______ **h** $\frac{35}{100}$ ______ **i** $\frac{5}{10}$ ______ **j** $\frac{1}{10}$ ______

2 Write the percentage equivalent for each decimal.

a 0·15 ______ **b** 0·45 ______ **c** 0·85 ______ **d** 0·55 ______ **e** 0·95 ______

f 0·8 ______ **g** 0·2 ______ **h** 0·7 ______ **i** 0·1 ______ **j** 0·05 ______

3 Write the decimal equivalent for each percentage.

a 25% ______ **b** 40% ______ **c** 90% ______ **d** 75% ______ **e** 10% ______

f 35% ______ **g** 25% ______ **h** 65% ______ **i** 5% ______ **j** 15% ______

4 For each square, colour and write the equivalent percentage.

a 0·3 = ______ %

b 0·15 = ______ %

c 0·6 = ______ %

d 0·95 = ______ %

e 0·5 = ______ %

f 0·89 = ______ %

g 0·46 = ______ %

h 0·7 = ______ %

i 0·82 = ______ %

j 0·05 = ______ %

1:07 Powers of Ten

CONCEPT

Ten thousands 10 000	Thousands 1000	Hundreds 100	Tens 10	Ones 1
10 × 10 × 10 × 10 10^4	10 × 10 × 10 10^3	10 × 10 10^2	10 10^1	1 1
6	4	7	3	8

$64\,738 = (6 \times 10\,000) + (4 \times 1000) + (7 \times 100) + (3 \times 10) + 8$
$= (6 \times 10^4) + (4 \times 10^3) + (7 \times 10^2) + (3 \times 10^1) + 8$

1 Write the numeral for:

a $(3 \times 10^4) + (7 \times 10^3) + (9 \times 10^2) + (5 \times 10^1) + 2$

b $(9 \times 10^4) + (6 \times 10^3) + (8 \times 10^2) + (3 \times 10^1) + 1$

c $(6 \times 10^4) + (2 \times 10^3) + (4 \times 10^2) + (7 \times 10^1) + 5$

d $(8 \times 10^4) + (9 \times 10^3) + (3 \times 10^2) + (5 \times 10^1) + 4$

2 Write the following in expanded notation using powers of ten.

a 6491

b 27 245

c 78 319

d 45 628

3 Write each number on the place-value chart.

a $(7 \times 10^4) + (9 \times 10^3) + (2 \times 10^2) + (3 \times 10^1) + 4$

b $(4 \times 10^4) + (6 \times 10^3) + (7 \times 10^2) + (9 \times 10^1) + 3$

c $(3 \times 10^4) + (5 \times 10^3) + (6 \times 10^2) + (8 \times 10^1) + 6$

d $(8 \times 10^4) + (3 \times 10^3) + (5 \times 10^2) + (6 \times 10^1) + 2$

Ten thousands	Thousands	Hundreds	Tens	Ones

4 Write the numeral for:

a 60 000 + 4000 + 900 + 50 + 8

b 90 000 + 6000 + 700 + 40 + 3

c 300 000 + 70 000 + 2000 + 500 + 90 + 8

d 700 000 + 80 000 + 5000 + 400 + 60 + 1

e 100 000 + 50 000 + 9000 + 300 + 50 + 6

1:08 Negative Numbers

1 Record the temperature shown on each thermometer.

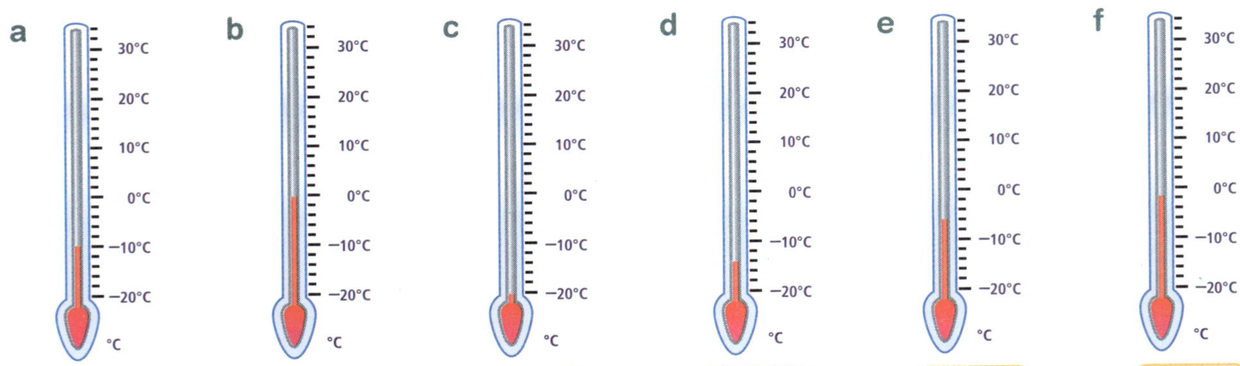

a ☐ °C b ☐ °C c ☐ °C d ☐ °C e ☐ °C f ☐ °C

2 Record the integers shown on the number line.

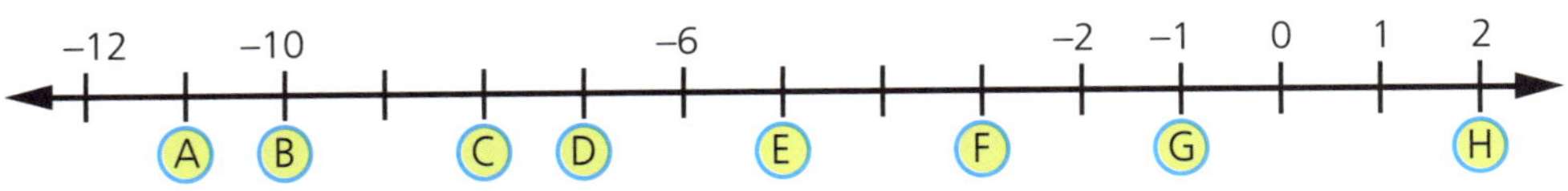

a H ☐ b G ☐ c F ☐ d A ☐

e D ☐ f B ☐ g C ☐ h E ☐

3 Record the integers shown on the number line.

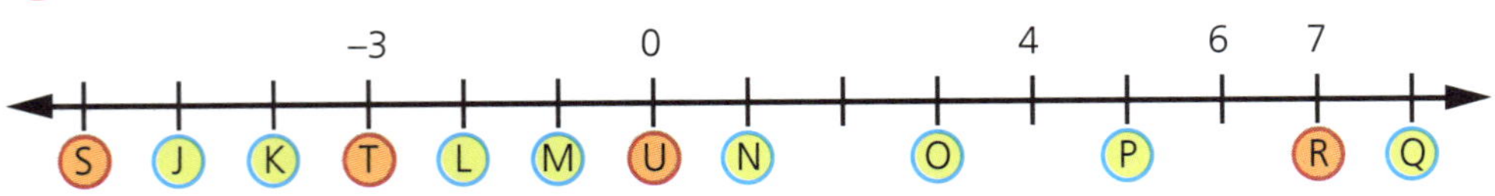

a K ☐ b L ☐ c Q ☐ d O ☐

e N ☐ f J ☐ g P ☐ h M ☐

i R ☐ j S ☐ k T ☐ l U ☐

Use the internet to find places that have had negative temperatures. Record your findings.

ICT

 • *AUSTRALIAN SIGNPOST MATHS 6* • ISBN 978 1 4886 2193 2

1:09 Positive and Negative Numbers

CONCEPT

We place negative numbers to the left of zero and positive numbers to the right.

–5 –4 –3 –2 –1 0 1 2 3 4 5

–30% –20% –10% 0 10% 20% 30%

1 Fill in the missing numbers on the number line.

a ☐ ☐ –1 0 ☐

b ☐ 0 1 ☐ ☐

c ☐ ☐ 0 ☐ 2

d ☐ –3 ☐ ☐ 0

2 Fill in the missing decimals or fractions on the number line.

a ☐ –2 0 2 ☐

b –20 –15 –10 ☐ ☐

c –8 ☐ –4 ☐ 0

d –30 ☐ ☐ 0 10

3 Fill in the missing percentages on the number line.

a ☐ –10% 0 10% ☐

b ☐ ☐ –25% 0 25%

c ☐ ☐ –10% –5% 0

d –15% 0 ☐ 30% ☐

Positive and Negative Integers Game

FUN SPOT

- The goal is to reach –20 on the number line.
- Each player starts at 4 on the number line. Take turns to throw two dice.
- An odd number can be used to move left, an even number to move right. The other dice represents how far you move along the number line.

  Example:  and a  can mean: 'move 4 to the left' or 'move 5 to the right'.

Finish (–20) Start (4)

–20 –18 –16 –14 –12 –10 –8 –6 –4 –2 0 2 4 6 8 10

 ISBN 978 1 4886 2193 2

1:10 Improper Fractions and Mixed Numbers

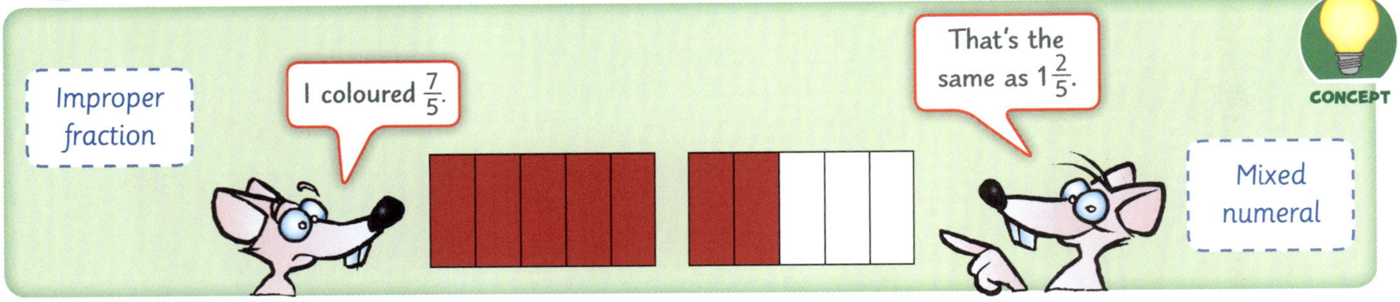

1 Write an improper fraction and mixed number for the coloured part in each model.

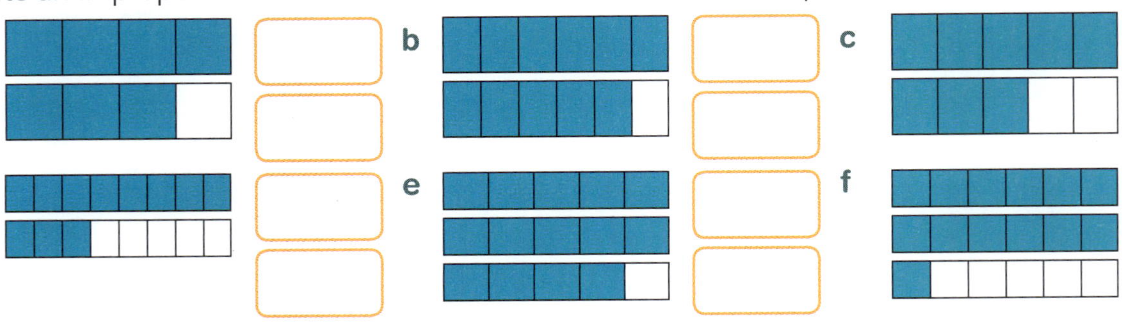

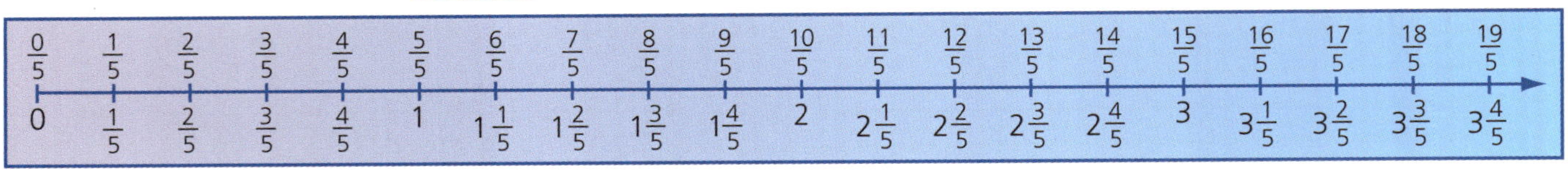

2 Use the number line above to write the mixed numeral for:

a $\frac{6}{5}$ b $\frac{11}{5}$ c $\frac{9}{5}$ d $\frac{13}{5}$ e $\frac{16}{5}$

f $\frac{12}{5}$ g $\frac{17}{5}$ h $\frac{7}{5}$ i $\frac{9}{5}$ j $\frac{8}{5}$

3 Use the number line to write the improper fraction for:

a $1\frac{4}{5}$ b $2\frac{3}{5}$ c $3\frac{2}{5}$ d $1\frac{1}{5}$ e $2\frac{2}{5}$

f $3\frac{1}{5}$ g $1\frac{2}{5}$ h $2\frac{1}{5}$ i $3\frac{4}{5}$ j $1\frac{3}{5}$

4 Write the mixed numeral for:

a $\frac{5}{4}$ b $\frac{13}{10}$ c $\frac{9}{8}$

d $\frac{7}{6}$ e $\frac{9}{4}$ f $\frac{17}{10}$

g $\frac{11}{8}$ h $\frac{13}{12}$ i $\frac{13}{8}$

j $\frac{11}{4}$ k $\frac{17}{6}$ l $\frac{17}{12}$

 ISBN 978 1 4886 2193 2

1:11 Patterns with Fractions and Decimals

CONCEPT

$\frac{7}{4}$ is called an improper fraction.

Did you know?

$\frac{7}{4}$ can also be written as $1\frac{3}{4}$. We call this a mixed number.

1 Write the mixed number for:

a $\frac{8}{5}$ ☐ b $\frac{11}{6}$ ☐ c $\frac{4}{3}$ ☐ d $\frac{162}{100}$ ☐

e $\frac{7}{3}$ ☐ f $\frac{13}{5}$ ☐ g $\frac{13}{6}$ ☐ h $\frac{124}{100}$ ☐

2 Write the improper fraction for:

a $1\frac{1}{2}$ ☐ b $2\frac{1}{4}$ ☐ c $1\frac{3}{4}$ ☐ d $2\frac{3}{4}$ ☐

e $3\frac{3}{5}$ ☐ f $5\frac{5}{8}$ ☐ g $9\frac{2}{5}$ ☐ h $10\frac{7}{8}$ ☐

i $15\frac{1}{10}$ ☐ j $15\frac{2}{10}$ ☐ k $15\frac{3}{10}$ ☐ l $15\frac{4}{10}$ ☐

3 Complete each pattern using decimals.

a 0·6, 0·8, 1, ☐, ☐, ☐ b 0·1, 0·3, 0·5, ☐, ☐, ☐

c 0·5, 1, 1·5, ☐, ☐, ☐ d 2·1, 2·2, 2·3, ☐, ☐, ☐

e 7·0, 7·3, 7·6, ☐, ☐, ☐ f 9·0, 9·5, 10·0, ☐, ☐, ☐

4 Complete each pattern using fractions.

a $\frac{1}{2}$, 1, $1\frac{1}{2}$, ☐, ☐, ☐ b $\frac{2}{8}$, $\frac{4}{8}$, $\frac{6}{8}$, ☐, ☐, ☐

c $\frac{1}{6}$, $\frac{3}{6}$, $\frac{5}{6}$, ☐, ☐, ☐ d $\frac{1}{5}$, $\frac{3}{5}$, 1, ☐, ☐, ☐

e $\frac{2}{10}$, $\frac{4}{10}$, $\frac{6}{10}$, ☐, ☐, ☐ f $\frac{1}{3}$, $\frac{2}{3}$, 1, ☐, ☐, ☐

Fraction Match

ACTIVITY

- Use two spinners (or dice) numbered 0 to 9.
- Spin each in turn and combine the two digits to make a fraction.
- On a prepared sheet, record each fraction as shown.
- Allow ten seconds to record each row.
- Mark answers before each new spin and score one point for each correct answer.

Fraction	Decimal	Percentage
$\frac{69}{100}$	0·69	69%

 • *AUSTRALIAN SIGNPOST MATHS 6* • ISBN 978 1 4886 2193 2

1:12 Using the Number Line

… –3, –2, –1, 0, 1, 2, 3, … are integers

We can extend our number line to the left of zero to include negative numbers.

–6 –4 –2 0 2 4 6 8

CONCEPT

1 Write the missing numbers.

a –3 –2 ☐ 0 1 ☐ 3

b ☐ –2 –1 0 ☐ 2 3

c –20 ☐ 0 10 20 ☐ 40

d –20 ☐ 0 ☐ 20 30 40

e ☐ –6 –4 –2 0 ☐ 4

f –20 –15 –10 ☐ 0 ☐ 10

g ☐ –10 –5 0 ☐ 10 15

h ☐ –2 0 ☐ 4 6 8

2 Complete each pattern.

a –3 ☐ ☐ ☐ 1 2 3

b –5 ☐ –3 ☐ –1 ☐ 1 ☐

c –3 –2 ☐ ☐ ☐ 2 3

d –8 ☐ –4 ☐ 0 ☐ 4 6

3 Which is the lower temperature:

a –2°C or 2°C? ☐ **b** –5°C or 0°C? ☐ **c** –5°C or 2°C? ☐

d –5°C or –2°C? ☐ **e** –5°C or –10°C? ☐ **f** 7°C or –7°C? ☐

4 Arrange these directed numbers in order from smallest to largest.

a 1, 5, –2, 4 ☐ **b** 0, –4, 1, –1 ☐

c –1, –2, –3, –4 ☐ **d** 2, –2, 3, –3 ☐

5 Write a number that stands for the lift floor given in each part.

a G ☐ **b** B2 ☐ **c** B1 ☐ **d** 3 ☐

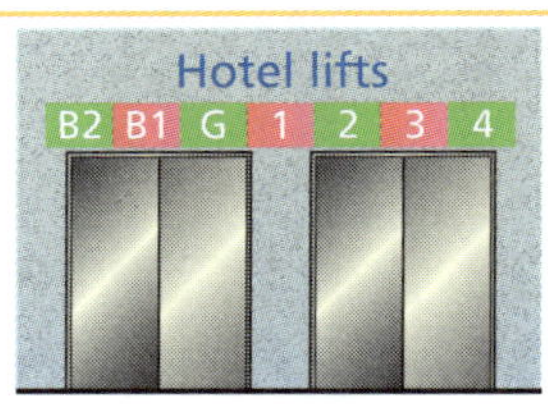

1:13 Positive and Negative Numbers

CONCEPT

Once a direction is chosen to be positive (+), the opposite direction is taken to be negative (–).

- Mount Everest is 8 000 metres above sea level, so its height could be written as +8 000 m or just 8 000 m.
- The Dead Sea is 100 metres below sea level, so its height could be written as –100 m.
- In cold countries the temperature is often below the freezing point of water (0°C). The temperature on a cold day could be –5°C.

- *–20 is read 'negative 20' or 'minus 20'.*
- *+20 is read 'positive 20' or just '20'.*

1 Write an integer (positive or negative number) for each statement.

a 50 m above sea level

b a loss of 10 kg

c 3 floors below ground

d a rise of 30 points

e a decrease of 14

f $100 withdrawal

2 Write the integer to show the change described in each sentence.

a Jan's mathematics mark was 8 marks higher this time.

b Today the temperature is 4°C below yesterday's temperature.

c John took 14 steps backwards.

d After playing marbles, Melissa had lost 15 marbles.

e The temperature fell from 37°C to 26°C.

f I bought shares for $23 per share and sold them for $19 per share.

g The level of water in the reservoir has dropped two metres.

h The population of our town has decreased by 150.

3 What temperature is shown on each thermometer below?

a

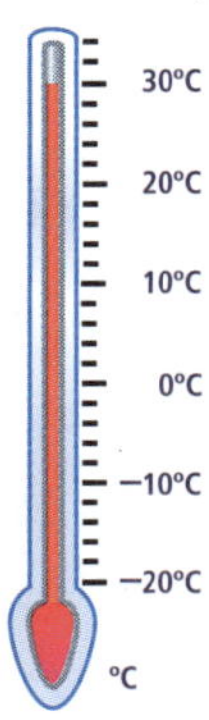

b

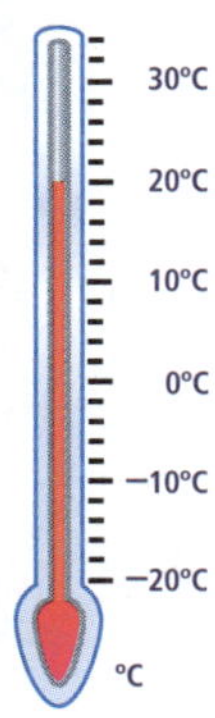

c

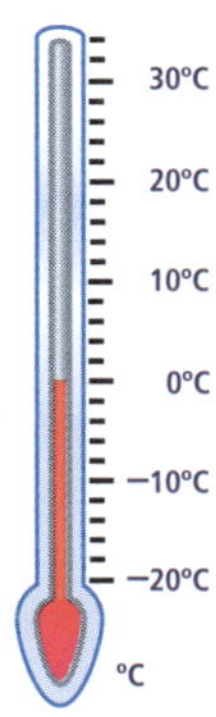

d

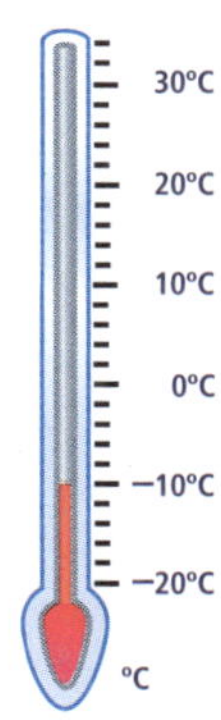

e

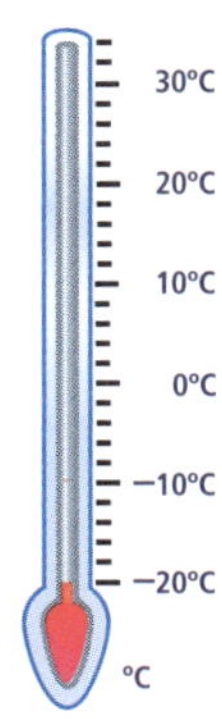

1:14 Using Negative Numbers

CONCEPT

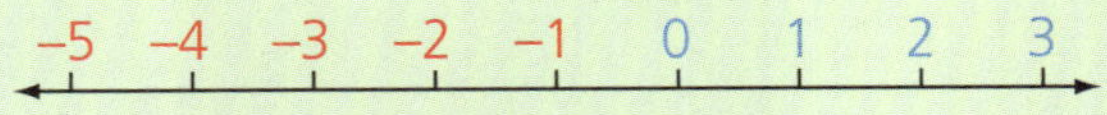

Negative numbers are numbers less than zero.

The temperature is minus ten degrees.

–10°C

1 Write the negative number from the number line at:

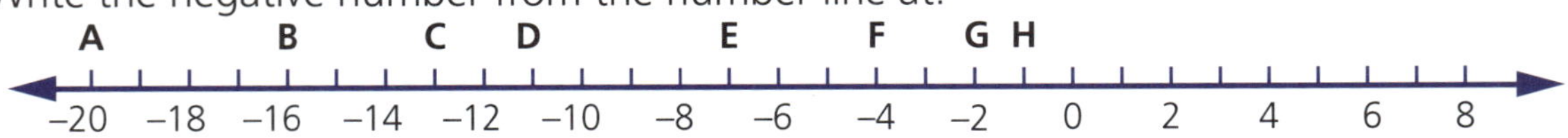

a G ☐ **b** A ☐ **c** F ☐ **d** B ☐ **e** C ☐
f E ☐ **g** D ☐ **h** H ☐

2 On Monday the temperature was 4 degrees Celsius (4°C). What would be the temperature on Tuesday if it was:

a 2 degrees colder? ☐ **b** 6 degrees colder? ☐ **c** 9 degrees colder? ☐
d 4 degrees colder? ☐ **e** 7 degrees colder? ☐ **f** 15 degrees colder? ☐

3 Sue had $11 in her bank account. What would be her bank balance if she wrote a cheque for:

a $6? ☐ **b** $15? ☐ **c** $11? ☐ **d** $28? ☐ **e** $31? ☐
f $43? ☐ **g** $70? ☐ **h** $16? ☐ **i** $50? ☐ **j** $100? ☐

4 Which numbers on the number line in Question 1 are 4 units from:

a 2? ☐ **b** –3? ☐ **c** –14? ☐

ICT

Example

[1] [0] [–] [2] [5] [=] ☐

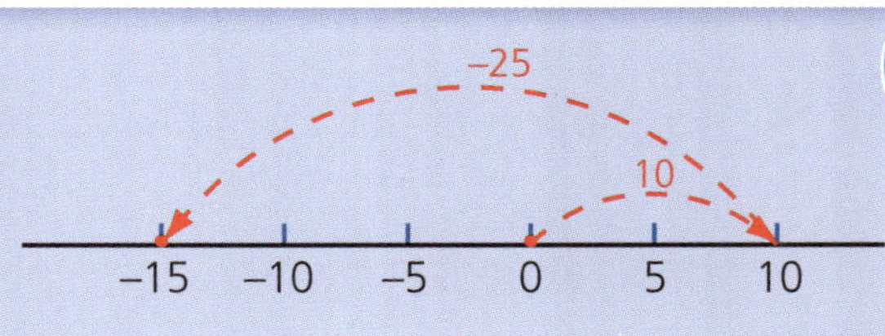

10 – 25 = –15

5 **a** 10 – 17 ☐ **b** 12 – 21 ☐
c 8 – 25 ☐ **d** 16 – 30 ☐
e 9 – 43 ☐ **f** 2 – 64 ☐
g 19 – 73 ☐ **h** 27 – 83 ☐
i 17 – 90 ☐ **j** 14 – 100 ☐

Use a number line to show the calculator sentences above.

10 – 25 gives a negative number for the answer.

1:15 Geometric Patterns

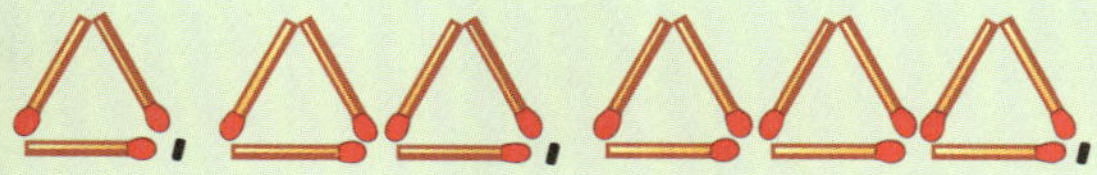

triangles × 3 = number of sides

CONCEPT

1 **a** Complete the table to show the number of sides needed to make the pattern of hexagons.

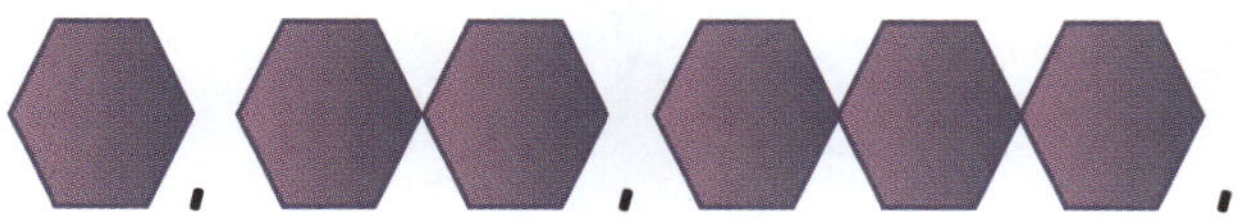

Number of Hexagons	1	2	3	4	5	6
Number of Sides	6					

b Write a rule to describe the pattern.

c How many sides would there be on 9 hexagons?

2 **a** Complete the table to show the number of stars needed to make the pattern of circles.

Number of Circles	1	2	3	4	5	6
Number of Stars	4					

b Write a rule to describe the pattern.

c How many stars would there be in 20 circles?

3 Adding counting numbers, starting from 1, gives triangular numbers.
Complete the table below to find the first six triangular numbers.

1	1 + 2	1 + 2 + 3	1 + 2 + 3 + 4	1 + 2 + 3 + 4 + 5	1 + 2 + 3 + 4 + 5 + 6

Patterns Using a Calculator

ICT

- Many calculators have a constant operator.

  9 × × 1 = 2 = 3 = 4 =

- Investigate the patterns made by the answers to these.

4 **a** 1 × 9 =
2 × 9 =
3 × 9 =
4 × 9 =

b 1 × 99 =
2 × 99 =
3 × 99 =
4 × 99 =

c 1 × 1 =
11 × 11 =
111 × 111 =
1111 × 1111 =

1:16 Fractions of a Group

CONCEPT

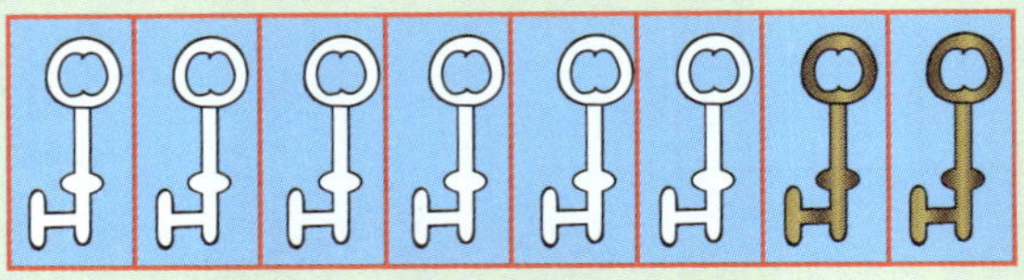

$\frac{1}{4}$ of 8 keys = 2 keys

$\frac{1}{4}$ of 8
$= 8 \div 4$
$= 2$

1 Find the fraction of each group.

a

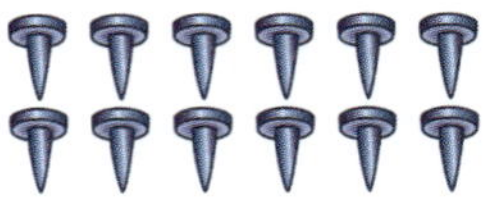

$\frac{1}{6}$ of 12 =

b

$\frac{1}{5}$ of 10 =

c

$\frac{1}{4}$ of 12 =

d

$\frac{1}{4}$ of 16 =

e

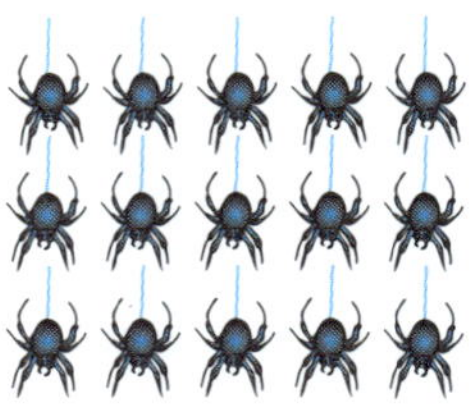

$\frac{1}{5}$ of 15 =

f

$\frac{1}{3}$ of 9 =

g

$\frac{1}{4}$ of 20 =

h

$\frac{1}{3}$ of 18 =

2 Use the array to find:

a $\frac{1}{2}$ of 24 =

b $\frac{1}{3}$ of 24 =

c $\frac{1}{4}$ of 24 =

d $\frac{1}{6}$ of 24 =

e $\frac{1}{8}$ of 24 =

f $\frac{1}{12}$ of 24 =

3 Find the fraction of each group.

a $\frac{1}{3}$ of 12

b $\frac{1}{10}$ of 30

c $\frac{1}{5}$ of 20

d $\frac{1}{3}$ of 15

e $\frac{1}{5}$ of 30

f $\frac{1}{8}$ of 32

g $\frac{1}{4}$ of 28

h $\frac{1}{10}$ of 70

- Use a calculator to check the answers to the questions above.

INVESTIGATION

4 **a** Colour $\frac{1}{2}$ of 10.

b Colour $\frac{1}{2}$ of 12.

c Colour $\frac{1}{2}$ of 14.

d Which is greater:

$\frac{1}{2}$ of 10, $\frac{1}{2}$ of 12 or $\frac{1}{2}$ of 14?

Explain your answer to a friend.

1:17 Fractions of a Group

CONCEPT

$\frac{1}{5}$ of 15

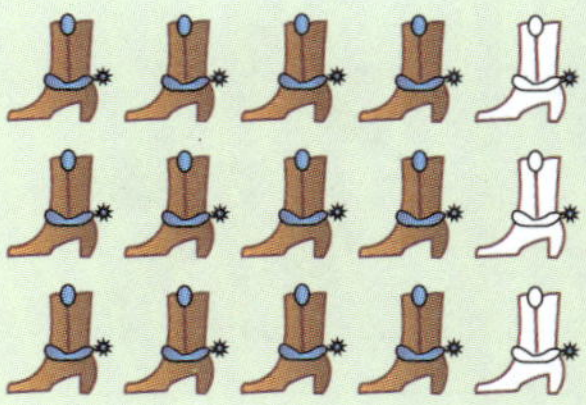

$\frac{1}{5}$ of 15 is 3.

So $\frac{4}{5}$ of 15 $= 4 \times 3$

$= 12$

$\frac{4}{5}$ of 15 $= \frac{4 \times 15}{5}$

$= \frac{60}{5}$

$= 12$

1 Find the fraction of each group.

a

$\frac{3}{5}$ of 10 = ☐

b

$\frac{2}{3}$ of 15 = ☐

c

$\frac{3}{4}$ of 12 ☐

d

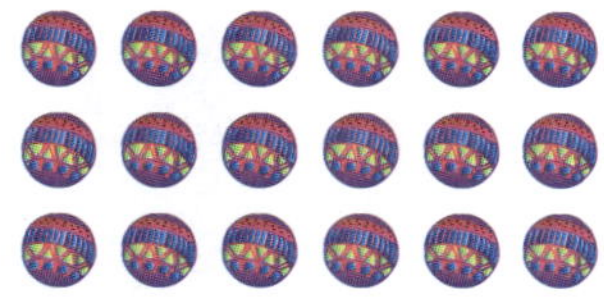

$\frac{2}{3}$ of 18 = ☐

2 Use the array to find:

a $\frac{3}{4}$ of 24 = ☐ **b** $\frac{3}{12}$ of 24 = ☐ **c** $\frac{2}{3}$ of 24 = ☐

d $\frac{5}{6}$ of 24 = ☐ **e** $\frac{2}{6}$ of 24 = ☐ **f** $\frac{5}{8}$ of 24 = ☐

g $\frac{5}{12}$ of 24 = ☐ **h** $\frac{3}{8}$ of 24 = ☐ **i** $\frac{7}{8}$ of 24 = ☐

3 Find the fraction of each group.

a $\frac{2}{3}$ of 30 = ☐ **b** $\frac{3}{8}$ of 16 = ☐ **c** $\frac{4}{5}$ of 20 = ☐ **d** $\frac{9}{10}$ of 60 = ☐

e $\frac{5}{12}$ of 36 = ☐ **f** $\frac{3}{5}$ of 25 = ☐ **g** $\frac{3}{10}$ of 70 = ☐ **h** $\frac{5}{8}$ of 40 = ☐

i $\frac{7}{10}$ of 120 = ☐ **j** $\frac{5}{6}$ of 42 = ☐ **k** $\frac{7}{8}$ of 32 = ☐ **l** $\frac{2}{5}$ of 60 = ☐

4

	Problem	Answer
a	Justin had 80 cricket cards but swapped $\frac{3}{8}$ of them. How many did he swap?	
b	The school canteen made 72 sandwiches and sold $\frac{5}{6}$ of them. How many were sold?	
c	Elle answered $\frac{9}{10}$ of her emails. If she had 130, how many did she answer?	
d	Con said that $\frac{4}{5}$ of 90 was 74. Was he right?	

Use either method to solve these problems.

© PEARSON AUSTRALIA 2018 • *AUSTRALIAN SIGNPOST MATHS 6* • ISBN 978 1 4886 2193 2

1:18 Addition and Subtraction of Fractions

CONCEPT

$$\frac{3}{8} + \frac{4}{8} = \frac{7}{8}$$

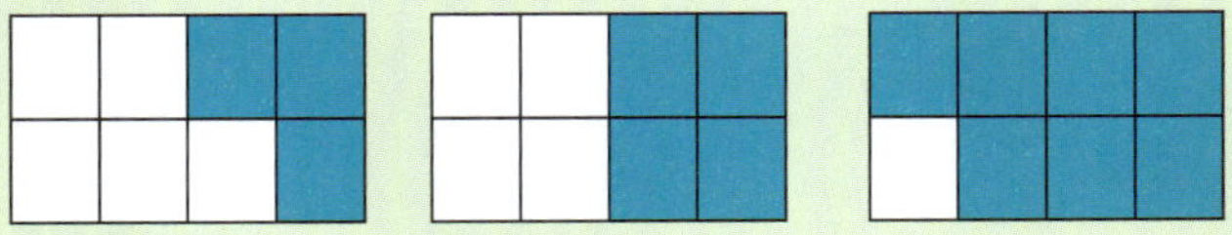

1 Use the array to find:

a $\frac{2}{12} + \frac{3}{12} =$ ☐ b $\frac{4}{12} + \frac{3}{12} =$ ☐ c $\frac{1}{12} + \frac{4}{12} =$ ☐

d $\frac{2}{12} + \frac{5}{12} =$ ☐ e $\frac{4}{12} + \frac{5}{12} =$ ☐ f $\frac{1}{12} + \frac{6}{12} =$ ☐

g $\frac{7}{12} + \frac{3}{12} =$ ☐ h $\frac{8}{12} + \frac{1}{12} =$ ☐ i $\frac{10}{12} + \frac{2}{12} =$ ☐

j $\frac{6}{12} - \frac{1}{12} =$ ☐ k $\frac{7}{12} - \frac{3}{12} =$ ☐ l $\frac{8}{12} - \frac{2}{12} =$ ☐

2 Add these fractions.

a $\frac{5}{8} + \frac{2}{8} =$ ☐ b $\frac{8}{10} + \frac{1}{10} =$ ☐ c $\frac{1}{4} + \frac{2}{4} =$ ☐ d $\frac{2}{6} + \frac{3}{6} =$ ☐

e $\frac{4}{10} + \frac{4}{10} =$ ☐ f $\frac{1}{6} + \frac{2}{6} =$ ☐ g $\frac{2}{8} + \frac{3}{8} =$ ☐ h $\frac{5}{10} + \frac{3}{10} =$ ☐

i 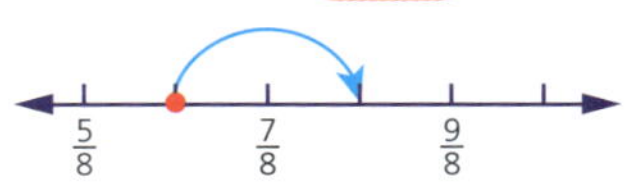

$\frac{6}{8} + \frac{2}{8} =$ ☐

j 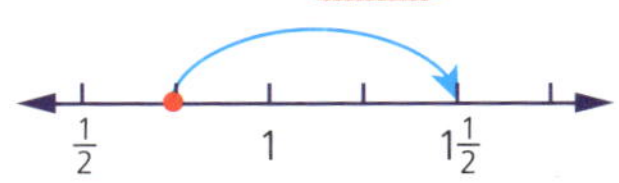

$\frac{3}{4} + \frac{3}{4} =$ ☐

3 Subtract these fractions.

a $\frac{2}{4} - \frac{1}{4} =$ ☐ b $\frac{7}{8} - \frac{3}{8} =$ ☐ c $\frac{4}{5} - \frac{1}{5} =$ ☐ d $\frac{4}{6} - \frac{3}{6} =$ ☐

e $\frac{9}{10} - \frac{2}{10} =$ ☐ f $\frac{4}{6} - \frac{2}{6} =$ ☐ g $\frac{5}{8} - \frac{3}{8} =$ ☐ h $\frac{5}{6} - \frac{1}{6} =$ ☐

i 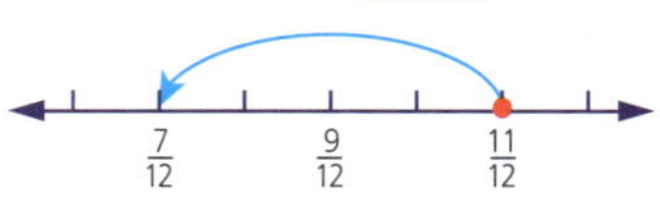

$\frac{11}{12} - \frac{4}{12} =$ ☐

j 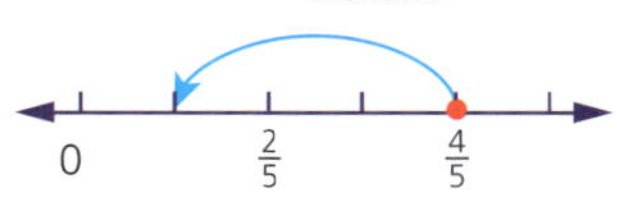

$\frac{4}{5} - \frac{3}{5} =$ ☐

4 Complete these.

a $\frac{1}{8} + \frac{4}{8} =$ ☐ b $\frac{3}{6} - \frac{1}{6} =$ ☐ c $\frac{5}{10} + \frac{3}{10} =$ ☐

d $\frac{7}{8} - \frac{4}{8} =$ ☐ e $\frac{3}{5} - \frac{1}{5} =$ ☐ f $\frac{3}{10} + \frac{5}{10} =$ ☐

g $\frac{3}{6} + \frac{1}{6} =$ ☐ h $\frac{6}{12} - \frac{4}{12} =$ ☐ i $\frac{4}{8} + \frac{1}{8} =$ ☐

1:19 Subtracting Fractions

CONCEPT

1 whole is $\frac{3}{3}$.

$1 - \frac{1}{3} = \frac{2}{3}$

1 whole is $\frac{4}{4}$.

$1 - \frac{1}{4} = \frac{3}{4}$

1 Complete these.

a $1 - \frac{1}{6} =$ ☐	**b** $1 - \frac{1}{10} =$ ☐	**c** $1 - \frac{1}{8} =$ ☐	**d** $1 - \frac{1}{12} =$ ☐
e $1 - \frac{1}{5} =$ ☐	**f** $1 - \frac{3}{4} =$ ☐	**g** $1 - \frac{7}{10} =$ ☐	**h** $1 - \frac{2}{5} =$ ☐
i $1 - \frac{2}{3} =$ ☐	**j** $1 - \frac{5}{6} =$ ☐	**k** $1 - \frac{3}{8} =$ ☐	**l** $1 - \frac{5}{12} =$ ☐

2 Complete these.

a $\frac{7}{8} - \frac{3}{8} =$ ☐	**b** $\frac{5}{6} - \frac{2}{6} =$ ☐	**c** $\frac{2}{3} - \frac{1}{3} =$ ☐	**d** $\frac{3}{4} - \frac{1}{4} =$ ☐
e $\frac{3}{6} - \frac{1}{6} =$ ☐	**f** $\frac{5}{8} - \frac{2}{8} =$ ☐	**g** $\frac{9}{10} - \frac{3}{10} =$ ☐	**h** $\frac{4}{5} - \frac{2}{5} =$ ☐
i $\frac{11}{12} - \frac{6}{12} =$ ☐	**j** $\frac{7}{10} - \frac{4}{10} =$ ☐	**k** $\frac{3}{5} - \frac{1}{5} =$ ☐	**l** $\frac{7}{12} - \frac{3}{12} =$ ☐

3 Complete these.

a $3 - \frac{1}{2} =$ ☐	**b** $2 - \frac{1}{3} =$ ☐	**c** $2 - \frac{1}{6} =$ ☐
d $3 - \frac{1}{5} =$ ☐	**e** $4 - \frac{1}{10} =$ ☐	**f** $4 - \frac{1}{12} =$ ☐
g $3 - \frac{1}{8} =$ ☐	**h** $4 - \frac{1}{4} =$ ☐	**i** $2 - \frac{3}{4} =$ ☐

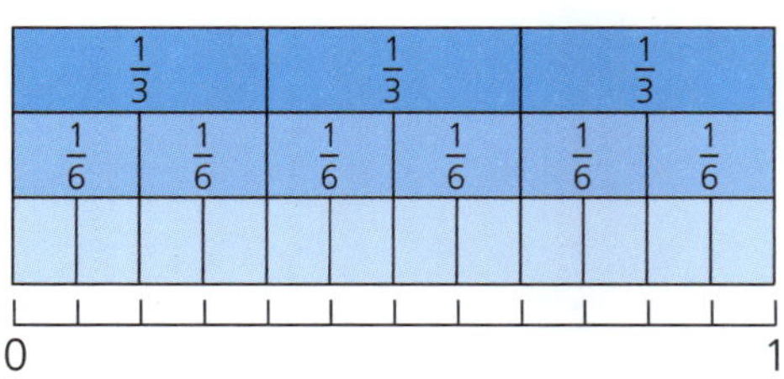

4 Use the diagram to show an equivalent fraction for each.

a $\frac{2}{12} =$ ☐	**b** $\frac{10}{12} =$ ☐	**c** $\frac{4}{6} =$ ☐
d $\frac{2}{3} =$ ☐	**e** $\frac{8}{12} =$ ☐	**f** $\frac{3}{6} =$ ☐

5 Use the diagram above to answer **true** or **false**.

a $\frac{1}{3} = \frac{4}{12}$ ☐	**b** $\frac{8}{12} = \frac{4}{6}$ ☐	**c** $\frac{2}{3} = \frac{6}{12}$ ☐	**d** $\frac{10}{12} = \frac{5}{6}$ ☐

1:20 Equivalent Fractions

CONCEPT

1									
$\frac{1}{2}$					$\frac{1}{2}$				
$\frac{1}{5}$		$\frac{1}{5}$		$\frac{1}{5}$		$\frac{1}{5}$		$\frac{1}{5}$	
$\frac{1}{10}$	$\frac{1}{10}$	$\frac{1}{10}$	$\frac{1}{10}$	$\frac{1}{10}$	$\frac{1}{10}$	$\frac{1}{10}$	$\frac{1}{10}$	$\frac{1}{10}$	$\frac{1}{10}$

1					
$\frac{1}{2}$			$\frac{1}{2}$		
$\frac{1}{3}$		$\frac{1}{3}$		$\frac{1}{3}$	
$\frac{1}{6}$	$\frac{1}{6}$	$\frac{1}{6}$	$\frac{1}{6}$	$\frac{1}{6}$	$\frac{1}{6}$

1 Write an equivalent fraction for each.

a $\frac{2}{10}$ ☐ **b** $\frac{4}{10}$ ☐ **c** $\frac{6}{10}$ ☐

d $\frac{4}{5}$ ☐ **e** $\frac{5}{5}$ ☐ **f** $\frac{1}{5}$ ☐

2 Write an equivalent fraction for each.

a $\frac{2}{6}$ ☐ **b** $\frac{4}{6}$ ☐ **c** $\frac{6}{6}$ ☐

d $\frac{1}{3}$ ☐ **e** $\frac{2}{3}$ ☐ **f** $\frac{3}{3}$ ☐

3 True or false?

a $\frac{1}{5} = \frac{2}{10}$ ☐ **b** $\frac{3}{5} = \frac{6}{10}$ ☐

c $\frac{4}{5} = \frac{9}{10}$ ☐ **d** $\frac{5}{10} = \frac{3}{5}$ ☐

e $\frac{8}{10} = \frac{4}{5}$ ☐ **f** $\frac{10}{10} = 1$ ☐

4 True or false?

a $\frac{2}{3} = \frac{4}{6}$ ☐ **b** $\frac{1}{3} = \frac{2}{6}$ ☐

c $\frac{3}{3} = \frac{6}{6}$ ☐ **d** $\frac{1}{6} = \frac{1}{3}$ ☐

e $\frac{3}{6} = \frac{2}{3}$ ☐ **f** $\frac{6}{6} = 1$ ☐

5 < or >?

a $\frac{1}{5}$ ☐ $\frac{1}{10}$ **b** $\frac{3}{5}$ ☐ $\frac{8}{10}$

c $\frac{6}{10}$ ☐ $\frac{2}{5}$ **d** $\frac{5}{10}$ ☐ $\frac{5}{5}$

6 < or >?

a $\frac{1}{3}$ ☐ $\frac{1}{6}$ **b** $\frac{3}{3}$ ☐ $\frac{4}{6}$

c $\frac{2}{6}$ ☐ $\frac{2}{3}$ **d** $\frac{5}{6}$ ☐ $\frac{1}{3}$

7 Use the array to find:

a $\frac{1}{3}$ of 12 ☐ **b** $\frac{1}{4}$ of 12 ☐ **c** $\frac{3}{4}$ of 12 ☐

d $\frac{1}{2}$ of 12 ☐ **e** $\frac{2}{3}$ of 12 ☐

8 Use the array to find the larger fraction.

a $\frac{1}{3}$ or $\frac{1}{4}$ ☐ **b** $\frac{3}{4}$ or $\frac{2}{3}$ ☐ **c** $\frac{1}{2}$ or $\frac{1}{3}$ ☐

d $\frac{3}{4}$ or $\frac{7}{12}$ ☐ **e** $\frac{2}{3}$ or $\frac{9}{12}$ ☐ **f** $\frac{5}{12}$ or $\frac{1}{2}$ ☐

ICT

Use a calculator to divide the numerator by the denominator.

Example: $\frac{1}{4}$ ←numerator ←denominator [1] [÷] [4] [=] 0·25

9 Repeat this activity for the following groups of fractions.

a $\frac{1}{4}$, $\frac{2}{8}$, $\frac{3}{12}$ **b** $\frac{3}{4}$, $\frac{6}{8}$, $\frac{9}{12}$ **c** $\frac{1}{2}$, $\frac{2}{4}$, $\frac{3}{6}$, $\frac{4}{8}$, $\frac{5}{10}$ **d** $\frac{1}{3}$, $\frac{2}{6}$, $\frac{3}{9}$

 • *AUSTRALIAN SIGNPOST MATHS 6* • ISBN 978 1 4886 2193 2

1:21 Equivalent Fractions

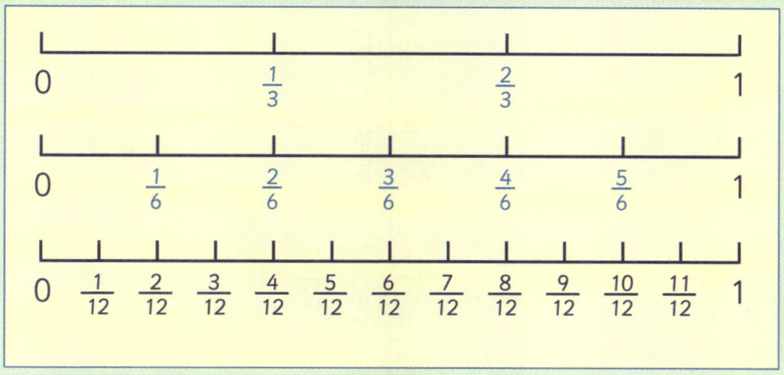

CONCEPT

1 Use the number lines to show an equivalent fraction for:

a $\frac{1}{6}$ ☐ **b** $\frac{5}{6}$ ☐ **c** $\frac{3}{6}$ ☐ **d** $\frac{2}{3}$ ☐

e $\frac{4}{12}$ ☐ **f** $\frac{4}{6}$ ☐ **g** ☐ **h** $\frac{6}{12}$ ☐

2 Use the number lines above to answer **true** or **false**.

a $\frac{2}{12} = \frac{1}{6}$ ☐ **b** $\frac{3}{6} = \frac{6}{12}$ ☐ **c** $\frac{2}{3} = \frac{9}{12}$ ☐ **d** $\frac{8}{12} = \frac{4}{6}$ ☐

e $\frac{1}{6} = \frac{1}{12}$ ☐ **f** $\frac{10}{12} = \frac{5}{6}$ ☐ **g** $\frac{1}{3} = \frac{4}{12}$ ☐ **h** $\frac{6}{12} = \frac{2}{3}$ ☐

3 Complete the number lines.

a

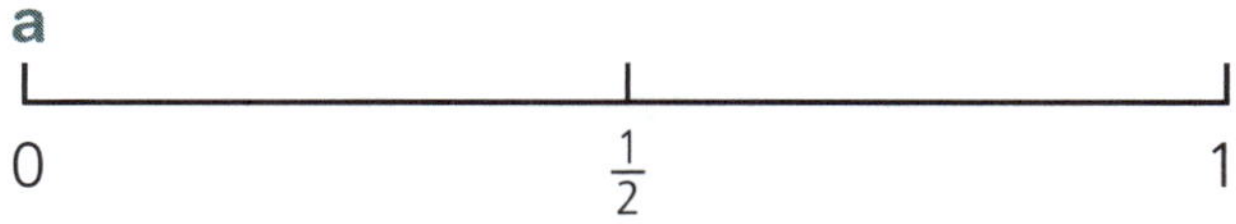

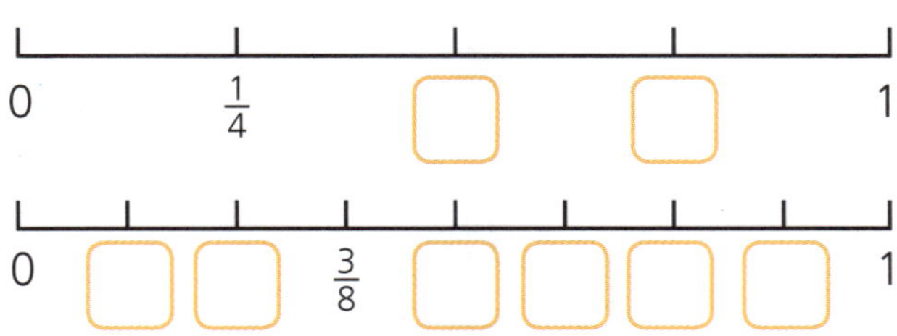

b

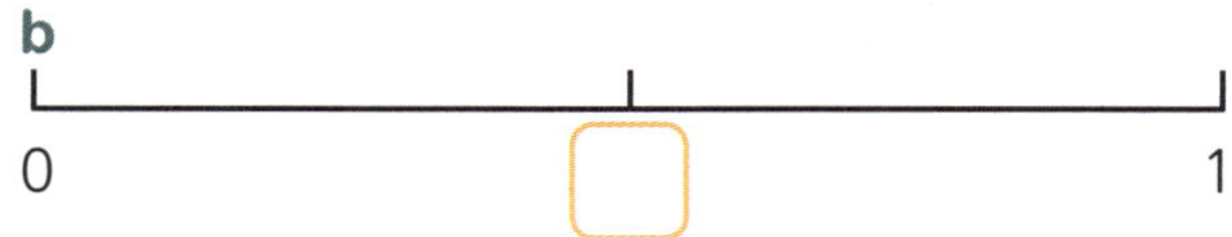

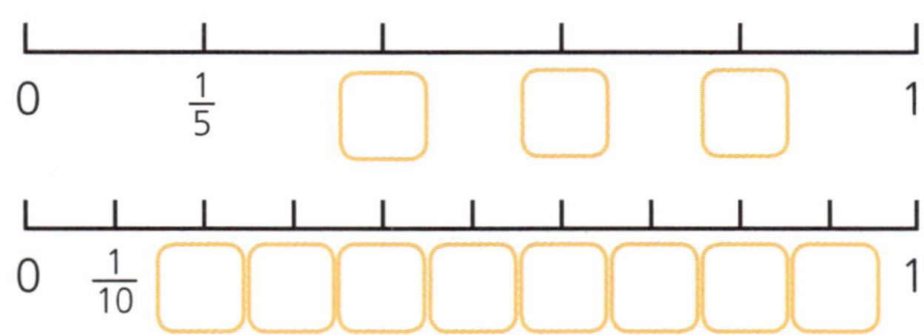

4 Use the number lines above to answer **true** or **false**.

a $\frac{4}{8} = \frac{1}{2}$ ☐ **b** $\frac{3}{4} = \frac{6}{8}$ ☐ **c** $\frac{1}{2} = \frac{5}{8}$ ☐ **d** $\frac{2}{4} = \frac{1}{2}$ ☐

e $\frac{7}{8} = \frac{3}{4}$ ☐ **f** $\frac{2}{10} = \frac{1}{5}$ ☐ **g** $\frac{4}{5} = \frac{8}{10}$ ☐ **h** $\frac{2}{5} = \frac{1}{2}$ ☐

i $\frac{3}{5} = \frac{6}{10}$ ☐ **j** $\frac{4}{10} = \frac{2}{5}$ ☐ **k** $\frac{5}{10} = \frac{1}{2}$ ☐ **l** $\frac{7}{10} = \frac{3}{5}$ ☐

5 Write an equivalent fraction for:

a $\frac{8}{10}$ ☐ **b** $\frac{2}{8}$ ☐ **c** $\frac{2}{5}$ ☐ **d** $\frac{6}{8}$ ☐

e $\frac{1}{5}$ ☐ **f** $\frac{6}{10}$ ☐ **g** $\frac{3}{6}$ ☐ **h** $\frac{3}{5}$ ☐

1:22 Equivalent Fractions

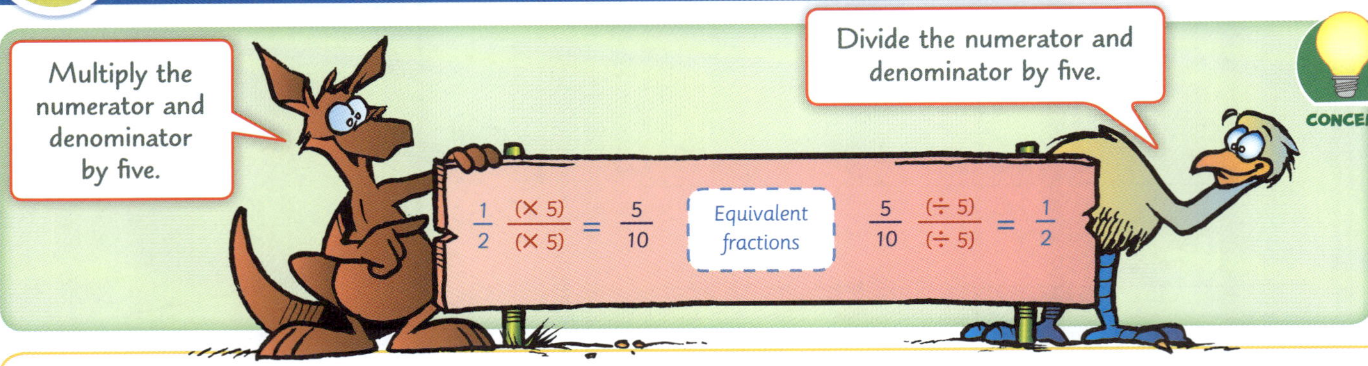

CONCEPT

1 Write an equivalent fraction for each.

a $\frac{3}{5}\,\frac{(\times 2)}{(\times 2)}$ $\frac{\square}{\square}$ **b** $\frac{1}{4}\,\frac{(\times 3)}{(\times 3)}$ $\frac{\square}{\square}$ **c** $\frac{1}{3}\,\frac{(\times 4)}{(\times 4)}$ $\frac{\square}{\square}$ **d** $\frac{1}{6}\,\frac{(\times 2)}{(\times 2)}$ $\frac{\square}{\square}$

e $\frac{2}{3}\,\frac{(\times 4)}{(\times 4)}$ $\frac{\square}{\square}$ **f** $\frac{4}{5}\,\frac{(\times 4)}{(\times 4)}$ $\frac{\square}{\square}$ **g** $\frac{3}{4}\,\frac{(\times 3)}{(\times 3)}$ $\frac{\square}{\square}$ **h** $\frac{5}{6}\,\frac{(\times 2)}{(\times 2)}$ $\frac{\square}{\square}$

2 Complete these to make a simpler fraction.

a $\frac{2}{4}\,\frac{(\div 2)}{(\div 2)}$ $\frac{\square}{\square}$ **b** $\frac{3}{6}\,\frac{(\div 3)}{(\div 3)}$ $\frac{\square}{\square}$ **c** $\frac{5}{20}\,\frac{(\div 5)}{(\div 5)}$ $\frac{\square}{\square}$ **d** $\frac{5}{10}\,\frac{(\div 5)}{(\div 5)}$ $\frac{\square}{\square}$

e $\frac{6}{8}\,\frac{(\div 2)}{(\div 2)}$ $\frac{\square}{\square}$ **f** $\frac{4}{10}\,\frac{(\div 2)}{(\div 2)}$ $\frac{\square}{\square}$ **g** $\frac{9}{12}\,\frac{(\div 3)}{(\div 3)}$ $\frac{\square}{\square}$ **h** $\frac{6}{10}\,\frac{(\div 2)}{(\div 2)}$ $\frac{\square}{\square}$

3 Multiply both the numerator and denominator by 4.

a $\frac{3}{5} = \frac{\square}{\square}$ **b** $\frac{1}{2} = \frac{\square}{\square}$ **c** $\frac{1}{10} = \frac{\square}{\square}$ **d** $\frac{2}{4} = \frac{\square}{\square}$ **e** $\frac{1}{5} = \frac{\square}{\square}$

4 Multiply both the numerator and denominator by 5.

a $\frac{1}{2} = \frac{\square}{\square}$ **b** $\frac{1}{3} = \frac{\square}{\square}$ **c** $\frac{4}{5} = \frac{\square}{\square}$ **d** $\frac{2}{3} = \frac{\square}{\square}$ **e** $\frac{3}{4} = \frac{\square}{\square}$

5 Divide both the numerator and denominator by 3.

a $\frac{6}{15} = \frac{\square}{\square}$ **b** $\frac{3}{9} = \frac{\square}{\square}$ **c** $\frac{6}{12} = \frac{\square}{\square}$ **d** $\frac{9}{18} = \frac{\square}{\square}$ **e** $\frac{3}{15} = \frac{\square}{\square}$

6 Simplify these fractions.

a $\frac{4}{20} = \frac{\square}{\square}$ **b** $\frac{10}{12} = \frac{\square}{\square}$ **c** $\frac{10}{15} = \frac{\square}{\square}$ **d** $\frac{5}{25} = \frac{\square}{\square}$ **e** $\frac{30}{100} = \frac{\square}{\square}$

1:23 Addition and Subtraction of Fractions

$\frac{8}{8}$ = 1 whole

To add fractions we change them so that they have the same denominator.

$\frac{3}{8} + \frac{1}{2} = \frac{3}{8} + \frac{4}{8} = \frac{7}{8}$

$\frac{7}{8} - \frac{1}{4} = \frac{7}{8} - \frac{2}{8} = \frac{5}{8}$

CONCEPT

1 Use the array above to answer these questions.

a $\frac{5}{8} + \frac{1}{4} = \frac{\square}{\square} + \frac{\square}{\square} = \frac{\square}{\square}$

b $\frac{7}{8} - \frac{1}{2} = \frac{\square}{\square} - \frac{\square}{\square} = \frac{\square}{\square}$

c $\frac{6}{8} - \frac{1}{4} = \frac{\square}{\square} - \frac{\square}{\square} = \frac{\square}{\square}$

d $\frac{3}{8} + \frac{1}{4} = \frac{\square}{\square} + \frac{\square}{\square} = \frac{\square}{\square}$

2 Use the array to answer these.

$\frac{10}{10}$ = 1 whole

a $\frac{4}{10} + \frac{2}{5} = \frac{\square}{\square} + \frac{\square}{\square} = \frac{\square}{\square}$

b $\frac{9}{10} - \frac{1}{2} = \frac{\square}{\square} - \frac{\square}{\square} = \frac{\square}{\square}$

c $\frac{10}{10} - \frac{3}{5} = \frac{\square}{\square} - \frac{\square}{\square} = \frac{\square}{\square}$

d $\frac{3}{10} + \frac{1}{2} = \frac{\square}{\square} + \frac{\square}{\square} = \frac{\square}{\square}$

3 Use the array to answer these.

a $\frac{3}{12} + \frac{2}{3} = \frac{\square}{\square} + \frac{\square}{\square} = \frac{\square}{\square}$

b $\frac{8}{12} - \frac{1}{2} = \frac{\square}{\square} - \frac{\square}{\square} = \frac{\square}{\square}$

c $\frac{3}{12} + \frac{1}{4} = \frac{\square}{\square} + \frac{\square}{\square} = \frac{\square}{\square}$

d $\frac{9}{12} - \frac{1}{4} = \frac{\square}{\square} - \frac{\square}{\square} = \frac{\square}{\square}$

4 Use the arrays above to calculate:

a $\frac{1}{5} + \frac{3}{10} = \square$ **b** $\frac{12}{12} - \frac{3}{4} = \square$ **c** $\frac{5}{12} + \frac{1}{3} = \square$ **d** $\frac{8}{10} - \frac{3}{5} = \square$

 • *AUSTRALIAN SIGNPOST MATHS 6* • ISBN 978 1 4886 2193 2

1:24 Addition and Subtraction of Fractions

CONCEPT

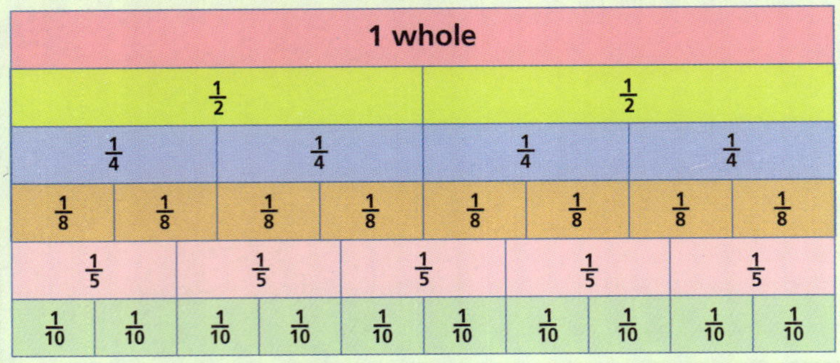

1 Add these fractions.

a $\frac{1}{3} + \frac{3}{6} = \frac{\square}{\square} + \frac{\square}{\square} = \square$

b $\frac{1}{4} + \frac{3}{8} = \frac{\square}{\square} + \frac{\square}{\square} = \square$

c $\frac{2}{5} + \frac{1}{10} = \frac{\square}{\square} + \frac{\square}{\square} = \square$

d $\frac{5}{8} + \frac{1}{4} = \frac{\square}{\square} + \frac{\square}{\square} = \square$

e $\frac{3}{10} + \frac{1}{2} = \frac{\square}{\square} + \frac{\square}{\square} = \square$

f $\frac{3}{5} + \frac{2}{10} = \frac{\square}{\square} + \frac{\square}{\square} = \square$

g $\frac{2}{5} + \frac{5}{10} = \frac{\square}{\square} + \frac{\square}{\square} = \square$

Add fractions

Subtract fractions

2 Subtract these fractions.

a $\frac{7}{8} - \frac{1}{2} = \frac{\square}{\square} - \frac{\square}{\square} = \square$

b $\frac{7}{10} - \frac{1}{5} = \frac{\square}{\square} - \frac{\square}{\square} = \square$

c $\frac{9}{12} - \frac{2}{3} = \frac{\square}{\square} - \frac{\square}{\square} = \square$

d $\frac{9}{10} - \frac{4}{5} = \frac{\square}{\square} - \frac{\square}{\square} = \square$

e $\frac{7}{12} - \frac{2}{6} = \frac{\square}{\square} - \frac{\square}{\square} = \square$

f $\frac{5}{6} - \frac{1}{3} = \frac{\square}{\square} - \frac{\square}{\square} = \square$

g $\frac{9}{10} - \frac{3}{5} = \frac{\square}{\square} - \frac{\square}{\square} = \square$

h $\frac{5}{12} - \frac{1}{4} = \frac{\square}{\square} - \frac{\square}{\square} = \square$

3 Solve these fraction problems.

a $\frac{5}{10} + \frac{2}{5} = \frac{\square}{\square} + \frac{\square}{\square} = \square$

b $\frac{5}{9} - \frac{1}{3} = \frac{\square}{\square} - \frac{\square}{\square} = \square$

c $\frac{7}{12} + \frac{2}{6} = \frac{\square}{\square} + \frac{\square}{\square} = \square$

d $\frac{5}{8} - \frac{1}{4} = \frac{\square}{\square} - \frac{\square}{\square} = \square$

1:25 Problems Using Fractions

1. If 18 basketball players were given $\frac{1}{2}$ an orange each at half-time, how many oranges would have to be purchased?
2. Greg had $\frac{3}{10}$ of his money invested in shares, $\frac{6}{10}$ was in the bank and the rest was in a safe at home. What fraction of his money was in the safe?
3. Maddy has 7 fish, 2 birds and 1 rabbit. What fraction of her pets are fish?
4. My mum and dad have 5 children: 3 boys and 2 girls. What fraction of the family are males?
5. There were 312 students who attended our school dance. This is exactly $\frac{1}{3}$ of my school. How many students belong to my school?
6. Our Student Council has 60 members. At our last meeting $\frac{4}{5}$ of the members were present. How many members were away?
7. Jack has a casual job and earns \$40 a week. Each pay he keeps $\frac{1}{4}$ of it and banks the rest. How much money does he keep?
8. Mum bought $\frac{1}{4}$ kg of cheese. It cost \$8.00 per kilogram. How much did she pay?
9. A petrol drum holds 100 litres when full. If the drum is now $\frac{3}{4}$ full, how many litres of petrol have been used?
10. On Saturday Tom stayed awake for 20 hours. If he worked for $\frac{3}{4}$ of this time, how many hours did he work?
11. The supermarket had 18 cartons of milk. We bought $\frac{1}{3}$ of the cartons. How many cartons were left?
12. If $\frac{1}{4}$ of my money is \$3.15, how much money do I have?
13. If $\frac{1}{10}$ of my footy cards is 23, how many cards do I have?
14. Sam decided to survey 75 Year 6 students. He found that $\frac{2}{3}$ of them liked maths the best. How many students liked maths best?
15. A bath is $\frac{3}{5}$ full. If the bath can hold 200 litres, how many litres are in the bath?
16. Fifty people were surveyed to find which coloured pencil they used the least. $\frac{1}{5}$ of them said green. How many people used green the least?

Fractions in the Environment

- Collect examples of fractions from newspapers and food packages.
- Discuss different ways in which fractions are used.

ACTIVITY

1:26 Finding Percentages

CONCEPT

To find a percentage of a quantity, write the percentage as a fraction. If we know $\frac{1}{10}$ of an amount we can multiply by 3 to find $\frac{3}{10}$ of the amount.

25% of $180

$= \frac{1}{4}$ of $180

= $180 ÷ 4

= $45

25% of $180 is $45.

30% of $1000

$\frac{1}{10}$ of $1000 = $100

$\frac{3}{10}$ of $1000 = 3 × $100

= $300

30% of $1000 is $300.

1 Find the discount on each item.

a sink

10% discount

Discount ______

b phone

10% off

Discount ______

c printer

25% discount

Discount ______

d computer

20% discount

Discount ______

2 Find the percentage of each quantity or amount.

a 10% of 40 ______ **b** 25% of 100 ______ **c** 50% of 90 ______

d 20% of 50 ______ **e** 40% of 80 ______ **f** 50% of 30 ______

g 20% of 200 ______ **h** 30% of 90 ______ **i** 25% of 300 ______

j 25% of $24 ______ **k** 20% of $40 ______ **l** 10% of $60 ______

m 50% of $18 ______ **n** 5% of $20 ______ **o** 40% of $70 ______

p 40% of $20 ______ **q** 30% of $50 ______ **r** 20% of $60 ______

Percentages on a Calculator

ICT

To find 25% of 20, enter [2] [0] [×] [2] [5] [%] [=] ______

25% = 0·25

(On some calculators the percentage will need to be entered as a decimal.)

3 Find:

a 30% of 180 ______ **b** 40% of 320 ______ **c** 15% of 500 ______

1:27 Finding Percentages

1 Find the percentage of each amount.

Remember to convert each percentage to a fraction.

a 10% of \$70		**b** 25% of \$32			
c 50% of \$30		**d** 20% of \$40			
e 50% of \$36		**f** 10% of \$50			
g 25% of \$24		**h** 10% of \$110		**i** 20% of \$35	

2 Find the percentage of each quantity.

a 20% of 45 boys		**b** 10% of 60 girls	
c 50% of 54 women		**d** 25% of 48 sheep	
e 10% of 140 pens		**f** 20% of 60 dogs	

3 Complete the table below.

a DVD

20% off

b bicycle

10% discount

c iron

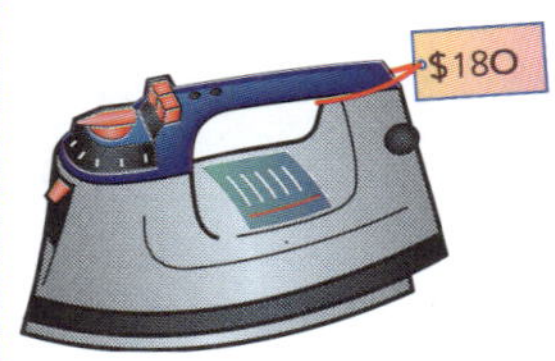

25% off

d CD player

5% discount

e kite

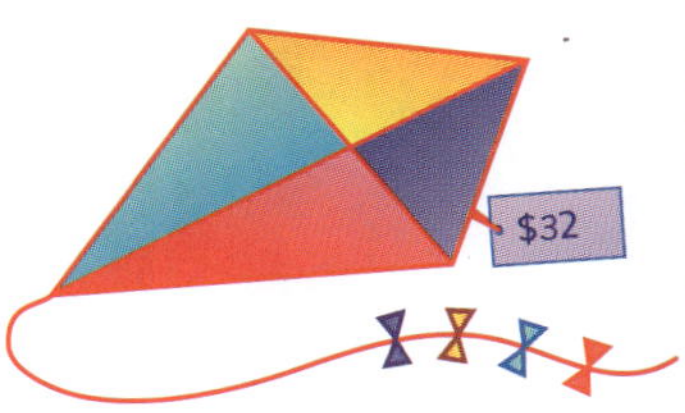

25% discount

f coffee maker

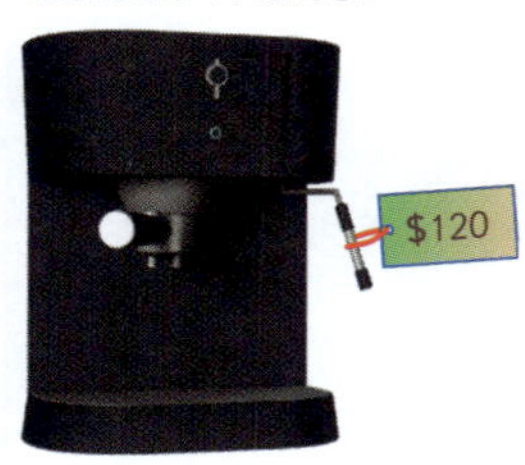

10% off

g camera

20% discount

h television

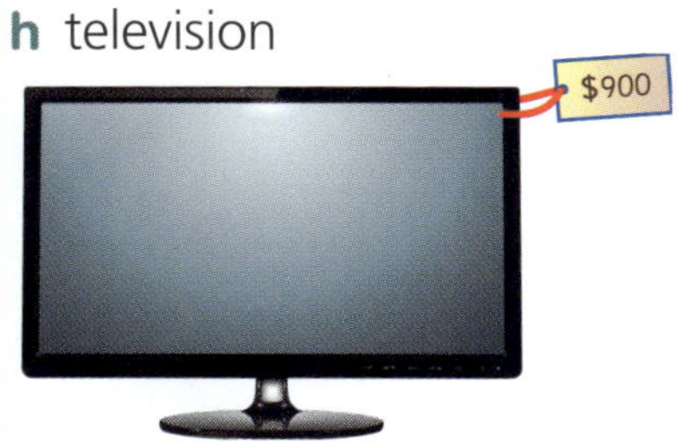

5% off

	a	b	c	d	e	f	g	h
Price	\$30							
Discount	\$6							
Discount Price	\$24							

Discounts on a Calculator

35% = 0·35

ICT

4 **a** To find 35% discount on \$90, enter [9] [0] [×] [3] [5] [%] [=]

b To find the discount price, enter [9] [0] [−] [9] [0] [×] [3] [5] [%] [=]

Make sure your calculator uses order of operations by checking answers to Question 3.

 ISBN 978 1 4886 2193 2

1:28 Prime and Composite Numbers

CONCEPT

- A **prime** number has only two factors, itself and 1.

  Example: 13 is a prime number. Its factors are 13 and 1.

- A **composite** number has more than two factors.

  Example: 9 is a composite number. Its factors are 9, 1 and 3.

In how many ways can I lay 12 tiles in rows and columns?

Here are all the possibilities:

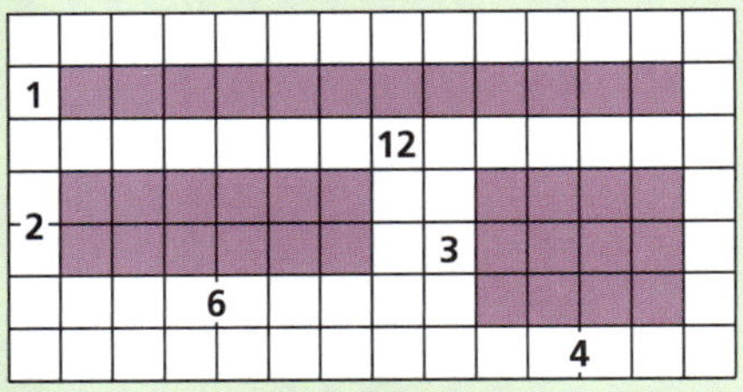

If only one pattern is possible, the number is a prime number.

The tiles above can be laid in three ways. The length of the sides are the factors of 12.
The factors of 12 are: 1, 12, 2, 6, 3 and 4.
12 is a multiple of each factor. 12 is a composite number.

1 The first pattern of tiles has been given for 6, 24, 20 and 7. Draw the other possible patterns.

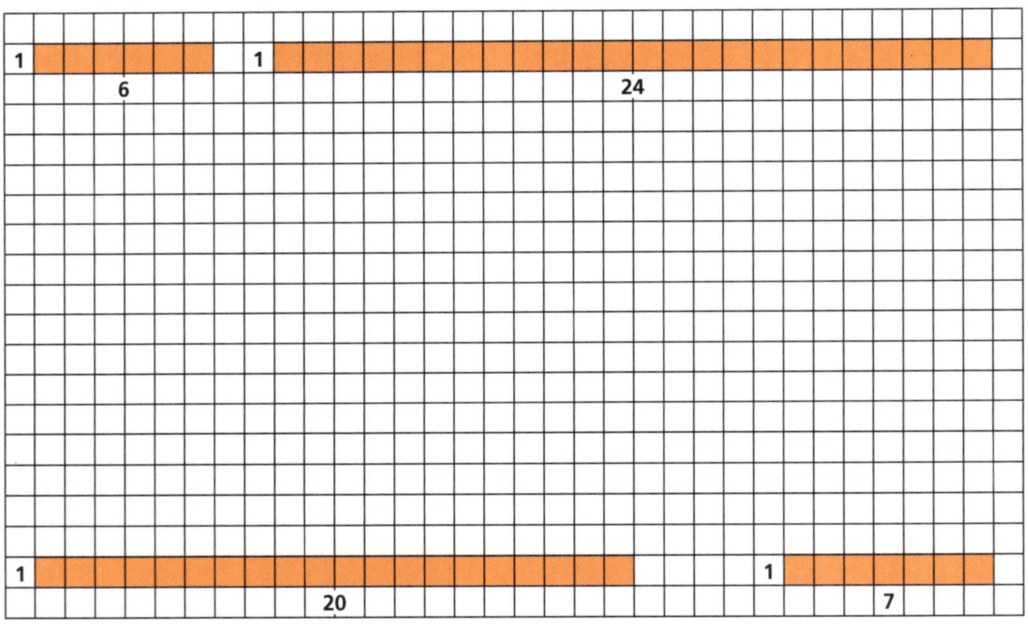

2 Use the patterns above to write all of the factors for:

a 6 ______ **b** 24 ______

c 7 ______ **d** 20 ______

3 **a** Which of the numbers in Question 2 is prime? ______

b Which of the numbers in Question 2 are composite? ______

c List the prime numbers that are less than 20. ______

 • *AUSTRALIAN SIGNPOST MATHS 6* • ISBN 978 1 4886 2193 2

1:29 Prime and Composite Numbers

A prime number has only 2 factors, itself and 1.

1 List all the factors of:

a 11 ______ **b** 8 ______ **c** 1 ______ **d** 9 ______

e 15 ______ **f** 13 ______ **g** 24 ______

Which of these numbers are:

h prime numbers? ______ **i** composite numbers? ______

2 List the factors of each number. Circle the prime numbers.

1				
2				
3				
4				
5				
6				

7						
8						
9						
10						
11						
12						

A composite number has more than two factors.

3 **a** List all of the prime numbers that are less than 30.

______

b List all of the composite numbers that are less than 30.

______

4 Complete these factor trees.

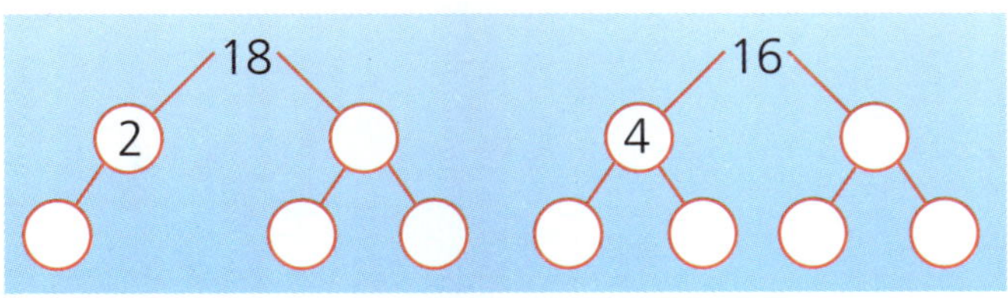

Factor trees help us write a number as a product of its prime factors.

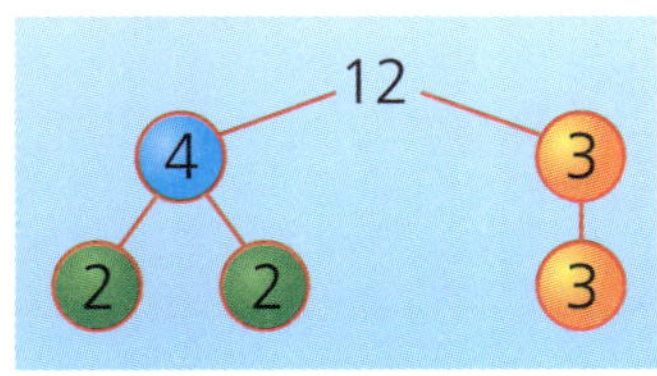

a 18 = ☐ × ☐ × ☐
= ☐

b 16 = ☐ × ☐ × ☐ × ☐
= ☐

INVESTIGATION

Writing numbers as products of prime factors can help us divide them.

A $210 \div 28 = \dfrac{\overset{1}{\cancel{2}} \times 3 \times 5 \times \overset{1}{\cancel{7}}}{\underset{1}{\cancel{2}} \times 2 \times \underset{1}{\cancel{7}}} = 15 \div 2 = 7\frac{1}{2}$

B $72 \div 48 = \dfrac{\overset{1}{\cancel{2}} \times \overset{1}{\cancel{2}} \times \overset{1}{\cancel{2}} \times \overset{1}{\cancel{3}} \times 3}{\underset{1}{\cancel{2}} \times \underset{1}{\cancel{2}} \times \underset{1}{\cancel{2}} \times 2 \times \underset{1}{\cancel{3}}} = 3 \div 2 = 1\frac{1}{2} = 1{\cdot}5$

Try these questions.

5 **a** $42 \div 24 = \dfrac{\square}{\square} = \square \div \square = \square$

b $78 \div 30 = \dfrac{\square}{\square} = \square \div \square = \square$

2:01 Multiplication Review

1 Complete these webs as quickly as you can. Learn any tables you get wrong.

a

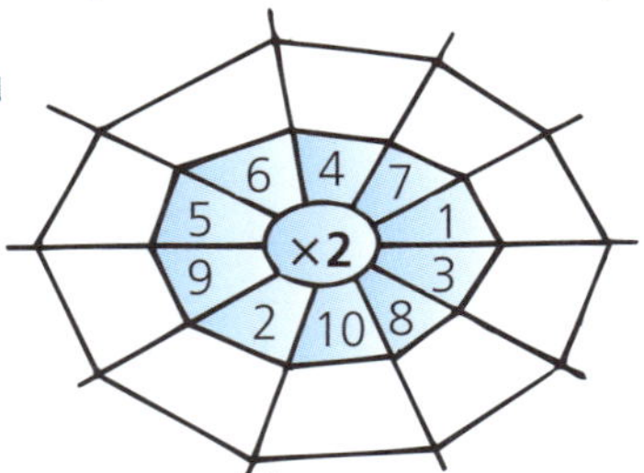

b

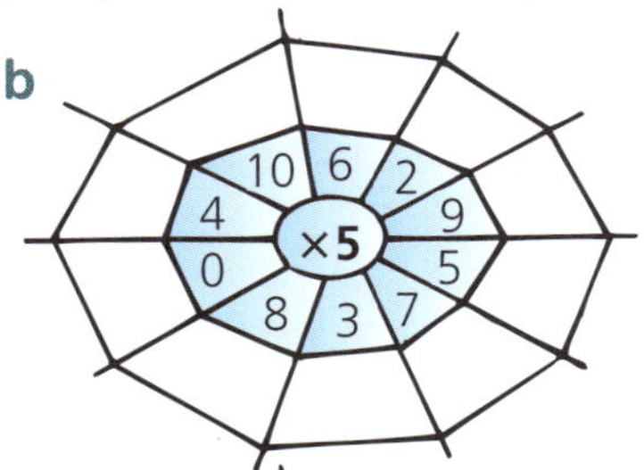

c

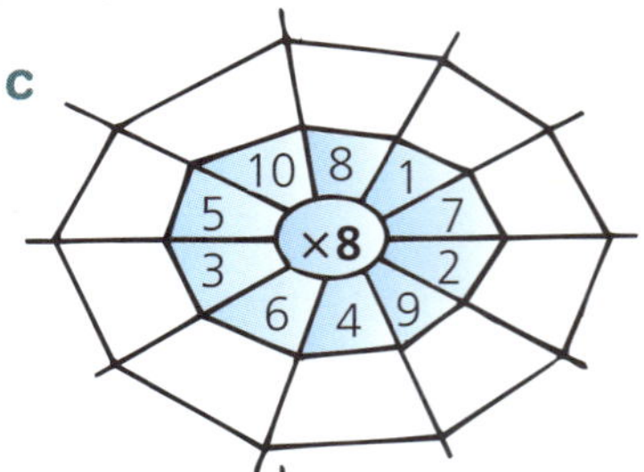

d

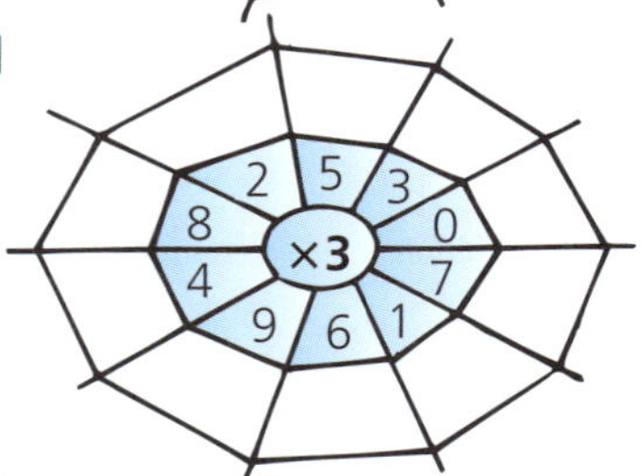

e

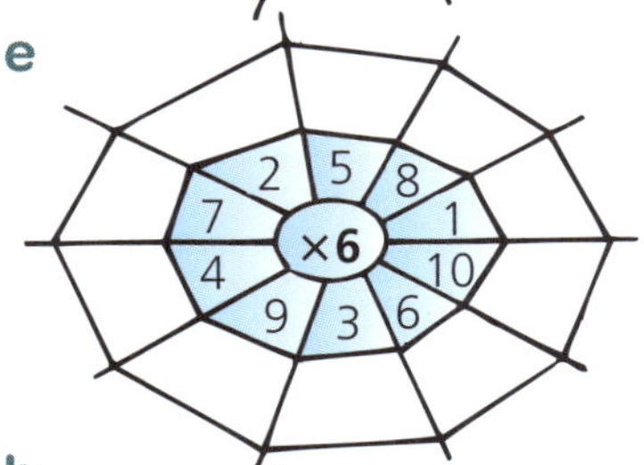

f

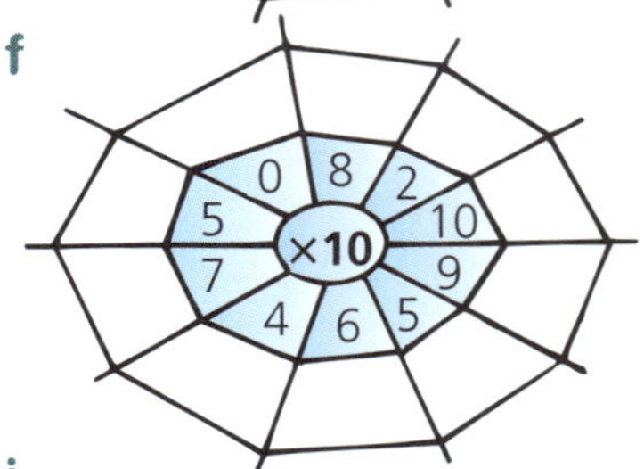

g

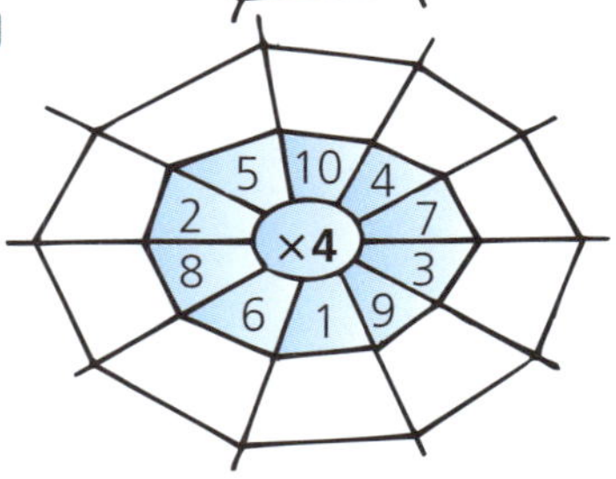

h

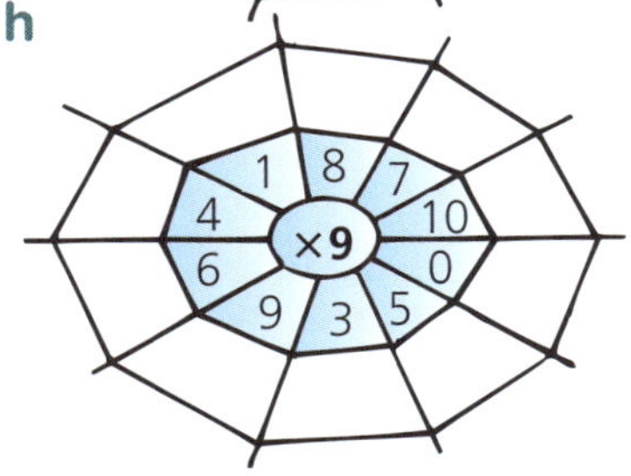

i

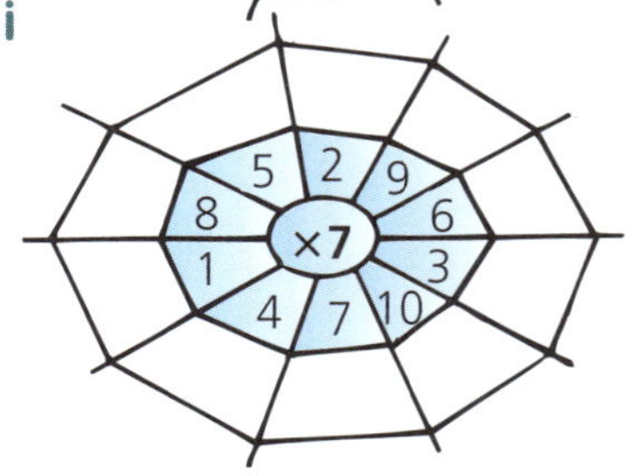

- A **multiple** is the answer when you multiply whole numbers.
- **Factors** of a number are whole numbers that multiply to give that number.
  The factors of 12 are 1, 12, 2, 6, 3 and 4.

CONCEPT

2 Write down all the factors of:

a 18 ______ **b** 49 ______

c 24 ______ **d** 13 ______

3 Write down the first eleven multiples of:

a 4 ______

b 9 ______

4 Complete:

a 3 squared = ___ **b** 8 squared = ___ **c** 10 squared = ___

d 49 = ___ squared **e** 1 = ___ squared **f** 25 = ___ squared

5 **True** or **false**?

a All the multiples of 3 that are less than 30 are odd numbers. ______

b All square numbers between 50 and 80 are even numbers. ______

 • *AUSTRALIAN SIGNPOST MATHS 6* • ISBN 978 1 4886 2193 2

2:02 Division

1. Find a fair share if these turtles were shared among:

   a 4 girls ☐ b 6 boys ☐ c 3 boys ☐ d 8 girls ☐

2. Find one share and the remainder if they are shared among:

   a 5 ponds ☐ b 7 ponds ☐

3. a How many groups of 3 turtles are there? ☐

   b How many groups of 4 turtles are there? ☐

4. How many groups of 13 are there and how many are left over?

   a 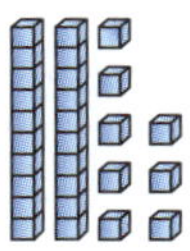 ☐ r ☐

   b 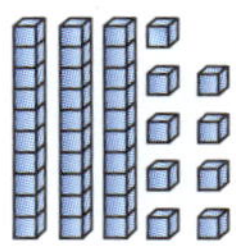 ☐ r ☐

   c 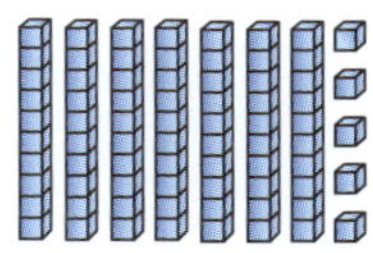 ☐ r ☐

5. Share each between four people. How many are there for each person and how many are left?

   a 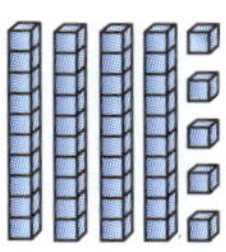 ☐ r ☐

   b 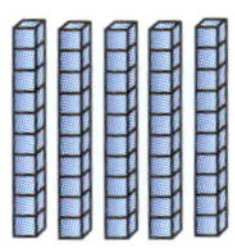 ☐ r ☐

   c 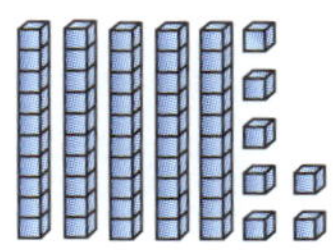 ☐ r ☐

6. Use the first number sentence to complete the other two.

a 7 × 8 = 56	b 9 × 6 = 54	c 8 × 6 = 48	d 7 × 9 = 63
56 ÷ 8 = ☐	54 ÷ 6 = ☐	48 ÷ 6 = ☐	63 ÷ 9 = ☐
56 ÷ 7 = ☐	54 ÷ 9 = ☐	48 ÷ 8 = ☐	63 ÷ 7 = ☐

7. a 3 × ☐ = 27 b 5 × ☐ = 35 c 4 × ☐ = 32 d 8 × ☐ = 16

   e ☐ × 7 = 28 f ☐ × 9 = 63 g ☐ × 8 = 64 h ☐ × 6 = 48

8. a 21 ÷ 3 = ☐ b 15 ÷ 5 = ☐ c 18 ÷ 6 = ☐

   d 16 ÷ 4 = ☐ e 40 ÷ 8 = ☐ f 90 ÷ 9 = ☐

   g 40 ÷ 10 = ☐ h 45 ÷ 9 = ☐ i 18 ÷ 3 = ☐

   j 25 ÷ 5 = ☐ k 20 ÷ 4 = ☐ l 36 ÷ 6 = ☐

   m 24 ÷ 4 = ☐ n 10 ÷ 10 = ☐ o 36 ÷ 4 = ☐

   p 27 ÷ 9 = ☐ q 24 ÷ 3 = ☐ r 81 ÷ 9 = ☐

   s 49 ÷ 7 = ☐ t 42 ÷ 6 = ☐ u 64 ÷ 8 = ☐

 • *AUSTRALIAN SIGNPOST MATHS 6* • ISBN 978 1 4886 2193 2

2:03 Addition Review

CONCEPT

Of light trucks registered in our study, 89 were blue and 9 193 were white. What was the total of these?

```
  1 1
9 1 9 3
+   8 9
-------
9 2 8 2
```

- Trade 10 ones for 1 ten.
- Trade 10 tens for 1 hundred.

Test your knowledge of last year's work.

1
- a) 14 + 550 + 248 + 93
- b) 245 + 76 + 193 + 157
- c) 214 + 106 + 97 + 233
- d) 9 + 20 + 188 + 624

2
- a) 6837 + 262
- b) 3086 + 2714
- c) 5307 + 1298
- d) 3846 + 2718

3
- a) 3076 + 297 + 4814
- b) \$20.08 + \$60.11 + \$18.91
- c) 6304 + 925 + 1045
- d) \$21.86 + \$19.37 + \$12.69

4
- a) 1086 + 193 + 2745 + 827
- b) 1843 + 3076 + 2184 + 1947
- c) 3827 + 938 + 1825 + 725
- d) 4386 + 827 + 99 + 2184

5
- a) 3074 + 6381
- b) 6038 + 2972
- c) 3186 + 5814

Use estimation to check all your answers.

6
- a) Of teenage drivers in our study, 1 573 were female and 2 679 were male. How many teenagers were there altogether?
- b) In the 80+ category there were 2 534 female drivers and 4 683 male drivers. How many drivers were 80 or older?

 • *AUSTRALIAN SIGNPOST MATHS 6* • ISBN 978 1 4886 2193 2

2:04 Subtraction Review

CONCEPT

Estimate before you calculate.

Of $7 000 received, $364 was paid in tax. How much was left?

$$
\begin{array}{r}
\ \ 9\ \ 9\ \ \\
6\ \ \not{1}\not{0}\ \not{1}\not{0}\ 10 \\
\$\ \ \not{7}\ \ \not{0}\ \ \not{0}\ \ \not{0} \\
-\ \$\ \ \ \ \ 3\ \ 6\ \ 4 \\
\hline
\$\ \ 6\ \ 6\ \ 3\ \ 6
\end{array}
\quad \text{OR} \quad
\begin{array}{r}
7000 = 6990 + 10 \\
6\ \ 9\ \ 9\ \ 10 \\
\$\ \ 7\ \ 0\ \ 0\ \ 0 \\
-\ \$\ \ \ \ \ 3\ \ 6\ \ 4 \\
\hline
\$\ \ 6\ \ 6\ \ 3\ \ 6
\end{array}
$$

$6636 was left.

1 a $875 - 38$ b $740 - 99$ c $482 - 49$ d $362 - 97$

2 a $600 - 193$ b $800 - 345$ c $300 - 93$ d $900 - 187$

3 a $6107 - 1836$ b $5318 - 967$ c $4093 - 3814$ d $3486 - 847$

4 a $2000 - 1084$ b $3000 - 915$ c $7000 - 3409$

2000 can be written as 1999 + 1.

5 a $1631 - 901$ b $7386 - 1836$ c $3543 - 667$

60 000 can be written as 59 999 + 1.

6 a $6000 - 1124$ b $3000 - 681$ c $7000 - 6936$

7 a Of 1 560 runners who began a race only 1 494 finished. How many dropped out? − ______

b I have run 3 285 m. How much further must I run to cover 4 200 m? − ______

 • *AUSTRALIAN SIGNPOST MATHS 6* • ISBN 978 1 4886 2193 2

2:05 Order of Operations

Do operations inside the brackets first.

$2 + (6 \times 3)$ ← brackets

We say:
$6 \times 3 = 18$ then
$2 + 18 = 20$.

CONCEPT

1 a $30 + (16 \div 4)$ b $23 - (14 + 6)$ c $4 + (5 \times 6)$
d $8 \div (16 - 12)$ e $30 - (30 - 10)$ f $46 - (19 + 11)$
g $(12 - 10) \times (20 - 5)$ h $(8 + 7) \div (13 - 8)$ i $6 \times (3 + 6 + 1)$

If only addition and subtraction are present, do the operations in order, going from left to right.

Example: $14 - 6 + 4 - 3$
$= 8 + 4 - 3$
$= 12 - 3$
$= 9$

CONCEPT

2 a $30 + 6 - 11$ b $21 - 12 + 8$ c $37 - 12 - 6$
d $60 + 14 - 4$ e $16 + 10 - 9$ f $33 - 10 + 3$
g $40 - 10 - 8 - 2$ h $42 + 8 + 9 - 7$ i $64 - 4 - 6 + 12$

If only multiplication and division are present, do the operations in order, going from left to right.

Example: $4 \times 5 \div 10 \times 3$
$= 20 \div 10 \times 3$
$= 2 \times 3$
$= 6$

CONCEPT

3 a $3 \times 4 \times 2$ b $9 \div 3 \times 5$ c $5 \times 6 \div 10$
d $12 \div 4 \times 6$ e $14 \div 7 \times 9$ f $20 \div 5 \times 3$
g $24 \div 6 \times 2 \div 4$ h $3 \times 8 \div 2 \div 3$ i $3 \times 2 \times 2 \div 6$

Order: 1 Brackets
2 × and ÷, left to right
3 + and –, left to right

Example: $15 - (12 - 2) \div 5$
$= 15 - 10 \div 5$
$= 15 - 2$
$= 13$

1 ( )
2 × ÷
3 + −

CONCEPT

4 a $3 + 7 \times 8$ b $20 - 3 \times 2$ c $9 - 18 \div 3$
d $15 + 4 \times 5$ e $16 - (13 - 8) \times 2$ f $6 \times 8 - (3 \times 7)$
g $6 + (8 - 3) \div 5$ h $18 + 20 \times 2 \div 5$ i $40 - 8 + 4 \times 2$

2:06 Order of Operations

CONCEPT

Use this order, when operations are mixed:

1. Do operations inside brackets.
2. Do × and ÷, going from left to right.
3. Do + and –, going from left to right.

$10 + 3 - (6 + 6) + 2 \times 2$
$= 10 + 3 - 12 + 2 \times 2$
$= 10 + 3 - 12 + 4$
$= 5$

1
- **a** $4 + 7 \times 3$
- **b** $(4 + 7) \times 3$
- **c** $10 - 6 - 4$
- **d** $18 - 6 \div 3$
- **e** $(18 - 6) \div 3$
- **f** $40 \div 4 - 2$
- **g** $8 + 23 + 7$
- **h** $8 + (23 + 7)$
- **i** $23 + 12 - 11$
- **j** $15 - 8 + 3$
- **k** $50 \div (5 \div 5)$
- **l** $6 \times 7 \times 10$

2
- **a** $56 - 20 \times 2$
- **b** $(56 - 20) \times 2$
- **c** $56 - (20 \times 2)$
- **d** $18 \times 9 \div 9$
- **e** $(18 \times 9) \div 9$
- **f** $18 \times (9 \div 9)$
- **g** $2 \times 49 - 15$
- **h** $2 \times (49 - 15)$
- **i** $60 \div 4 + 11$
- **j** $60 \div (4 + 11)$

3
- **a** $5 \times 6 + 4 \times 5$
- **b** $5 \times (6 + 4) \times 5$
- **c** $(5 \times 6 + 4) \times 5$
- **d** $48 \div 4 + 4 \div 2$
- **e** $48 \div (4 + 4 \div 2)$
- **f** $48 \div (4 + 4) \div 2$
- **g** $50 - (5 + 5 \div 5)$
- **h** $50 - 5 - (5 - 5)$
- **i** $15 + 30 \div (9 - 4)$
- **j** $(15 + 30) \div 9 - 4$

4
- **a** $80 - 5 - 5 - 5 - 5 - 5 - 5$
- **b** $80 - (5 + 5 + 5 + 5 + 5 + 5)$
- **c** $48 - 6 - 6 - 6 - 6 - 6 - 6$
- **d** $48 - (6 + 6 + 6 + 6 + 6 + 6)$
- **e** $64 \div 2 \div 2 \div 2 \div 2 \div 2 \div 2$
- **f** $64 \div (2 \times 2 \times 2 \times 2 \times 2 \times 2)$
- **g** $6 \times 5 + 3 \times 8 + 6 \times 8 - 100$
- **h** $40 \div 5 - 36 \div (8 + 1) - 8 \div 4$

Triangular Numbers

FUN SPOT

5 Chemists count pills in triangular pill trays. If the pills form a triangle, the number of pills is called a triangular number.

a How many pills are in each of these pill trays?

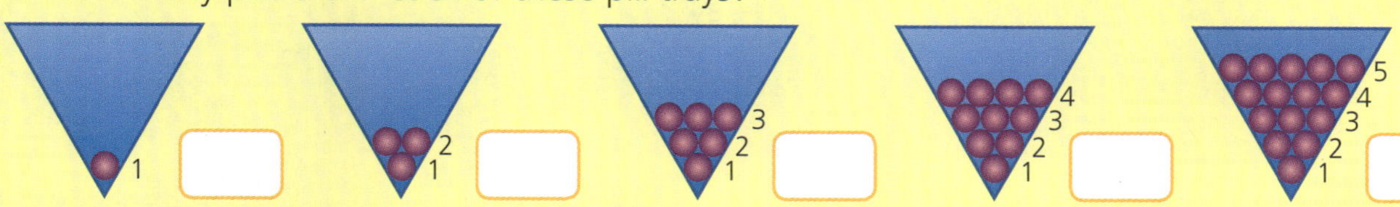

Each layer has one more than the layer below it.

b You have found the first five triangular numbers. Now find the next five.

2:07 Multiplying 10s, 100s and 1000s

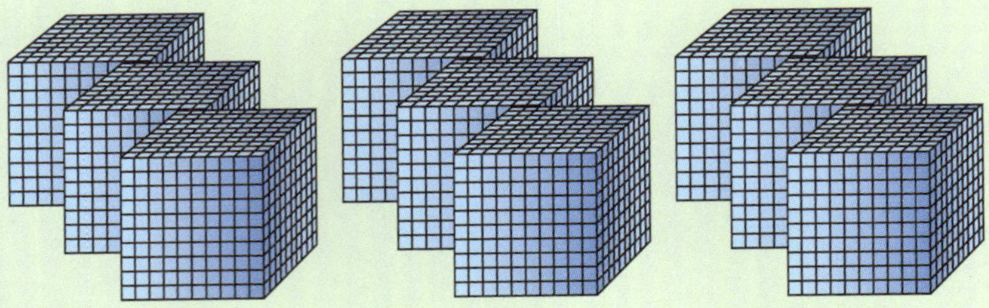

Th	H	T	U
3	0	0	0
×			3
9	0	0	0

1

a 4 × 2 tens = ___ tens

b 7 × 4 tens = ___ tens

c 9 × 5 tens = ___ tens

d 28 tens = ___ hundreds ___ tens

e 45 tens = ___ hundreds ___ tens

f 5 × 7 hundreds = ___ hundreds

g 8 × 4 hundreds = ___ hundreds

h 32 hundreds = ___ thousands ___ hundreds

i 7 × 6 thousands = ___ thousands

2

a 80 × 3

b 40 × 9

c 30 × 10

d 70 × 7

e 90 × 6

f 200 × 3

g 500 × 4

h 600 × 8

i 700 × 6

j 300 × 8

k 8000 × 7

l 6000 × 5

m 3000 × 6

n 5000 × 10

o 8000 × 9

3

a 4 packets of 50 toothpicks = ___ toothpicks

b 8 lots of 600 stamps = ___ stamps

c 7 stories of 3000 words = ___ words

d 7 days work, $6000 paid each day. Total paid = ___

e One computer costs $4000. The cost of 7 computers = ___

f One load of milk is 7000 L. Six loads of milk = ___

g Each night Alan used 6000 L of water on his new lawn. How much water did he use in 10 days? ___

h Alan used 8 packets of seed to plant his lawn. Each packet contained 5000 seeds. How many seeds did he use? ___

If 4000 seeds in each of 8 packets grew, how many grew altogether? ___ How many seeds did not grow? ___

 • *AUSTRALIAN SIGNPOST MATHS 6* • ISBN 978 1 4886 2193 2

2:08 Multiplication of Larger Numbers

CONCEPT

437 ties were sold for $5 each. How much was received for the ties altogether?

Extended Form

	4	3	7
×			5
		3	5
	1	5	0
2	0	0	0
2	1	8	5

Contracted Form

Th	H	T	U
	4	3	7
×	1	3	5
2	1	8	5

What is traded goes here.

- 5 × 7 ones make 35 ones. Write 5 in the units column and put a small 3 in the tens column.
- 5 × 3 tens plus 3 traded makes 18 tens. Write 8 in the tens column and a small 1 in the hundreds column.
- 5 × 4 hundreds plus the 1 traded is 21 hundreds. Write 21 hundreds as 2 thousands, 1 hundred.

$2185 was received altogether.

1

a 13 × 7   b 18 × 5   c 15 × 7   d 43 × 3   e 27 × 5

f 53 × 4   g 55 × 8   h 67 × 9   i 48 × 6

2

a 128 × 5   b 274 × 3   c 208 × 4   d 119 × 7   e 228 × 4

f 523 × 4   g 813 × 3   h 614 × 7   i 504 × 6   j 714 × 5

k $2.37 × 8   l $3.65 × 4   m $8.32 × 6   n $4.97 × 3   o $8.40 × 9

Investigating the Greatest Product

ACTIVITY

1 Throw four dice. Make a 1-digit and a 3-digit number using the numbers on the dice. Multiply these to get the product.

2 By rearranging, find all of the possible products.

3 How was the greatest product made?

4 Test your answer to **3** by throwing the dice again.

2 × 345	3 × 245
2 × 354	3 × 254
2 × 435	3 × 425
2 × 453	3 × 452
2 × 534	3 × ☐
2 × 543	3 × ☐

2:09 Mental Strategies for Subtraction

1 Use counting on to answer these.

a 39 – 35		**b** 40 – 38	
c 88 – 84		**d** 97 – 93	
e 71 – 68		**f** 79 – 71	
g 68 – 63		**h** 45 – 39	
i 60 – 56		**j** 52 – 49	
k 91 – 86		**l** 63 – 56	

Counting On

75 – 68 = ☐

Count on from 68:

"69, 70 and 5 more."

The answer is 7.

We could say 68 + ☐ = 75.

Counting Back

81 – 32 = ☐

Count back 30:

"71, 61, 51."

Count back 2:

"50, 49."

The answer is 49.

This is the jump strategy.

2 Use counting back to answer these.

a 78 – 22		**b** 95 – 45	
c 56 – 14		**d** 78 – 14	
e 344 – 31		**f** 365 – 31	
g 169 – 23		**h** 186 – 36	
i 281 – 12		**j** 102 – 4	
k 414 – 25		**l** 636 – 27	

3

a $\begin{array}{r} 84 \\ -\ 28 \\ \hline \end{array}$ **b** $\begin{array}{r} 71 \\ -\ 58 \\ \hline \end{array}$ **c** $\begin{array}{r} 62 \\ -\ 39 \\ \hline \end{array}$ **d** $\begin{array}{r} 45 \\ -\ 19 \\ \hline \end{array}$

e $\begin{array}{r} 60 \\ -\ 27 \\ \hline \end{array}$ **f** $\begin{array}{r} 42 \\ -\ 19 \\ \hline \end{array}$ **g** $\begin{array}{r} 37 \\ -\ 18 \\ \hline \end{array}$ **h** $\begin{array}{r} 87 \\ -\ 49 \\ \hline \end{array}$

164 – 38
Subtract 40 and then add 2.

4

a $\begin{array}{r} 137 \\ -\ 9 \\ \hline \end{array}$ **b** $\begin{array}{r} 514 \\ -\ 98 \\ \hline \end{array}$ **c** $\begin{array}{r} 815 \\ -\ 97 \\ \hline \end{array}$ **d** $\begin{array}{r} 246 \\ -\ 99 \\ \hline \end{array}$

e $\begin{array}{r} 665 \\ -\ 38 \\ \hline \end{array}$ **f** $\begin{array}{r} 987 \\ -\ 59 \\ \hline \end{array}$ **g** $\begin{array}{r} 534 \\ -\ 119 \\ \hline \end{array}$ **h** $\begin{array}{r} 560 \\ -\ 58 \\ \hline \end{array}$

Use rounding to check your answers.

5

a There were 368 people in the hall. 98 were dancing. How many were not dancing? ☐

b We had 352 fireworks. We used 246. How many were not used? ☐

c Heather has 170 birds. How many finches has she if 134 birds are not finches? ☐

2:10 Problem Solving

At the beginning of the year Spencer was 109 cm tall. At the end of the year he was 120 cm tall. How much had he grown during the year?

Setting out

Find: How much had he grown?

Number sentence:

120 – 109 = ☐

Answer:

He had grown 11 cm.

Working:

$$\begin{array}{rrr} & \overset{1}{\not{2}} & \overset{10}{\not{0}} \\ 1 & 2 & 0 \\ -\;1 & 0 & 9 \\ \hline & 1 & 1 \end{array}$$

CONCEPT

1 We had 37 jugs of soft drink at the party. Each one held 4 L. How much soft drink did we have?

Find: ☐
Number sentence: ☐
Answer: ☐
Working:

2 On Monday I spent $108, on Tuesday $272, on Wednesday $133 and on Thursday $95. What was the average spent each day?

Find: ☐
Number sentence: ☐
Answer: ☐
Working:

3 A carpet shop sold 230 m of carpet in the first week, 315 m in the second, 302 m in the third and 261 m in the fourth week of May. How many metres of carpet were sold during May?

Find: ☐
Number sentence: ☐
Answer: ☐
Working:

4 I went shopping with $450. When I returned I had $68. How much had I spent?

Find: ☐
Number sentence: ☐
Answer: ☐
Working:

5 How many tins would fit into 8 cartons if 64 tins fit into one carton?

Find: ☐
Number sentence: ☐
Answer: ☐
Working:

6 We travelled for three days. On the first day we travelled 413 km, on the second 380 km, and on the third day we travelled 123 km. How far did we travel?

Find: ☐
Number sentence: ☐
Answer: ☐
Working:

2:11 Division Review

In 9 litres of sea water we found 216 g of salt. How much salt would have been in one litre of this sea water?

$$\begin{array}{r} 24 \\ 9\overline{)2\,1\,^{3}6} \end{array}$$

In one litre of sea water there was 24 g of salt.

CONCEPT

Test your knowledge of last year's work.

1 a $5\overline{)44}$ b $4\overline{)27}$ c $7\overline{)40}$ d $6\overline{)37}$

2 a $3\overline{)36}$ b $6\overline{)66}$ c $2\overline{)84}$ d $3\overline{)63}$

3 a $3\overline{)57}$ b $4\overline{)56}$ c $5\overline{)85}$ d $8\overline{)96}$

4 a $3\overline{)642}$ b $6\overline{)528}$ c $5\overline{)825}$ d $8\overline{)440}$

5 a $7\overline{)849}$ b $9\overline{)574}$ c $4\overline{)717}$ d $3\overline{)280}$

6 a $10\overline{)370}$ b $10\overline{)583}$ c $10\overline{)800}$ d $10\overline{)614}$

7 a $3\overline{)612}$ b $5\overline{)535}$ c $4\overline{)828}$ d $6\overline{)642}$

8 a $4\overline{)601}$ b $7\overline{)842}$ c $4\overline{)900}$ d $6\overline{)842}$

9 a Paul found three shells of lengths 38 mm, 57 mm and 19 mm. What was the average length of the shells? ☐

b Mia raised $315 for charity, Josh raised $186, Zach $115, Naomi $220 and Erin $114.

What was the average amount raised? ☐

2:12 Division

Eight people can be seated at each table.
How many tables are needed for 263 people?

$8\overline{)26^{2}3}$ = 32 r 7

33 tables will be needed. A table will be needed to seat the remaining 7.

CONCEPT

1

a 7)40 b 5)35 c 9)77 d 4)34 e 10)93

f 5)75 g 6)84 h 5)90 i 6)90 j 3)87

2

a 2)484 b 3)360 c 5)555 d 6)606 e 4)844

f 4)504 g 5)665 h 7)938 i 3)702 j 8)976

3

a 6)939 b 4)700 c 5)752 d 7)900 e 8)264

f 10)600 g 10)350 h 10)740 i 10)300 j 10)720

k 4)123 l 8)856 m 3)902 n 5)535 o 9)273

4

a 357 eggs are put into cartons of six. How many cartons are filled?

b 357 cars are taken to town by truck. How many trips are needed if a truck can carry six cars?

c Six spark plugs are needed to service a car. How many cars can be serviced using 155 spark plugs?

d I used 938 mL of milk to fill seven identical containers. How much did I pour into each?

e Callum had $960. This is ten times as much as Holly. How much does she have?

2:13 Division Involving Fractions

Four students are to share 13 slices of toast. How many should each receive?

$\begin{array}{r} 3\text{ r }1 \\ 4\overline{)13} \end{array}$ **or** $\begin{array}{r} 3\frac{1}{4} \\ 4\overline{)13} \end{array}$

Each student should receive $3\frac{1}{4}$ (or 3·25) slices of toast.

CONCEPT

1 How much would each person receive if 4 students shared:

a 5 slices of toast? ☐ **b** 6 slices of toast? ☐ **c** 7 slices of toast? ☐

2 Each month 3 painters shared the painting of homes. How many homes should each person paint if altogether there were:

a 7 homes? ☐ **b** 11 homes? ☐ **c** 23 homes? ☐

3 Write each answer as a mixed number.

a $5\overline{)23}$ **b** $5\overline{)19}$ **c** $2\overline{)35}$ **d** $3\overline{)44}$

e $10\overline{)57}$ **f** $4\overline{)31}$ **g** $7\overline{)40}$ **h** $9\overline{)71}$

i $8\overline{)90}$ **j** $6\overline{)88}$ **k** $5\overline{)93}$ **l** $4\overline{)95}$

If the remainder can be shared, write the answer as a mixed number.

3 pies were shared among 10 teachers. How much did each teacher eat?

$\begin{array}{r} 0\frac{3}{10} \\ 10\overline{)3} \end{array}$ Each teacher ate $\frac{3}{10}$ of a pie.

4 **a** $5\overline{)3}$ **b** $4\overline{)2}$

c $10\overline{)7}$ **d** $3\overline{)1}$ **e** $8\overline{)5}$

5 Write each answer as a mixed number.

a $3\overline{)281}$ **b** $5\overline{)423}$ **c** $9\overline{)187}$ **d** $10\overline{)299}$

e $6\overline{)745}$ **f** $8\overline{)903}$ **g** $7\overline{)912}$ **h** $4\overline{)700}$

© PEARSON AUSTRALIA 2018 • *AUSTRALIAN SIGNPOST MATHS 6* • ISBN 978 1 4886 2193 2

2:14 Averages

- An average is a fair share.
- **Mean** is another name for average.
- To find the average **add** the scores then **divide** by the number of scores.

CONCEPT

Find the average of 12, 21, 19, 13 and 20.
Average = (12 + 21 + 19 + 13 + 20) ÷ 5
= 85 ÷ 5
= 17

17
5)8³5

Average = 17

Find the average of 18 cm, 50 cm and 28 cm.
Average = (18 + 50 + 28) ÷ 3
= 96 ÷ 3
= 32

32
3)96

Average = 32 cm

1 Find the average of each group of numbers.

a 38, 15, 22   3)

b 6, 5, 14, 8, 7   5)

c 4, 7, 3, 6, 10, 8, 4, 2, 7, 9

d 73, 29, 65, 43, 40, 52, 20

e 437, 863, 248, 675, 382

f 3 810, 1 280, 1 130, 1 250, 2 370

g 15, 25, 30, 29, 9, 18, 31, 28, 11, 29

h 12, 15, 18, 9, 11, 15, 6, 12

i 1 865, 4 017, 96, 843, 209

j 936, 405, 763, 804, 95, 327

k $38.40, $10.00, $6.20

l $7.30, $12.50, $3.90

CONCEPT

To find the average speed, divide the total distance travelled by the time taken.

$$\text{Speed} = \frac{\text{distance}}{\text{time}}$$

Example
I travelled from my home to Melbourne, a distance of 186 km, in 3 hours. What is my average speed?

$$\text{Average speed} = \frac{186\text{ km}}{3\text{ h}}$$

62
3)186

= 62 km/h

2 Find the average speed in each case.

a I travelled 450 km in 5 hours. ☐ km/h   5)450

b It took Ben 6 hours to travel 426 km. ☐ km/h

c We flew 822 km in 3 hours. ☐ km/h

d The snail moved 8·4 cm in 4 minutes. ☐ cm/min

e In 5 days we travelled 4 500 km. ☐ km/day

f Sound travels 1 675 m in 5 seconds. ☐ m/s

2:15 Addition to 999 999

At the last count Jock's two sheep stations had 374 295 and 439 015 sheep. How many sheep did he have altogether?

$$\begin{array}{r} {\scriptstyle 1\ 1\ \ \ 1\ 1\ \ } \\ 374295 \\ +\,439015 \\ \hline 813310 \end{array}$$

Altogether Jock had 813310 sheep.

374 thousands
295 ones
374295

CONCEPT

1

a 42300 + 19200

b 35321 + 11800

c 56684 + 9066

d 18957 + 61846

e 175800 + 94250

f 405312 + 6918

g 125094 + 609887

h 354500 + 267600

2

a 35256 + 9184 + 4215

b 8645 + 26314 + 83021

c 43400 + 48915 + 8335

d 75340 + 10659 + 10936

e 340000 + 75900 + 471550

f 87366 + 135344 + 218900

g 9560 + 28537 + 696315

h 650000 + 97000 + 165000

3

a 386915 + 9084 + 16121 + 103514

b 55950 + 4831 + 130090 + 7346

c 246176 + 164308 + 127351 + 263867

Use rounding to check your answers.

4

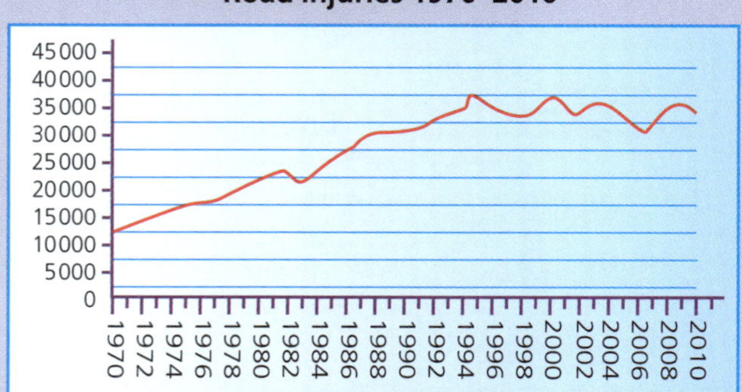

ICT

- This graph shows the number of road injuries in each year. (It could be drawn as a line of dots.)
- Find an approximation for the total number of injuries that occurred from:

a 1971 to 1980 ______

b 1991 to 2000 ______

c 2001 to 2010 ______

2:16 Subtraction to 999 999

HTh means hundred thousands.
TTh means ten thousands.

$^{1}2$ means 12.

CONCEPT

986 270 people left a country during a war, but 146 523 did not return. How many returned?

HTh	TTh	Th	H	T	U
		15			
	7	~~5~~	12	6	10
9	~~8~~	~~6~~	~~2~~	~~7~~	~~0~~
− 1	4	6	5	2	3
8	3	9	7	4	7

Another setting out:

9 ^{7}~~8~~ ^{15}~~6~~ ^{1}2 ^{6}~~7~~ ^{1}0
− 1 4 6 5 2 3
8 3 9 7 4 7

839 747 returned.

1

a 56 340 − 9160
b 87 115 − 33 485
c 64 009 − 21 333
d 32 457 − 8098
e 96 000 − 81 000
f 89 693 − 65 355
g 65 250 − 9346
h 22 760 − 897

2

a 193 583 − 27 638
b 345 350 − 18 181
c 560 499 − 20 716
d 783 615 − 9999
e 620 650 − 136 177
f 750 000 − 385 000
g 684 326 − 232 895

Ask: Does the answer make sense?

3

a Mr Rich wanted to buy a house advertised for sale at $895 000. He offered $819 900 and the offer was accepted. How much less than the advertised price did he pay?

b Mr Rich earned $555 360 last year but had to pay $257 285 in tax. How much did he have left?

c The house Mr Rich bought was on 12 hectares of land. 45 000 m^2 of the land was covered with trees. How many square metres were not covered by trees? (1 ha = 10 000 m^2)

d On this land, Mr Rich grew apple trees. This year he produced 112 000 apples. He sold 85 000 apples. How many were not sold?

e In February 2012 there were 842 300 part-time workers in Victoria. If there were 626 800 workers in Melbourne, how many were outside of Melbourne?

2:17 5-Digit Subtraction from Thousands

CONCEPT

Can you explain how these methods work?

Method 1 (trading):

```
      9  9
   8 10 10 10
 1 9  0  0  0
-   3  6  7  3
 1 5  3  2  7
```

or

$19\,000 = 18\,999 + 1$

```
  1 8 9 9 9 +1
-   3 6 7 3
  1 5 3 2 6 +1
```

→

```
   8 9 9 10
 1 9 0 0 0
-  3 6 7 3
 1 5 3 2 7
```

1

a $10\,000 - 4136$

b $10\,000 - 1423$

c $10\,000 - 8096$

d $10\,000 - 2401$

e $30\,000 - 7046$

f $20\,000 - 8016$

g $60\,000 - 2955$

h $50\,000 - 3641$

2

a $90\,000 - 18\,073$

b $80\,000 - 36\,724$

c $50\,000 - 19\,334$

d $80\,000 - 79\,424$

e $70\,000 - 847$

f $40\,000 - 943$

g $30\,000 - 29\,948$

Use rounding to check your answers.

3

a My odometer shows 34 000 km. When I left Alice Springs it showed 32 845 km. How far have I driven?

b My odometer now shows 36 385 km. How far must I travel before it shows 37 000 km?

c Mrs Cadwallader ordered 3 000 ice blocks for the canteen. When they arrived, 421 were missing. How many ice blocks arrived at the canteen?

d Our town ran a race of 30 km. When the leader had run 17 000 m, I had run only 15 175 m. How far was the leader in front of me?

e Claire and Marcus are saving to buy a house. Claire has saved $16 390 while Marcus has saved $18 000. How much more has Marcus saved?

How much money do they have altogether?

f An aircraft at a height of 37 000 m above sea level passed over a high altitude balloon that was floating at a height of 34 472 m. How far above the balloon was the aircraft?

 • *AUSTRALIAN SIGNPOST MATHS 6* • ISBN 978 1 4886 2193 2

2:18 Money

INVESTIGATION

Before money was used, people traded for things they needed.
Over time, most societies developed a form of money.
Research one of the following topics:

- When were coins first used?
- Where was paper money first used?
- What materials have been used for coins?

1

a
```
   $19.65
   $12.40
 + $16.32
```

b
```
   $19.43
   $77.60
   $ 8.35
 + $19.09
```

c
```
   $131.27
   $ 96.94
   $ 18.29
 + $  2.75
```

d
```
   $142.55
   $216.97
 + $ 35.62
```

2

a
```
   $39.60
 − $27.85
```

b
```
   $83.35
 − $64.90
```

c
```
   $132.50
 − $ 96.78
```

d
```
   $392.40
 − $187.65
```

e
```
   $402.05
 − $275.63
```

f
```
   $371.50
 − $183.76
```

g
```
   $502.31
 − $497.99
```

h
```
   $1316.40
 − $ 975.80
```

3 When we shop, the total of each bill is rounded off to the nearest five cents. Find the totals. Complete the table to show the change from $30 for each bill.

a	b	c	d	e
$4.97	$9.81	$3.26	$4.35	$4.93
$3.86	$4.37	$8.59	$3.67	$7.64
$5.73	$5.26	$6.96	$6.39	$5.89
$2.99	$7.95	$7.95	$4.26	$5.03
$6.57		$1.23	$7.18	$6.26

	a	b	c	d	e
Total					
Rounded off					
Change	$30.00 −	−	−	−	−

 • *AUSTRALIAN SIGNPOST MATHS 6* • ISBN 978 1 4886 2193 2

2:19 Travel Maths

Read this map for information. The numbers on the roads show the distances (in kilometres) between places or road junctions. Highway numbers are shown on shields.

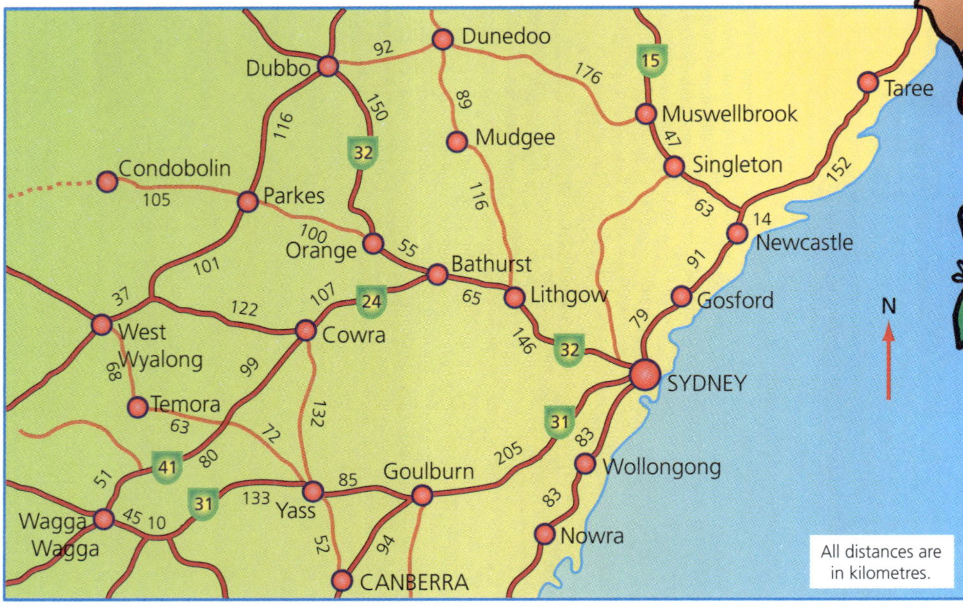

Estimate to check your answers.

1 Steven travelled to Dubbo from Sydney via Highway 32. His sister Wendy travelled to Dubbo from Newcastle passing through Singleton, Muswellbrook and Dunedoo. How far did each travel and what was the difference between these distances?

Steven ______ km Wendy ______ km Difference ______ km

2 What is the shortest distance from:

a Parkes to Cowra? ______ **b** Dubbo to Cowra? ______

c Wagga Wagga to Sydney? ______ **d** Dubbo to Lithgow? ______

e Parkes to Yass? ______ **f** Canberra to Lithgow? ______

3 **a** A traveller drives directly to Cowra from Wagga Wagga on Mondays, Wednesdays and Fridays, returning home to Wagga Wagga each night. How far does he travel on these trips in one week? ______

b If his car uses 1 litre of petrol for each 10 kilometres covered, how much petrol will be used in a week of travelling to Cowra and back? ______

c What would be the cost if each litre of petrol costs $1.60 cents? ______

4 During our last holiday, we travelled from Newcastle to Nowra. The trip took 6 hours. How far did we travel and what was the average speed? ______ ______

5 Between which towns is the distance 162 km? ______

2:20 Order of Operations

START

Do operations inside brackets.

↓

Do multiplication and division going from left to right.

↓

Do addition and subtraction going from left to right.

STOP

Example

8 × (3 + 4) – 12 ÷ (1 + 2)

Do brackets.

= 8 × 7 – 12 ÷ 3

Do multiplication and division.

= 56 – 4

Do addition and subtraction.

= 52

CONCEPT

1 a 3 × (5 + 5) ☐ b 6 – 3 × 2 ☐ c 7 + 16 ÷ 4 ☐

d 10 – (8 – 4) ☐ e (16 – 6) × 8 ☐ f (15 + 3) ÷ 6 ☐

g 14 – 6 × 2 ☐ h 18 + 10 ÷ 2 ☐ i 50 ÷ 5 × 5 ☐

j 20 ÷ 5 – 4 ☐ k 16 × 3 + 40 ☐ l 16 + 3 × 40 ☐

2 a 2 × 3 + 2 × 10 ☐ b 4 + 6 × 7 + 3 ☐

c 6 × 8 – 7 × 4 – 4 ☐ d 15 × (6 + 11 – 7) ☐

e (6 + 3) × 7 – 15 + 1 ☐ f 80 – (63 – 13) + 7 ☐

g 8 – 5 + 5 – 8 + 8 + 2 ☐ h 16 + 4 × 18 ÷ 2 – 4 ☐

i 150 + (0 ÷ 7) × 10 – 50 ☐ j 74 – 8 – 8 – (8 – 8) ☐

k 15 ÷ (3 × 5) + (3 × 5) ☐ l 50 + 16 ÷ (8 ÷ 8) – (6 × 8) ☐

Order:
1 ( )
2 × and ÷
3 + and –

3 Put in brackets to make each sentence true.

a 16 – 3 × 2 = 26 b 6 – 3 × 2 = 6 c 10 × 6 + 4 = 100

d 6 × 3 + 5 = 48 e 50 ÷ 5 + 5 = 5 f 70 ÷ 10 – 3 = 10

g 27 – 14 – 8 = 21 h 4 + 3 × 12 – 2 – 10 = 72

i 100 × 40 – 20 – 10 = 1 000

Example:
6 + 12 ÷ 3 + 1 = 9
Guess and check:
(6 + 12) ÷ 3 + 1 = 7 No
6 + 12 ÷ (3 + 1) = 9 Yes

A Bridge of Numbers

FUN SPOT

4 a Fill in these rectangles as if the solid black arrows mean ÷ 4 and the solid red arrows mean × 8.

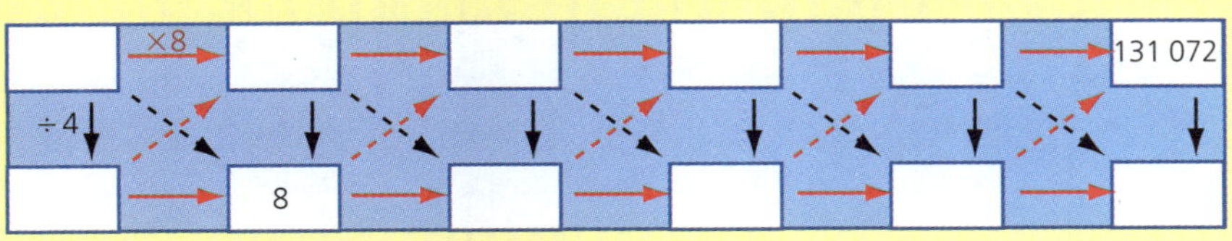

What operation is needed to link the boxes along:

b the black broken arrows? ☐

c the red broken arrows? ☐

2:21 Strategies with Decimals

- 1 = 10 tenths = 100 hundredths = 9 tenths and 10 hundredths
- 1 = 1·0 = 1·00 = 0·9 + 0·10

5·3 + 0·75 = 5 + 0·3 + 0·7 + 0·05
= 5 + 1 + 0·05
= 6·05

CONCEPT

1 Complete:

a 0·7 + 0·3 = ☐	**b** 0·07 + 0·93 = ☐	**c** 0·08 + 0·02 = ☐
d 0·1 + 0·9 = ☐	**e** 0·05 + 0·95 = ☐	**f** 0·05 + 0·05 = ☐
g 0·6 + 0·4 = ☐	**h** 0·09 + 0·01 = ☐	**i** 0·09 + 0·91 = ☐

2 Try these.

a 1·03 + 0·97 = ☐	**b** 1·6 + 0·4 = ☐	**c** 1·06 + 0·94 = ☐
d 1·71 + 0·29 = ☐	**e** 1·09 + 0·91 = ☐	**f** 1·4 + 0·60 = ☐

3 Complete these.

a 2·4 + 0·6 = ☐	**b** 3·7 + 0·3 = ☐	**c** 2·1 + 0·9 = ☐
d 3·71 + 0·39 = ☐	**e** 2·3 + 0·7 = ☐	**f** 2·05 + 0·05 = ☐

ICT

4 Using a calculator, press these keys:

100 [÷] 10 [=] ☐ [÷] 10 [=] ☐ [÷] 10 [=] ☐ [÷] 10 [=] ☐ [÷] 10 [=] ☐

5 Check your answers to Questions 1, 2 and 3 using a calculator.

6 Use a calculator to answer these.

a $8.50 + $9.00 ☐
b $19.20 + $0.80 ☐
c $8.00 + $3.70 ☐
d $4.85 + $2.05 ☐
e $9.63 + $4.17 ☐

- $2.00 + $2.70 is entered as
  [2] [+] [2] [·] [7] [=]
  = $4.70.
- $2.30 + $4.70 is entered as
  [2] [·] [3] [+] [4] [·] [7] [=]
  = $7.

4·7 = 4·70

2:22 Addition of Decimals

CONCEPT

Felicity drank 2·857 L of water. James drank 3·375 L.
How much did they drink altogether?

	U		Tth	Hth	Thth	
	1		1	1		
	2	·	8	5	7	L
+	3	·	3	7	5	L
	6	·	2	3	2	L

We trade 10 thousandths for 1 hundredth, then trade 10 hundredths for 1 tenth, then trade 10 tenths for a one.

1

a

U	Tth	Hth	Thth
2 ·	4	1	3
+ 1 ·	3	7	8

b

U	Tth	Hth	Thth
3 ·	5	9	2
+ 0 ·	1	3	5

c

U	Tth	Hth	Thth
6 ·	8	4	2
+ 1 ·	3	0	5

d

U	Tth	Hth	Thth
4 ·	6	9	3
+ 7 ·	2	0	5

e

U	Tth	Hth	Thth
1 ·	7	6	1
+ 4 ·	8	0	9

f

U	Tth	Hth	Thth
6 ·	0	3	7
+ 3 ·	3	8	5

g

U	Tth	Hth	Thth
9 ·	5	0	8
+ 3 ·	6	8	1

h

U	Tth	Hth	Thth
7 ·	3	9	2
+ 2 ·	0	5	7

i

U	Tth	Hth	Thth
6 ·	9	8	3
+ 7 ·	5	4	2

j

U	Tth	Hth	Thth
1 ·	6	0	2
+ 3 ·	7	7	5

k

U	Tth	Hth	Thth	
7 ·	9	1	1	L
+ 3 ·	6	8		L

l

U	Tth	Hth	Thth	
4 ·	6	9	9	m
+ 6 ·	7	3		m

m

U	Tth	Hth	Thth	
5 ·	9	9	9	km
+ 3 ·	7	2	1	km

n

U	Tth	Hth	Thth	
6 ·	8	3	9	L
+ 3 ·	5	8	2	L

o

U	Tth	Hth	Thth	
7 ·	8	0	4	m
+ 0 ·	9	9	9	m

2 Estimate to check all your answers.

FUN SPOT

1 whole = $\frac{1000}{1000}$

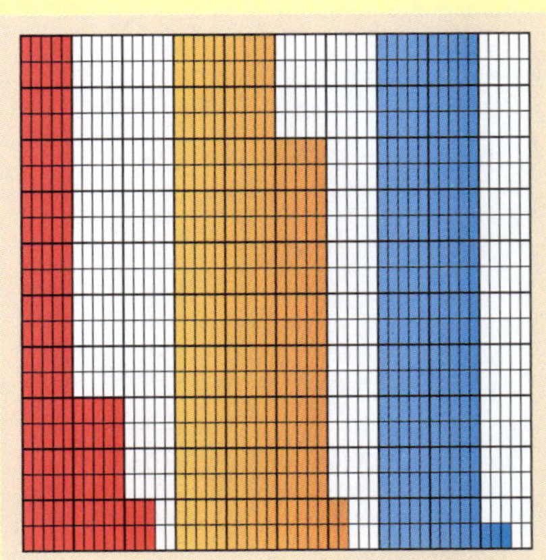

3 What decimal is represented on this decimal square by the colour:

a red? ☐

b blue? ☐

c orange? ☐

What decimal is represented by the total of:

d red and orange? ☐

e red and blue? ☐

f orange and blue? ☐

 • *AUSTRALIAN SIGNPOST MATHS 6* • ISBN 978 1 4886 2193 2

2:23 Addition of Decimals

CONCEPT

Keep the points under points.

```
  3·19          1
  7·5       3·190
  9·237  →  7·500
  ____      9·237
           ______
           19·927
```

Zeros at the end of decimals don't change values.

3·14 = 3·140

1

a	b	c	d	e
5·6 2·91 + 7·35	4·71 8·39 + 6·5	7·3 10·42 + 6·7	9·25 8·4 + 6·92	27·64 48·39 + 39·8

f	g	h	i	j
3·92 7·6 +18·39	5·4 6·95 + 7·87	7·62 12·34 + 48·5	13·4 9·68 + 4·97	63·75 9·6 + 15·96

2

a	b	c	d	e
6·39 2·715 + 8·68	9·6 8·347 + 7·694	7·306 8·59 + 4·7	10·4 3·965 + 25·73	38·5 9·639 + 12·58

f	g	h	i	j
7·245 13·4 + 6·98	11·09 6·3 + 9·467	9·06 35·307 + 10·865	73·015 9·6 + 28·39	67·49 14·7 + 2·643

1

a Emily bought 2·5 L of juice. Sophie bought 1·125 L.
How much did they buy altogether?

b Rahul ran 3·4 km on Tuesday and 6·75 km on Thursday.
How far did he run?

Estimate to check your answers.

c Samson joined 2 pieces of wood 34·72 cm and 25·69 cm long.
What is the total length?

d Joy bought 1·15 kg of turkey and 3·725 kg of lamb.
What was the total weight of the meat?

e On one roll, we have 3·85 m of material. On another we have 7·5 m.
How much material do we have altogether?

2:24 Subtraction of Decimals

$$\begin{array}{r} 16.1 \\ -\ 9.643 \\ \hline \end{array} \rightarrow \begin{array}{r} 16.100 \\ -\ 9.643 \\ \hline 6.\ \ \ 7 \end{array}$$

CONCEPT

1

a $16.6 - 8.4$

b $31.46 - 25.99$

c $72.54 - 56.75$

d $99.214 - 86.831$

e $82.3 - 79.6$

f $96.5 - 38.72$

g $102.3 - 99.68$

h $47.326 - 29.6$

i $462.786 - 186.53$

j $275.413 - 196.8$

k $35.621 - 19.835$

l $95.204 - 67.638$

2

a Sarah ran 6·752 km. Casey ran 8·259 km. How much further did Casey run?

b Penny had 35·729 L of water in the bath. She let out 12·53 L. How much water was left?

c Tim had 5·942 kg of flour. He used 3·753 kg. How much does he have left?

d Jake had a piece of wood 1·723 m long. He cut off 0·79 m. How much did he have left?

ACTIVITY

- Subtract 2 decimal numbers from the number 1. The aim is to reach a final answer that is as close to 0 as possible.
- Each time, the number you take away will be made by rolling 3 dice. Use the 3 numbers you roll to create each decimal number. If you roll 2, 5 and 6 you could make 0·256, 0·265, 0·526, 0·562, 0·625 or 0·652.

 • *AUSTRALIAN SIGNPOST MATHS 6* • ISBN 978 1 4886 2193 2

2:25 Multiplying Thousands

	TTh	Th	H	T	U
		1	6	0	9	5
×	3		5	3	6
	9	6	5	7	0

CONCEPT

Use the method above to answer these.

1
- a 1346 × 6
- b 1460 × 5
- c 2516 × 3
- d 2307 × 4
- e 1175 × 8

2
- a 8250 × 3
- b 8417 × 4
- c 5117 × 4
- d 6209 × 4
- e 4237 × 3
- f 9350 × 7
- g 4965 × 8
- h 7385 × 3
- i 3385 × 4
- j 5980 × 6

3
- a 11200 × 5
- b 28350 × 3
- c 30125 × 2
- d 13617 × 6
- e 20735 × 4
- f 18399 × 4
- g 15000 × 6
- h 13847 × 5
- i 25000 × 3
- j 18978 × 5

4
- a There are 3600 seconds in one hour. How many are in four hours?
- b There are 1440 minutes in a day. How many minutes are in one week?
- c There are 1461 days in 4 years. How many days had Rachel lived on her twelfth birthday?
- d On average, there are 8766 hours in a year. How many hours will Heather have lived on her eighth birthday?

FUN SPOT

- Calculate the number of years, months, weeks, days, hours and seconds that you have lived so far. Today my age was:

 ____ years, ____ months, ____ days, ____ hours, ____ minutes

ISBN 978 1 4886 2193 2

2:26 Multiplication of Decimals

Multiplying by a Whole Number

6 × 0·3 is 6 lots of three tenths
= 18 tenths
= 1·8

The same number of figures are after the decimal point in the answer as in the question.

$$\begin{array}{r} 0\cdot3 \\ \times\ {}^{1}\ \ 6 \\ \hline 1\cdot8 \end{array}$$

CONCEPT

1
- **a** 3 × 0·2 ☐ **b** 2 × 0·4 ☐ **c** 3 × 0·3 ☐ **d** 4 × 0·2 ☐
- **e** 6 × 0·5 ☐ **f** 5 × 0·3 ☐ **g** 9 × 0·6 ☐ **h** 7 × 0·4 ☐
- **i** 8 × 0·7 ☐ **j** 9 × 0·9 ☐ **k** 8 × 0·8 ☐

3·0 = 3

2
- **a** 0·5 × 3 ☐ **b** 0·1 × 7 ☐ **c** 0·2 × 4 ☐
- **d** 0·7 × 2 ☐ **e** 0·9 × 3 ☐ **f** 0·8 × 7 ☐
- **g** 0·4 × 9 ☐ **h** 0·6 × 5 ☐ **i** 0·8 × 5 ☐

3

a $\begin{array}{r} 1\cdot6 \\ \times\quad 2 \\ \hline \end{array}$ **b** $\begin{array}{r} 2\cdot4 \\ \times\quad 3 \\ \hline \end{array}$ **c** $\begin{array}{r} 6\cdot1 \\ \times\quad 5 \\ \hline \end{array}$ **d** $\begin{array}{r} 2\cdot7 \\ \times\quad 4 \\ \hline \end{array}$

e $\begin{array}{r} 2\cdot8 \\ \times\quad 5 \\ \hline \end{array}$ **f** $\begin{array}{r} 7\cdot4 \\ \times\quad 5 \\ \hline \end{array}$ **g** $\begin{array}{r} 9\cdot2 \\ \times\quad 5 \\ \hline \end{array}$ **h** $\begin{array}{r} 4\cdot5 \\ \times\quad 4 \\ \hline \end{array}$

i $\begin{array}{r} 7\cdot3 \\ \times\quad 8 \\ \hline \end{array}$ **j** $\begin{array}{r} 6\cdot5 \\ \times\quad 7 \\ \hline \end{array}$ **k** $\begin{array}{r} 4\cdot7 \\ \times\quad 6 \\ \hline \end{array}$ **l** $\begin{array}{r} 8\cdot3 \\ \times\quad 9 \\ \hline \end{array}$

$$\begin{array}{r} 7\cdot5 \\ \times\ {}^{1}\quad 3 \\ \hline 22\cdot5 \end{array} \qquad \begin{array}{r} 8\cdot6 \\ \times\ {}^{3}\quad 5 \\ \hline 43\cdot0 \end{array}$$
or 43

4
- **a** How far did I travel if I drove 8·5 km each day for 5 days? ☐
- **b** What is the total length of rope if each of 5 ropes is 9·2 m long? ☐
- **c** Alan drinks 0·8 L of milk each day. How much is drunk in 7 days? ☐
- **d** What is the cost of 5 yachts if each costs 2·8 million dollars? ☐

5 Use a calculator to complete these questions.
- **a** 7 × 165·3 L ☐ **b** 9 × 27·8 m ☐
- **c** 6 × 111·9 kg ☐ **d** 76·5 × 5 L ☐
- **e** 8 × 44·4 kg ☐ **f** 4 × 156·8 t ☐

ICT

2:27 Multiplication of Decimals

CONCEPT

$$\begin{array}{r} 3\cdot17 \\ \times\quad {}_{2}4 \\ \hline 12\cdot68 \end{array}$$

1
- **a** $2\cdot13 \times 3$
- **b** $6\cdot72 \times 4$
- **c** $12\cdot96 \times 6$
- **d** $23\cdot07 \times 8$
- **e** $35\cdot84 \times 9$

2
- **a** $10.46 × 5
- **b** $12.37 × 6
- **c** $18.96 × 8
- **d** $132.59 × 7
- **e** $356.37 × 9

3 Find the cost of:
- **a** 5 litres of milk
- **b** 4 boxes of cereal
- **c** 8 jars of jam
- **d** 12 litres of ice cream
- **e** 3 kg cheese and 4 L milk
- **f** 3 jars of jam and 5 kg of cheese
- **g** 6 kg of grapes

$5.90

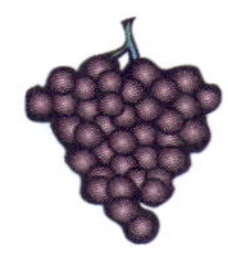
$4.90/kg

$2.70

$3.75

$3.75

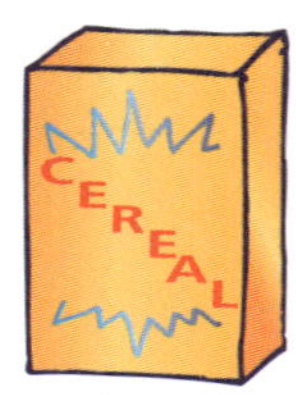

$6.50

CONCEPT

Find the total length of 6 pencils if each is 12·5 cm long.

$$\begin{array}{r} 12\cdot5 \\ \times\ {}_{1}\ {}_{3}\quad 6 \\ \hline 75\cdot0 \end{array}$$

Total length is 75 cm.

Find the total mass of 9 chairs if each has a mass of 4·125 kg.

$$\begin{array}{r} 4\cdot125 \\ \times\ {}_{1}\ \ {}_{2}\ {}_{4}\quad 9 \\ \hline 37\cdot125 \end{array}$$

Total mass is 37·125 kg.

4
- **a** $1\cdot2 \times 5$
- **b** $1\cdot537 \times 4$
- **c** $2\cdot638 \times 3$
- **d** $1\cdot7 \times 6$
- **e** $2\cdot3 \times 8$
- **f** $2\cdot6 \times 7$

5
- **a** Find the total length of 8 pencils if each pencil has a length of 13·8 cm.
- **b** Find the total mass of 7 chairs if each chair has a mass of 3·845 kg.
- **c** A glass holds 0·325 L. How much would 6 of these glasses hold?

 • *AUSTRALIAN SIGNPOST MATHS 6* • ISBN 978 1 4886 2193 2

2:28 Multiplication of Decimals

1
a $3 \cdot 1 \times 4$
b $6 \cdot 2 \times 7$
c $9 \cdot 7 \times 8$

Example: $2 \cdot 3 \times 4 = 9 \cdot 2$ (carry 1)

d $27 \cdot 64 \times 5$
e $36 \cdot 59 \times 9$
f $42 \cdot 75 \times 6$

g $65 \cdot 04 \times 7$
h $53 \cdot 6 \times 6$
i $85 \cdot 03 \times 9$
j $37 \cdot 93 \times 8$
k $93 \cdot 2 \times 4$

2
a $3 \cdot 246 \times 4$
b $1 \cdot 953 \times 3$
c $2 \cdot 869 \times 5$
d $4 \cdot 126 \times 7$
e $5 \cdot 073 \times 6$

f $2 \cdot 305 \times 4$
g $4 \cdot 096 \times 7$
h $1 \cdot 738 \times 5$
i $2 \cdot 007 \times 9$
j $3 \cdot 105 \times 6$

3
a $13·65 × 7
b $27.05 × 5
c $45.80 × 6

d $120.63 × 4
e $195.07 × 3
f $201.36 × 9

Multiply money in the same way.

3
a $310 \cdot 4$ L $\times 5$
b $167 \cdot 2$ m $\times 8$
c $21 \cdot 625$ kg $\times 7$

d $328 \cdot 3$ cm $\times 4$
e $512 \cdot 6$ g $\times 6$
f $80 \cdot 148$ km $\times 5$

Make $10

FUN SPOT

- Players take turns to roll a dice and multiply the number shown by 30.
- That number of cents is collected from a bank.
- The game continues until a player has $10.

$4 \times 30 = 120$

 • *AUSTRALIAN SIGNPOST MATHS 6* • ISBN 978 1 4886 2193 2

2:29 Division of Numbers Larger than 999

How many sets of six quoits can be made from 3891 quoits?

$$\begin{array}{r} 6\,4\,8 \text{ r } 3 \\ 6\overline{)3\,8^{2}9^{5}1} \end{array}$$

648 sets can be made.

CONCEPT

1 These have no remainder.

a 3)1584 **b** 5)3615

c 6)4434 **d** 4)1836

e 8)9496 **f** 6)8634

1 Once you begin, put a figure above each number. Show remainders.

a 4)2 3 1 8 **b** 5)3 3 8 6 **c** 7)6 1 4 5 **d** 9)1 4 8 6

e 4)7 3 1 0 8 **f** 3)8 4 6 7 2 **g** 6)8 1 4 7 4 **h** 4)6 9 2 6 0

2 **a** $7185 was shared among five workers. How much did each receive?

b Don Bradman scored 2429 runs in one season of cricket, for a batting average of 115·66 runs per innings. Don scored seven times as many runs as George. How many did George score?

c How many six-page books can be made from 2946 pages?

d Four sheets of paper were handed to each person in the hall. If 2128 sheets were given out, how many people were present?

e Five truck tyres cost $1835. What is the cost of one tyre?

f Eight people shared 3 L of orange juice. What is a fair share?

g Mia placed 1365 books on five bookshelves, with the same number of books on each shelf. How many books were on each shelf?

3 Find the missing digits by multiplying, then check to justify your answer.

a 1 8 0 1 over 4) **b** 3 1 5 6 over 2) **c** 6 1 8 over 5)

d 1 4 2 5 over 3) **e** 1 7 3 2 over 5) **f** 2 1 3 8 over 3)

g 9 2 5 over 4) **h** 6 2 4 over 7) **i** 1 0 8 6 over 9)

 • AUSTRALIAN SIGNPOST MATHS 6 • ISBN 978 1 4886 2193 2

2:30 Division with Zeros in the Answer

CONCEPT

Harry paid $8 136 for nine tickets. How much did he pay for one ticket?

```
    9 0 4
9)8 1 3³6
```

Harry paid $904 for one ticket.

How many gift packs of four belts can Fred make from 8 123 belts?

```
   2 0 3 0 r 3
4)8 1¹2 3
```

Fred can make 2 030 gift packs. 3 belts remain.

1 Don't forget the zeros as you do these.

a 2)1 2 1 2	**b** 4)3 2 0 8	**c** 3)2 1 2 4	**d** 8)1 6 5 6
e 5)5 0 2 5	**f** 3)6 1 2 0	**g** 4)8 3 6 0	**h** 7)7 1 4 7
i 7)9 4 5 0	**j** 4)8 2 8 4	**k** 6)7 2 5 4	**l** 5)5 0 3 5

2 Some of these will have remainders.

a 2)1 0 8 1	**b** 3)1 2 0 1	**c** 5)3 5 0 4	**d** 4)2 4 0 2
e 7)8 4 0 6	**f** 4)4 1 4 1	**g** 6)9 0 0 2	**h** 5)7 0 1 4
i 9)9 0 1 8 0	**j** 6)7 2 1 3 2	**k** 5)5 1 5 2 5	**l** 8)9 0 4 1 6

3 **a** The same number of sheep was put on six trains. How many sheep were put on each train if 13 098 sheep were sent altogether?

b $8416 was earned by eight students working together. How much did each earn if the money was shared fairly?

c How much water would be given to each person if eight people shared 5 L?

d Old John has lived for 18 928 days. How many weeks is that? How many years has he lived? (Take 52 weeks = 1 year.)

 • *AUSTRALIAN SIGNPOST MATHS 6* • ISBN 978 1 4886 2193 2

2:31 Division of Large Numbers by 10

CONCEPT

Brendan places 10 stamps on each page. How many pages will 2856 stamps fill?

$$\begin{array}{r} 285 \text{ r } 6 \\ 10\overline{)2\,8^{8}5^{5}6} \end{array}$$

285 pages will be filled. 6 stamps remain.

Change 3457 mm into centimetres. (Note: 10 mm = 1 cm)

$$\begin{array}{r} 345\tfrac{7}{10} \\ 10\overline{)3\,4^{4}5^{5}7} \end{array}$$

$345\frac{7}{10}$ cm (or 345·7 cm)

1

a $10\overline{)8352}$	**b** $10\overline{)1865}$	**c** $10\overline{)3648}$	**d** $10\overline{)9014}$
e $10\overline{)7840}$	**f** $10\overline{)2380}$	**g** $10\overline{)5080}$	**h** $10\overline{)4110}$
i $10\overline{)8145}$	**j** $10\overline{)9000}$	**k** $10\overline{)3806}$	**l** $10\overline{)7200}$

2

a $10\overline{)16380}$	**b** $10\overline{)45600}$	**c** $10\overline{)37240}$	**d** $10\overline{)86000}$
e $10\overline{)36353}$	**f** $10\overline{)47864}$	**g** $10\overline{)10056}$	**h** $10\overline{)71346}$
i $10\overline{)216320}$	**j** $10\overline{)130056}$	**k** $10\overline{)186500}$	**l** $10\overline{)371039}$

3 Change to centimetres.

a 8000 mm ☐ **b** 56850 mm ☐ **c** 18680 mm ☐ **d** 7840 mm ☐

4 Write the number of tens in:

a 6930 ☐ **b** 5100 ☐ **c** 61270 ☐ **d** 56000 ☐

5 Ten Christmas decorations fill one box. How many boxes can be filled with 36000 decorations? ☐

To divide a whole number by 10, remove the last digit and make it the remainder.

6 Heather has ten times as much money as Sandy. How much has Sandy if Heather has $8165? ☐

$816.5 is $816.50.

 • *AUSTRALIAN SIGNPOST MATHS 6* • ISBN 978 1 4886 2193 2

2:32 Multiplying and Dividing by Powers of 10

$$\begin{array}{r} 0\cdot831 \\ \times\ 100 \\ \hline 83\cdot100 \end{array}$$

Place decimal **under** decimal for multiplication.

The decimal point has moved 2 places to the right.

Place decimal **over** decimal for division.

$$100\,\overline{)\,70\cdot300} \quad = 0\cdot703$$

The decimal point has moved 2 places to the left.

CONCEPT

1 Solve these on a calculator and try to find a rule that links the questions to the answers.

a 64·72 × 10	**b** 25·231 × 100	**c** 86·5092 × 1 000
d 44·2933 × 10	**e** 17·39 × 10	**f** 50·9461 × 100
g 37·5277 × 1 000	**h** 29·805 × 100	**i** 37·001 × 100

When we multiply by 10, 100 or 1 000, we move the decimal point 1, 2 or 3 places to the right.

2 Use the rule to answer these.

a 77·2914 × 1 000	**b** 15·73 × 10	**c** 7·7931 × 100
d 84·2913 × 1 000	**e** 91·0033 × 10	**f** 8·7935 × 1 000
g 21·068 × 100	**h** 64·3375 × 10	**i** 40·3571 × 1 000
j 35·113 × 10	**k** 9·6121 × 1 000	**l** 94·274 × 100

3 Solve these on a calculator and try to find a rule that links the questions to the answers.

a 27·14 ÷ 10	**b** 563·36 ÷ 100	**c** 7245·6 ÷ 1 000
d 906·7 ÷ 100	**e** 877·94 ÷ 10	**f** 630·4 ÷ 100
g 6311·9 ÷ 1 000	**h** 462·731 ÷ 100	**i** 856·094 ÷ 100

When we divide by 10, 100 or 1 000, we move the decimal point 1, 2 or 3 places to the left.

4 Use the rule to answer these.

a 1534·83 ÷ 1 000	**b** 97·46 ÷ 10	**c** 725·63 ÷ 100
d 884·366 ÷ 100	**e** 735·08 ÷ 10	**f** 83·251 ÷ 10
g 805·603 ÷ 100	**h** 844·9 ÷100	**i** 5623·9 ÷ 1 000
j 72·634 ÷ 10	**k** 3621·7 ÷ 1 000	**l** 836·73 ÷ 10
m 8044·6 ÷ 1 000	**n** 723·8 ÷ 100	**o** 846·9 ÷ 100

2:33 Dividing by a Multiple of 10

CONCEPT

- To divide by 10, cross off the last zero.
  $160\not{0} \div 10 = 160$
  $382\not{0} \div 10 = 382$
- To divide by 70, divide by 10 and then divide by 7.
  $4200 \div 70 = 7\not{0}\overline{)420\not{0}}$ = 60
  To divide by 10, we cross out the last zero.

1 a $10\overline{)35000}$ b $10\overline{)9600}$ c $10\overline{)7360}$ d $10\overline{)24000}$

2 a $100\overline{)35000}$ b $100\overline{)9600}$ c $100\overline{)7000}$ d $100\overline{)24000}$

3 Divide by the multiple of 10 by first dividing by 10.

a $20\overline{)8000}$ b $40\overline{)160}$ c $30\overline{)2700}$ d $70\overline{)35000}$

e $60\overline{)1320}$ f $50\overline{)1050}$ g $80\overline{)1600}$ h $90\overline{)10710}$

CONCEPT

- To divide by 10, move the decimal point one place to the left.
  $1385 \div 10 = 138{\cdot}5$
  $\$857 \div 10 = \$85.7$
  $= \$85.70$
- To divide by 60, divide by 10 and then divide by 6.
  $31056 \div 60$
  $= 3105{\cdot}6 \div 6$
  $= 6\overline{)31^{1}0^{4}5{\cdot}^{3}6}$ = 517·6
  $6\not{0}\overline{)31^{1}0^{4}5{\cdot}^{3}6{\cdot}}$ = 517·6

4 a $20\overline{)412}$ b $20\overline{)7104}$ c $30\overline{)1011}$ d $30\overline{)4020}$ e $40\overline{)728}$

f $50\overline{)7385}$ g $40\overline{)1044}$ h $60\overline{)1122}$ i $50\overline{)9145}$ j $60\overline{)3222}$

k $70\overline{)\$4907}$ l $90\overline{)\$3033}$ m $80\overline{)\$7632}$ n $90\overline{)\$7101}$ o $70\overline{)\$3689}$

5 Round off each answer to the nearest whole number.

a $70\overline{)387}$ b $40\overline{)3817}$ c $50\overline{)8156}$ d $60\overline{)3407}$ e $30\overline{)4186}$

 ISBN 978 1 4886 2193 2

2:34 Constructing Number Sentences

The sum of a number and 6 is 8·7.
■ + 6 = 8·7

Subtract 6 from 8·7.
■ = 2·7

CONCEPT

Increase means **add on**.
Decrease means **take away**.
Difference is the answer when **subtracting**.
Product is the answer when you **multiply**.
Quotient is the answer when you **divide**.
To **square** a number, multiply it by **itself**.

1 Find the value of each missing number.

a $6 + ■ = 7\frac{1}{2}$ ∴ ■ = ☐
b ■ + 11 = 12·2 ∴ ■ = ☐
c $▲ - 8 = \frac{1}{4}$ ∴ ▲ = ☐
d ▲ ÷ 3 = 1·1 ∴ ▲ = ☐
e 7 × ■ = 7·7 ∴ ■ = ☐
f $■ - \frac{3}{4} = \frac{1}{4}$ ∴ ■ = ☐

∴ means therefore.

2 Check the given solution by substituting, then write T (**true**) or F (**false**).

a ■ − 3·2 = 1·8 ∴ ■ = 4 ☐
b ■ + 109 = 824 ∴ ■ = 720 ☐
c ▲ × 3 = 4·5 ∴ ▲ = 1·5 ☐
d ▲ ÷ 7 = 36 ∴ ▲ = 252 ☐
e 27 + ■ = 270 ∴ ■ = 10 ☐
f 150 ÷ ■ = 25 ∴ ■ = 6 ☐

3 Write each expression if ■ is the missing number.

a 3 more than the number ☐
b 7 less than the number ☐
c 5 times the number ☐
d the sum of the number and 2 ☐
e the square of the number ☐
f one fifth of the number ☐

4 Construct a number sentence to match the problem, then find the answer.

a A number is increased by 4 to make 11. ☐
b A number is decreased by 11 to give 9. ☐
c The difference between a number and 7 is 6. ☐
d The product of 9 and a number is 27. ☐
e The quotient of a number and 4 is 7. ☐
f The square of a number is 25. ☐
g The sum of a number and 8 is 14. ☐
h A number minus 34 is equal to 82. ☐
i When a number is divided by 7 it gives 9. ☐
j The average of 15 and a number is 16. ☐

2:35 Division of Decimals

Dividing by a Whole Number

0·045 ÷ 3 is 45 thousandths ÷ 3
= 15 thousandths
= 0·015

The decimal point in the answer is placed above the decimal point in the question.

$3\overline{)0\cdot04^{1}5}$ = 0·015

$3\overline{)4^{1}5}$ = 15

CONCEPT

1 a 0·8 ÷ 2 b 0·9 ÷ 3 c 0·7 ÷ 7
d 1·6 ÷ 4 e 2·4 ÷ 8 f 2·8 ÷ 4
g 3·5 ÷ 5 h 8·1 ÷ 9 i 3·6 ÷ 6

$2\overline{)0\cdot6}$ = 0·3
$5\overline{)2\cdot0}$ = 0·4
Look at the pattern.

2 a $2\overline{)8\cdot4}$ b $3\overline{)9\cdot3}$ c $5\overline{)5\cdot5}$ d $4\overline{)6\cdot4}$
e $2\overline{)0\cdot84}$ f $3\overline{)0\cdot93}$ g $5\overline{)0\cdot55}$ h $4\overline{)0\cdot64}$

3 a $2\overline{)0\cdot804}$ b $3\overline{)0\cdot804}$ c $4\overline{)0\cdot804}$
d $2\overline{)0\cdot024}$ e $3\overline{)0\cdot024}$ f $4\overline{)0\cdot024}$

Place a figure above every figure in the question.
$2\overline{)0\cdot03^{1}4}$ = 0·017

4 a $5\overline{)15\cdot5}$ b $5\overline{)123\cdot0}$ c $5\overline{)1\cdot75}$ d $8\overline{)0\cdot824}$
e $3\overline{)2\cdot142}$ f $9\overline{)2\cdot142}$ g $8\overline{)19\cdot84}$ h $8\overline{)0\cdot024}$

5 Use a calculator to complete these questions.
a 131·57 m ÷ 5 b 7855·2 cm ÷ 8
c 3·976 km ÷ 7 d 81·862 km ÷ 11
e 10·272 L ÷ 4 f 118·781 L ÷ 13

ICT

2:36 Division of Decimals

CONCEPT

When dividing decimals we can place extra zeros at the end as we are dividing.

$8\overline{)5 \cdot 4} = 8\overline{)5 \cdot 40} = 8\overline{)5 \cdot 4^{6}0^{4}0}$ with answer $0 \cdot 675$

We place a figure above each figure after the decimal.

1 True (**T**) or false (**F**)?

a 5 = 5·0	**b** 5 = 5·00	**c** 5 = 0·5
d 6 = 6·000	**e** 3·4 = 3·40	**f** 2·1 = 2·10
g 46 = 46·0	**h** 5·11 = 5·110	**i** 8·000 = 8

2

a $5\overline{)5 \cdot 60}$ **b** $5\overline{)5 \cdot 6}$ **c** $5\overline{)1 \cdot 8}$

d $4\overline{)9 \cdot 00}$ **e** $4\overline{)7}$ **f** $4\overline{)1 \cdot 8}$

3

a $8\overline{)11 \cdot 400}$ **b** $8\overline{)26 \cdot 16}$ **c** $8\overline{)130 \cdot 4}$

d $\frac{1}{5}$ of 21 kg **e** $\frac{1}{4}$ of 83 km

f $\frac{1}{2}$ of 37 L **g** $\frac{1}{8}$ of 10·2 m

Find $\frac{1}{4}$ of 21 L.

$4\overline{)21 \cdot {}^{1}0{}^{2}0}$ with answer $5 \cdot 25$

Answer: 5·25 L

4

a $8\overline{)3 \cdot 000}$ **b** $4\overline{)27 \cdot 1}$ **c** $5\overline{)25 \cdot 6}$ **d** $8\overline{)1}$

e $4\overline{)399}$ **f** $2\overline{)137 \cdot 1}$ **g** $6\overline{)8 \cdot 1}$ **h** $6\overline{)126 \cdot 9}$

ICT

5 Use a calculator to complete these questions.

a 150·72 m ÷ 12 **b** 129·015 L ÷ 15

c 0·546 km ÷ 21 **d** 176·415 km ÷ 57

2:37 Extending Multiplication Facts

CONCEPT

2 × 8	16
2 × 80	160
2 × 800	1 600

20 × 8	160
20 × 80	1 600
20 × 800	16 000

Complete each pattern.

1 a

3 × 8	
3 × 80	
3 × 800	

b

2 × 3	
2 × 30	
2 × 300	

c

6 × 4	
6 × 40	
6 × 400	

d

4 × 7	
4 × 70	
4 × 700	

e

8 × 6	
8 × 60	
8 × 600	

f

7 × 9	
7 × 90	
7 × 900	

2 a

30 × 8	
30 × 80	
30 × 800	

b

20 × 4	
20 × 40	
20 × 400	

c

50 × 7	
50 × 70	
50 × 700	

d

80 × 7	
80 × 70	
80 × 700	

e

60 × 3	
60 × 30	
60 × 300	

f

90 × 9	
90 × 90	
90 × 900	

3 a

8 × 20	
80 × 20	
800 × 20	
8 000 × 20	

b

70 × 3	
70 × 30	
70 × 300	
70 × 3 000	

c

4 × 80	
40 × 80	
400 × 80	
4 000 × 80	

d

3 × 90	
30 × 90	
300 × 90	
3 000 × 90	

e

50 × 6	
50 × 60	
50 × 600	
50 × 6 000	

Constant Multipliers on a Calculator

ICT

4 Press the keys below, and write the answer on the display for each empty box.

a 40 [×] [×] 20 [=] [ ] 200 [=] [ ] 2 000 [=] [ ] 20 000 [=] [ ]

Here 20, 200, 2 000 and 20 000 have all been multiplied by 40.

b 30 [×] [×] 30 [=] [ ] 300 [=] [ ] 3 000 [=] [ ] 30 000 [=] [ ]

Here 30, 300, 3 000 and 30 000 have all been multiplied by 30.

© PEARSON AUSTRALIA 2018 • *AUSTRALIAN SIGNPOST MATHS 6* • ISBN 978 1 4886 2193 2

2:38 Multiplying Numbers Ending in Zeros

CONCEPT

10 × 23
= 230

20 × 23
= 2 × 230

30 × 23
= 3 × 230

70 × 23
= 7 × 230

1
a 10 × 35 ___ b 15 × 10 ___ c 10 × 78 ___ d 51 × 10 ___
e 20 × 35 ___ f 15 × 20 ___ g 20 × 78 ___ h 51 × 20 ___
i 30 × 35 ___ j 15 × 30 ___ k 30 × 78 ___ l 51 × 30 ___

20 × 10 = 200 | 20 × 20 = 400 | 20 × 30 = 600 | 20 × 70 = 1400

2
a 30 × 10 ___ b 30 × 20 ___ c 30 × 30 ___
d 10 × 40 ___ e 20 × 40 ___ f 30 × 40 ___
g 10 × 80 ___ h 20 × 80 ___ i 30 × 80 ___

To multiply numbers ending in zeros, write down the zeros and multiply the numbers that remain, e.g. 600 × 80 = 48 000.

3
a 40 × 20 ___ b 10 × 90 ___ c 20 × 30 ___ d 40 × 10 ___
e 30 × 50 ___ f 40 × 30 ___ g 70 × 20 ___ h 60 × 50 ___
i 80 × 30 ___ j 60 × 60 ___ k 80 × 70 ___ l 20 × 90 ___

4
a 500 × 60 ___ b 800 × 40 ___ c 500 × 70 ___
d 80 × 800 ___ e 40 × 700 ___ f 70 × 500 ___
g 60 × 400 ___ h 90 × 800 ___ i 200 × 60 ___

Check answers.

5

×	60	90	300	800	400	600	500
50							
200							
700							
900							

 • *AUSTRALIAN SIGNPOST MATHS 6* • ISBN 978 1 4886 2193 2

2:39 Multiplication by 2-Digit Numbers

CONCEPT

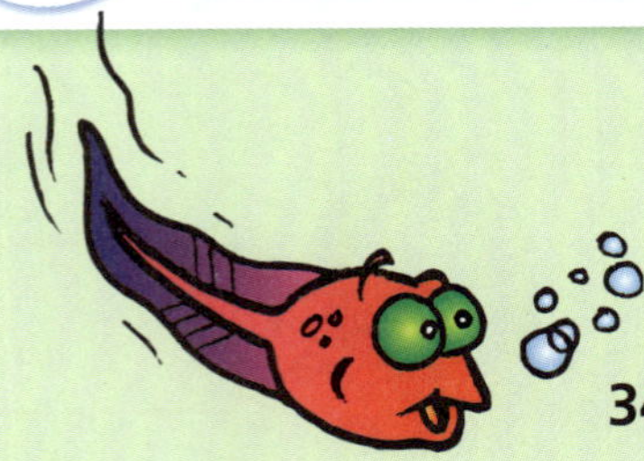

$34 \times 86 = (30 \times 86) + (4 \times 86)$

$$\begin{array}{r} 86 \\ \times\ {}^{1}30 \\ \hline 2580 \end{array} \qquad \begin{array}{r} 86 \\ \times\ {}^{2}4 \\ \hline 344 \end{array}$$

$34 \times 86 = 2580 + 344 = 2924$

To multiply by 30, put down the 0 and multiply by 3.

1 a $\begin{array}{r} 42 \\ \times 30 \\ \hline \end{array}$ b $\begin{array}{r} 56 \\ \times 20 \\ \hline \end{array}$ c $\begin{array}{r} 81 \\ \times 40 \\ \hline \end{array}$ d $\begin{array}{r} 43 \\ \times 30 \\ \hline \end{array}$

e 40×65 ☐ f 70×92 ☐ g 90×27 ☐

2 a $17 \times 61 = (10 \times 61) + (☐ \times ☐)$

b $16 \times 95 = (☐ \times ☐) + (6 \times 95)$

c $31 \times 53 = (30 \times ☐) + (1 \times ☐)$

d $46 \times 66 = (40 \times ☐) + (6 \times ☐)$

e $72 \times 37 = (☐ \times 37) + (☐ \times 37)$

f $29 \times 48 = (☐ \times 48) + (☐ \times 48)$

g $49 \times 59 = (☐ \times ☐) + (☐ \times ☐)$

h $72 \times 64 = (☐ \times ☐) + (☐ \times ☐)$

3 a $\begin{array}{r} 63 \\ \times 10 \\ \hline \end{array}$ $\begin{array}{r} 63 \\ \times 5 \\ \hline \end{array}$ $15 \times 63 = \_\_\_ + \_\_\_ = ☐$

b $\begin{array}{r} 74 \\ \times 20 \\ \hline \end{array}$ $\begin{array}{r} 74 \\ \times 8 \\ \hline \end{array}$ $28 \times 74 = \_\_\_ + \_\_\_ = ☐$

c $\begin{array}{r} 92 \\ \times 40 \\ \hline \end{array}$ $\begin{array}{r} 92 \\ \times 6 \\ \hline \end{array}$ $46 \times 92 = \_\_\_ + \_\_\_ = ☐$

d $\begin{array}{r} 81 \\ \times 50 \\ \hline \end{array}$ $\begin{array}{r} 81 \\ \times 2 \\ \hline \end{array}$ $52 \times 81 = \_\_\_ + \_\_\_ = ☐$

e $\begin{array}{r} 75 \\ \times 70 \\ \hline \end{array}$ $\begin{array}{r} 75 \\ \times 5 \\ \hline \end{array}$ $75 \times 75 = \_\_\_ + \_\_\_ = ☐$

f $\begin{array}{r} 88 \\ \times 60 \\ \hline \end{array}$ $\begin{array}{r} 88 \\ \times 9 \\ \hline \end{array}$ $69 \times 88 = \_\_\_ + \_\_\_ = ☐$

g $\begin{array}{r} 52 \\ \times 30 \\ \hline \end{array}$ $\begin{array}{r} 52 \\ \times 5 \\ \hline \end{array}$ $35 \times 52 = \_\_\_ + \_\_\_ = ☐$

h $\begin{array}{r} 96 \\ \times 40 \\ \hline \end{array}$ $\begin{array}{r} 96 \\ \times 2 \\ \hline \end{array}$ $42 \times 96 = \_\_\_ + \_\_\_ = ☐$

i $\begin{array}{r} 28 \\ \times 50 \\ \hline \end{array}$ $\begin{array}{r} 28 \\ \times 7 \\ \hline \end{array}$ $57 \times 28 = \_\_\_ + \_\_\_ = ☐$

j $\begin{array}{r} 84 \\ \times 60 \\ \hline \end{array}$ $\begin{array}{r} 84 \\ \times 3 \\ \hline \end{array}$ $63 \times 84 = \_\_\_ + \_\_\_ = ☐$

4 Each night for two weeks, Luke's dad read 35 pages of *Watership Down*. How many pages were read in the two weeks? ☐

 • *AUSTRALIAN SIGNPOST MATHS 6* • ISBN 978 1 4886 2193 2

2:40 Multiplication by 2-Digit Numbers

CONCEPT

36 students paid $72 each to go to camp. How much was paid altogether?

36 × 72 =

```
    7 2          7 2
×   3 0   +   ×    6
  2 1 6 0        4 3 2
```

```
        7 2
  ×     3 6
      4 3 2   (6 × 72)
    2 1 6 0   (30 × 72)
    2 5 9 2
```

$2592 was paid altogether.

1

a
```
  3 2
× 1 4
       (4 × 32)
____   (10 × 32)
```

b
```
  5 3
× 3 2
       (2 × 53)
____   (30 × 53)
```

c
```
  9 3
× 5 1
       (1 × 93)
____   (50 × 93)
```

d
```
  6 2
× 3 4
       (4 × ☐)
____   (30 × ☐)
```

e
```
  7 1
× 4 4
       (4 × ☐)
____   (40 × ☐)
```

f
```
  4 3
× 3 1
       (☐ × 43)
____   (☐ × 43)
```

g
```
  9 4
× 2 2
       (☐ × 94)
____   (☐ × 94)
```

2 Write out the algorithm and use it to find the answer.

a 12 × 89
```
  8 9
× 1 2
       (☐ × ☐)
____   (☐ × ☐)
```

b 53 × 27
```
  2 7
× 5 3
       (☐ × ☐)
____   (☐ × ☐)
```

c 63 × 56
```
  5 6
×
       (☐ × ☐)
____   (☐ × ☐)
```

d 18 × 49
```
×
       (☐ × ☐)
____   (☐ × ☐)
```

e 37 × 95
```
×
       (☐ × ☐)
____   (☐ × ☐)
```

f 72 × 33
```
×
       (☐ × ☐)
____   (☐ × ☐)
```

g 25 × 73
```
×
       (☐ × ☐)
____   (☐ × ☐)
```

h 46 × 53
```
×
       (☐ × ☐)
____   (☐ × ☐)
```

3

a One small bottle of correction fluid contains 18 mL.
How much correction fluid would be contained in 56 small bottles? ☐

b In a box that weighed 200 g, Glen sent 36 Christmas balls to Brett.
If each ball weighed 14 g, what was the gross weight of the package? ☐

2:41 Multiplication by 2-Digit Numbers

CONCEPT

34 members of our club booked one-way tickets to Brisbane. Each ticket cost $88. How much did the tickets cost altogether?

The tickets cost $2992.

Working

```
    8 8
×   3 4
  3 5 2
2 6 4 0
2 9 9 2
```

1

a 46×12

b 93×11

c 81×24

d 75×32

e 67×12

f $53 \times 41 = 41 \times 53$

g $62 \times 27 = 27 \times 62$

h $14 \times 99 = 99 \times 14$

i $25 \times 18 = 18 \times 25$

j $32 \times 94 = 94 \times 32$

k 16×52

l 33×69

m 28×31

n 95×42

2

a Each month Rhonda sends $38 to World Vision. How much does she send in a year?

b A gift shop purchased 36 teapots for $54 each and sold them for $87 each.
How much did the teapots cost the gift shop altogether?
What was the total amount the shop received by selling all the teapots?
How much profit did the gift shop make?

c Complete this table for other items sold by the gift shop.

Item	Number of Items	Cost of One	Total Cost	Selling Price of One	Total from Sales	Total Profit
sugar bowl	65	$14		$28		
oval tray	43	$22		$35		
cup/saucer	76	$28		$47		

2:42 Multiplication by 2-Digit Numbers

CONCEPT

Brad bragged that his stamp collection was 28 times larger than his brother's. His brother had 168 stamps. How many stamps did Brad say he had?

```
    1 6 8
×     2 8
  1 3 4 4   ← (8 × 168)
  3 3 6 0   ← (20 × 168)
  4 7 0 4
```

Brad said he had 4704 stamps.

1

a
```
  4 2 9
×   1 3
```

b
```
  1 5 3
×   2 7
```

c
```
  3 6 2
×   4 5
```

d
```
  2 8 7
×   7 2
```

e
```
  8 0 3
×   7 5
```

f
```
  7 4 0
×   5 6
```

g
```
  9 0 4
×   3 4
```

h
```
  6 1 0
×   8 9
```

i
```
  3 1 1
×   6 8
```

j 17 × 952
```
  9 5 2
×   1 7
```

k 38 × 750
```
×
```

l 26 × 5800
```
×
```

2 a Through air, sound travels at 330 metres each second. How far would it travel in 35 seconds?

Through sea water, sound travels 4 times faster. How far does it travel through sea water in 35 seconds?

b The platform of the Assyrian palace of Sargon covered about 11 hectares and was 15 metres high. It would have taken 8600 people 14 years to construct the platform. If one person could do all this work, how long would it take them?

Estimating the Number of Blocks in a Jar

FUN SPOT

- Take turns to fill (or partly fill) a jar with place-value ones.
- Estimate the number of layers of ones in the jar.
- Look underneath to estimate the number in each layer.
- Multiply, to estimate the number of ones in the jar.
- See whose estimate is closest.

2:43 Number Sentences

We can use opposite operations to find a missing number.

+ 6 is the opposite of – 6
× 4 is the opposite of ÷ 4

■ – 16 = 7
The opposite of – 16 is + 16.
So ■ = 7 + 16
= 23

▲ ÷ 3 = 13
The opposite of ÷ 3 is × 3.
So ▲ = 13 × 3
= 39

CONCEPT

1 Find the value of the missing number, then write **T** (**true**) or **F** (**false**).

a ■ + 6 = 10 So ■ =
Does 10 – 6 equal ■?

b ■ – 1 = 99 So ■ =
Does 99 + 1 equal ■?

c ▲ + 83 = 91 So ▲ =
Does 91 – 83 equal ▲?

d ▲ – 37 = 28 So ▲ =
Does 28 + 37 equal ▲?

e ■ ÷ 2 = 41 So ■ =
Does 41 × 2 equal ■?

f ■ × 5 = 40 So ■ =
Does 40 ÷ 5 equal ■?

g ▲ ÷ 7 = 11 So ▲ =
Does 11 × 7 equal ▲?

h ▲ × 6 = 126 So ▲ =
Does 126 ÷ 6 equal ▲?

2 Find the value of each missing number.

a 5 + ■ = 12 – 4 So ■ =

b ■ × 4 = 100 – 16 So ■ =

c ▲ – 7 = 6 × 3 So ▲ =

d ▲ ÷ 3 = 15 – 8 So ▲ =

e ■ – 136 = 9 × 7 So ■ =

f ▲ × 5 = 36 + 14 So ▲ =

We can check an answer by substituting it into the number sentence.

If ■ = 23 is a solution to ■ + 98 = 121, then 23 + 98 = 121.

If ▲ = 8 is a solution to 7 × ▲ = ▲ + 48, then 7 × 8 = 8 + 48.

CONCEPT

3 Check the given solution by substituting, then write **T** (**true**) or **F** (**false**).

a If ■ + 14 = 30, then ■ = 10

b If ■ – 37 = 13, then ■ = 50

c If ▲ × 4 = 420, then ▲ = 15

d If ▲ ÷ 7 = 20, then ▲ = 140

e ■ + 8 + 10 = 32, ■ = 14

f ■ × 3 – 4 = 16, ■ = 7

g 16 – ▲ = 8 + 7, ▲ = 1

h 100 ÷ ▲ + 5 = 30, ▲ = 4

i ■ + 8 = 2 × ■, ■ = 8

j 2 × ■ – 35 = ■ – 15, ■ = 20

k (▲ + 4) × 3 + 7 = 22, ▲ = 1

l 4 × ▲ + 1 = 2, ▲ = $\frac{1}{4}$

 ISBN 978 1 4886 2193 2

2:44 Divisibility and Factors

Divisor	Divisibility Test	Example
2	The number must be even, i.e. it must end in 0, 2, 4, 6 or 8.	73 916 is divisible by 2 as it is even (ends in 6).
3	The sum of the digits is divisible by 3.	10 203 is divisible by 3 as it has a digit sum divisible by 3 (1 + 0 + 2 + 0 + 3 = 6).
4	The number made by the last two digits must be divisible by 4.	111 024 is divisible by 4 as 24 ÷ 4 = 6 (ie 24 is divisible by 4).
5	The last digit must be 5 or 0.	64 015 is divisible by 5 as it ends in 5.
8	The number made by the last three digits must be divisible by 8.	11 160 is divisible by 8 as 160 ÷ 8 = 20 (i.e. 160 is divisible by 8).
9	The sum of the digits is divisible by 9.	88 110 is divisible by 9 as 8 + 8 + 1 + 1 + 0 = 18.
10	The last digit must be 0.	27 370 is divisible by 10 as it ends in 0.

CONCEPT

A number is divisible by 6 if it is divisible by 2 and 3.

1 Underline the numbers that are:

a	divisible by 2:	154	387	4441	37 902	89 366	25 819
b	divisible by 3:	184	732	1092	36 304	67 313	111 111
c	divisible by 4:	614	812	4308	17 224	83 906	111 110
d	divisible by 5:	307	415	8400	81 194	55 504	111 110
e	divisible by 8:	11 008	23 000	65 832	614 016	821 104	7184
f	divisible by 9:	10 070	14 301	38 246	91 422	128 700	66 811
g	divisible by 10:	37 015	38 400	75 830	415 004	111 010	41 875

2 Write down all the factors of:

a 24: ☐ ☐ ☐ ☐ ☐ ☐ ☐ ☐

b 40: ☐ ☐ ☐ ☐ ☐ ☐ ☐ ☐

c 105: ☐ ☐ ☐ ☐ ☐ ☐ ☐ ☐

d 81: ☐ ☐ ☐ ☐ ☐

e 120: ☐ ☐ ☐ ☐ ☐ ☐ ☐ ☐ ☐ ☐ ☐ ☐ ☐ ☐ ☐ ☐

The factors of 36
36 ÷ **1** = **36**
36 ÷ **2** = **18**
36 ÷ **3** = **12**
36 ÷ **4** = **9**
36 ÷ 5 = 7 r 1
36 ÷ **6** = 6

If a number is divisible by a number, then it is also divisible by the factors of that number.

3 If a number is:

a divisible by 30, it is also divisible by ☐ ☐ ☐ ☐ ☐ ☐ ☐.

b divisible by 24, it is also divisible by ☐ ☐ ☐ ☐ ☐ ☐ ☐.

c divisible by 100, it is also divisible by ☐ ☐ ☐ ☐ ☐ ☐ ☐ ☐.

  ISBN 978 1 4886 2193 2

2:45 Estimation with Decimals

CONCEPT

4·12 + 84·691 → Rounding → 4 + 85 = 89 ← 89 is our estimate.

Both rounded up will give an **overestimate**.

e.g. 2·942 × 8·5

Rounding both up: 3 × 9 = 27

27 is an **overestimate**.

Both rounded down will give an **underestimate**.

e.g. 2·41 × 12·3

Rounding both down: 2 × 12 = 24

24 is an **underestimate**.

1 Round each decimal to the nearest whole number. Write whether you have rounded up (**U**) or rounded down (**D**).

a 84·3 ☐ ☐ **b** 13·9 ☐ ☐ **c** 28·2 ☐ ☐

d 47·8 ☐ ☐ **e** 5·6 ☐ ☐ **f** 26·4 ☐ ☐

g 19·9 ☐ ☐ **h** 21·7 ☐ ☐ **i** 33·3 ☐ ☐

≑ means 'is approximately equal to'.

2 Mary obtained these answers using a calculator. Use estimation to discover which 3 answers are incorrect. Put a cross next to the wrong answers.

a 4·186 × 5·1 ≑ 41·6 ☐ **b** 3·421 × 2·2 ≑ 7·5 ☐

c 2·178 × 7·3 ≑ 22·7 ☐ **d** 6·492 × 3·3 ≑ 21·4 ☐

e 5·512 × 3·5 ≑ 19·29 ☐ **f** 7·122 × 5·7 ≑ 40·6 ☐

g 2·32 × 6·1 ≑ 14·15 ☐ **h** 4·341 × 3·2 ≑ 19·2 ☐

3 Mentally round each number to the nearest whole number, then estimate the answer.

a 29·3 + 8·99 ≑ ☐ **b** 30·812 + 6·1 ≑ ☐

c 100·65 – 23·4 ≑ ☐ **d** 17·3 – 7·012 ≑ ☐

e 1·863 × 6·714 ≑ ☐ **f** 2·375 × 3·41 ≑ ☐

g 9·243 + 5·341 ≑ ☐ **h** 8·66 × 4·73 ≑ ☐

i 18·657 – 11·62 ≑ ☐ **j** 22·831 + 9·22 ≑ ☐

Rounding

If the next digit is 5 or more, round up.
e.g. 7·514 → 8.

If the next digit is 4 or less, round down.
e.g. 13·49 → 13.

Use a calculator to check the answers to Questions 2 and 3.

ICT

© PEARSON AUSTRALIA 2018 • *AUSTRALIAN SIGNPOST MATHS 6* • ISBN 978 1 4886 2193 2

2:46 Using Rounding to Check Answers

We can estimate answers using whole numbers to check the reasonableness of our answers.

	Round	
3 2 · 3 0 1	3 2	↓ Rounded down
+ 1 4 · 4 4 0	+ 1 4	↓ Rounded down
4 6 · 7 4 1	4 6	← **underestimate**

Both numbers were rounded down so this will be an underestimate.

CONCEPT

1 Round these numbers to the nearest whole number to estimate. Complete the written algorithms to check that your answers are reasonable.

a

	Round
5 1 · 3 9 2	☐
+ 3 4 · 0 4 3	+ ☐
____	____

b

	Round
4 3 · 6 5 3	☐
+ 2 7 · 4 0 2	+ ☐
____	____

c

	Round
6 3 · 9 7 2	☐
− 5 8 · 6 0 1	− ☐
____	____

d

	Round
8 0 · 3 5 9	☐
− 4 2 · 0 9 3	− ☐
____	____

Will your estimate be an underestimate or an overestimate?

We can use our knowledge of decimals to estimate mentally.

$0{\cdot}4 \times 36$ is almost $0{\cdot}5 \times 36$.

$\frac{1}{2}$ of $36 = 18$

$0{\cdot}4 \times 36 \doteqdot 18$

18 is a good estimate.

$\doteqdot$ means 'is approximately equal to'.

$83{\cdot}125 \div 3{\cdot}9$

$\doteqdot 80 \div 4$

$= 20$

20 is a good estimate.

Use a calculator to check how close your estimates were.

CONCEPT

In Questions 2 and 3, round numbers to estimate the answer.

2

a $0{\cdot}6 \times 12 \doteqdot$ ☐ **b** $42 \times 1{\cdot}843 \doteqdot$ ☐ **c** $22 \times 1{\cdot}2 \doteqdot$ ☐

d $5{\cdot}3 \times 8 \doteqdot$ ☐ **e** $28 \times 0{\cdot}482 \doteqdot$ ☐ **f** $3{\cdot}9 \times 7 \doteqdot$ ☐

g $10{\cdot}8 \times 7 \doteqdot$ ☐ **h** $100 \times 3{\cdot}31 \doteqdot$ ☐ **i** $7{\cdot}12 \times 6 \doteqdot$ ☐

3

a $59{\cdot}83 \div 3 \doteqdot$ ☐ **b** $33{\cdot}812 \div 8 \doteqdot$ ☐ **c** $20{\cdot}9 \div 7 \doteqdot$ ☐

d $8{\cdot}205 \div 2 \doteqdot$ ☐ **e** $15{\cdot}08 \div 5 \doteqdot$ ☐

f $17{\cdot}938 \div 2 \doteqdot$ ☐ **g** $19{\cdot}79 \div 4 \doteqdot$ ☐

Ask: Does the answer make sense?

 • *AUSTRALIAN SIGNPOST MATHS 6* • ISBN 978 1 4886 2193 2

2:47 A Strategy for Multiplying Decimals

1 Complete this table.

Question	Number of Decimal Places in Question	Calculator Answer	Number of Decimal Places in Answer
0·3 × 5	1	1.5	1
0·4 × 0·7	2	0.28	2
0·41 × 2	2		
0·103 × 3	3		
0·121 × 0·2	4		

INVESTIGATION

When multiplying decimals, the number of figures after the decimal point in the answer must be the same as the total number of figures that come after the decimal points in the question.

2 How many decimal places has each?

a 8·45 ☐ **b** 3613·9 ☐ **c** 74·267 ☐ **d** 7245·1 ☐
e 794·02 ☐ **f** 600·111 ☐ **g** 5·87 ☐ **h** 56·609 ☐
i 10·571 ☐ **j** 5513·8 ☐ **k** 21·6917 ☐ **l** 6·0143 ☐

3 Try to do these in your head. Multiply as though there are no decimal points, then use the rule above to include the correct number of decimal points.

a 0·2 × 8 ☐ **b** 0·8 × 3 ☐ **c** 9 × 0·03 ☐
d 7 × 0·06 ☐ **e** 0·9 × 4 ☐ **f** 5 × 0·3 ☐
g 0·04 × 9 ☐ **h** 0·03 × 2 ☐ **i** 0·2 × 2 ☐
j 0·03 × 5 ☐ **k** 0·08 × 6 ☐ **l** 0·02 × 0·7 ☐
m 0·07 × 0·1 ☐ **n** 0·04 × 0·2 ☐ **o** 0·7 × 0·3 ☐

4 Use the rule above to complete these.

a 0·2 × 4·3 ☐ **b** 0·11 × 10 ☐ **c** 2 × 0·53 ☐
d 1·07 × 0·02 ☐ **e** 3·1 × 0·4 ☐ **f** 0·6 × 0·6 ☐
g 8·24 × 0·01 ☐ **h** 1·5 × 0·3 ☐ **i** 4·2 × 5 ☐

5 **a** Each metre length of a pole has a mass of 0·8 kg. What is the mass of 0·3 metres of a pole? ☐

b Find the area of a square of length 0·6 cm. ☐

c Find the area of a rectangle of length 0·4 cm and width 0·3 cm. ☐

d The mill produces 0·8 kg of flour each minute. How much is produced in 1·5 minutes? ☐

2:48 Division Strategies

CONCEPT

Multiply both numbers by the same amount to make an equivalent division calculation.

300 ÷ 60 (÷10, ÷10) → 30 ÷ 6 = 5

3·5 ÷ 0·5 (×10, ×10) → 35 ÷ 5 = 7

60 ÷ 4 (÷2, ÷2) → 30 ÷ 2 = 15

Divide by a factor of the divisor, or multiply by 10 or 100 to remove decimals.

$135 \div 15 = \frac{135}{15} = \frac{135 \div 5}{15 \div 5} = \frac{27}{3} = 9$ or $135 \div 15 = \frac{135 \times 2}{15 \times 2} = \frac{270}{30} = 9$

We get the same answer if we divide or multiply each term in 135 ÷ 15 by the same number.

Divide or multiply each part by a number to make the question easier, then find the answer.

1 **a** 400 ÷ 50 (÷10, ÷10) ☐ ÷ ☐ = ☐
b 4·5 ÷ 0·9 (×10, ×10) ☐ ÷ ☐ = ☐
c 48 ÷ 12 (÷2, ÷2) ☐ ÷ ☐ = ☐
d 2·1 ÷ 0·7 (×10, ×10) ☐ ÷ ☐ = ☐
e 360 ÷ 90 (÷10, ÷10) ☐ ÷ ☐ = ☐
f 120 ÷ 15 (÷3, ÷3) ☐ ÷ ☐ = ☐
g 60 ÷ 15 (÷3, ÷3) ☐ ÷ ☐ = ☐
h 810 ÷ 90 (÷10, ÷10) ☐ ÷ ☐ = ☐
i 5·6 ÷ 0·8 (×10, ×10) ☐ ÷ ☐ = ☐

2 **a** 180 ÷ 20 ☐ ÷ ☐ = ☐
b 1·2 ÷ 0·6 ☐ ÷ ☐ = ☐
c 7·2 ÷ 0·9 ☐ ÷ ☐ = ☐
d 108 ÷ 12 ☐ ÷ ☐ = ☐
e 540 ÷ 60 ☐ ÷ ☐ = ☐
f 112 ÷ 14 ☐ ÷ ☐ = ☐
g 126 ÷ 14 ☐ ÷ ☐ = ☐
h 300 ÷ 50 ☐ ÷ ☐ = ☐
i 1·4 ÷ 0·2 ☐ ÷ ☐ = ☐

3 **a** 240 ÷ 60 ☐
b 4·5 ÷ 0·5 ☐
c 8·8 ÷ 1·1 ☐
d 280 ÷ 40 ☐
e 1·2 ÷ 0·6 ☐
f 810 ÷ 90 ☐
g 400 ÷ 80 ☐
h 550 ÷ 50 ☐
i 2·4 ÷ 0·8 ☐

2:49 Addition of Large Numbers

This table shows the number of cars registered.

NSW	3,258,552
Vic	2,882,452
SA	891,137
Tas	259,068

How many cars were registered altogether in these states?

```
  2 2 2 1 2 1
3 2 5 8 5 5 2
2 8 8 2 4 5 2
  8 9 1 1 3 7
  2 5 9 0 6 8
-------------
7 2 9 1 2 0 9
```

There were 7,291,209 cars registered.

CONCEPT

1

a
```
  380186
 1242346
+  94685
```

b
```
 6342694
  987486
+2156071
```

c
```
 1890600
 2477300
+8243100
```

d
```
  946215
 5093614
+1420817
```

e
```
  304218
   99324
 2460814
+   3967
```

f
```
 23814211
  2608694
  5214414
+16935062
```

g
```
 3835000000
  910000000
 1049000000
+2340000000
```

2

a
```
 $ 65314.50
 $182708.25
+$ 99314.75
```

b
```
 $684319.74
 $ 96250.08
+$145619.09
```

c
```
 $423800.00
 $ 98500.00
+$125600.00
```

d
```
 $246153.60
 $ 93146.00
+$563463.40
```

3 Use the table below to find the population at the end of September 2010.

NSW	7,253,400	Vic	5,567,100	Qld	4,532,300
WA	2,306,200	SA	1,647,800	Tas	508,500
NT	230,200	ACT	359,700		
Other Territories (Jervis Bay, Christmas Island and the Cocos Islands)					2,000

Commas are often used in large numbers.

ACTIVITY

4 At the end of September 2010, the estimated resident population of Australia was 22 407 200 people. This was an increase of 345 500 since the end of September 2009. Estimate when Australia's population reached 22 200 000.

Show your method here.

 • *AUSTRALIAN SIGNPOST MATHS 6* • ISBN 978 1 4886 2193 2

2:50 Subtraction of Large Numbers

CONCEPT

By how much is the area of the Atlantic Ocean (76 800 000 km^2) larger than the area of the Indian Ocean (68 600 000 km^2)?

```
   6 16
   7 6 8 0 0 0 0 0
 − 6 8 6 0 0 0 0 0
     8 2 0 0 0 0 0
```

1

a 4 302 600 − 125 300	**b** 8 526 480 − 1 847 145	**c** 10 475 000 − 9 934 000	**d** 15 680 000 − 1 236 340
e 84 390 000 − 36 488 000	**f** 60 000 000 − 9 186 000	**g** 59 600 000 − 27 800 000	**h** 16 000 000 − 1 843 712

2

a The Pacific Ocean has a coastline of 135 663 km and a maximum depth of 10 924 m, while the Atlantic Ocean has a coastline of 111 866 km and a maximum depth of 8 380 m. What is their difference in:

i maximum depth? ______ **ii** coastline? ______

b By how much area is the largest ocean, the Pacific Ocean (166 229 000 km^2), greater than the largest sea, the South China Sea (2 590 600 km^2)? ______

c In December 1999 in France, the biggest tree loss by a windstorm occurred. 360 million trees were destroyed. If 3 475 800 trees were planted in the following year, how much was this less than the number lost? ______

d In 2010 China's population was estimated to be 1 381 000 000. How much less than 2 billion (2 000 000 000) is this? ______

e In 2010 the world's population was 6 900 000 000 (6·9 billion). How many people did not live in China? ______

f The average distance from the Earth to the Sun is 150 000 000 km. If the Earth's present position is 1 860 000 km less than the average distance, how far is it from the Sun at present? ______

3 Estimate or use a calculator to check your answers.

 • *AUSTRALIAN SIGNPOST MATHS 6* • ISBN 978 1 4886 2193 2

2:51 Problem Solving

CONCEPT

What two numbers have a difference of 2 and a product of 728?

Calculator Solution

Guess and check.

Guess 1:	32	and	34			
Check:	32	×	34 =	1 088	This is too big.
Guess 2:	22	and	24			
Check:	22	×	24 =	528	This is too small.
Guess 3:	26	and	28			
Check:	26	×	28 =	728	This is just right!

Answer: The numbers are **26** and **28**.

ICT

1 For each sum or difference, find two numbers that have the given product.

You can use a calculator.

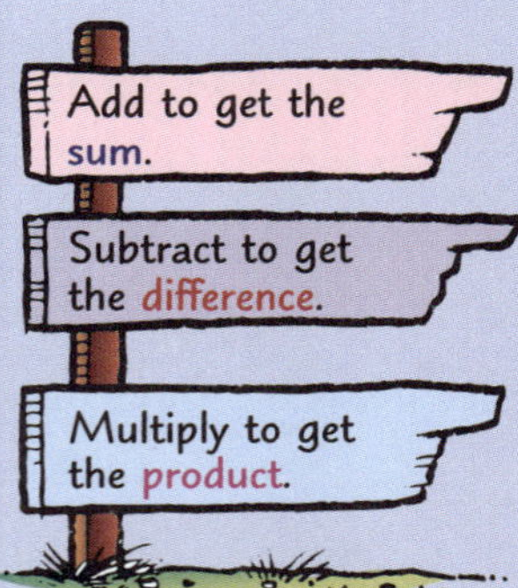

	Difference	Product	Numbers
a	3	270	
b	1	210	
c	6	567	
d	4	1 020	
e	10	231	
f	5	500	
g	7	408	
h	9	220	
i	2	840	

	Sum	Product	Numbers
j	40	300	
k	33	242	
l	75	1 400	
m	30	125	
n	70	1 216	
o	104	2 700	
p	98	2 400	
q	55	750	
r	91	2 050	

2 Rohan used his calculator to obtain this result: 6·5 ÷ 4 = 1·625.
How would he write the answer if the question was:

a What is the length of one part if 6·5 km is divided into 4 parts?

b Four melons cost $6.50. What should be the price of one melon?

Calculator Crossword

FUN SPOT

3 To fill in the clues, follow these steps.

- Do each question on a calculator.
- Turn your calculator upside down to change the number into a word.
- Write in the word.

Across

1 33 514 + 21 864
5 900 (read backwards)
6 10 000 – 4 063
7 50 (read backwards)
9 10 more than 24
11 6 680 – 6 629
13 70 (read backwards)

Down

1 55 762 + 1 956
2 9 988 – 3 995
3 7 412 – 6 907
4 50 (backwards)
8 142 × 5
10 1 000 – 267
12 50 (backwards)

2:52 Using Strategies to Solve Problems

1 Yesterday I bought plants for my garden. I bought five azaleas; a white one for $4.15, a red one for $3.80, an orange one for $3.25 and a double azalea for $5. My favourite purchase was a variegated one which cost $5.85. My taxi home cost $8.50.

a Which azalea cost the most? ______ **b** How much did I spend on azaleas? ______

c What three azaleas cost a total of $13.25? ______

Draw a diagram.

Example

A girl places marbles along a line 30 cm long. If she places them 10 cm apart, how many marbles does she need?

CONCEPT

2 Posts are placed in a straight line, one metre apart. If the line of posts extends for 10 metres, how many posts are used? ______

How many posts would be used if they were placed in a circle? ______

3 Five children link hands to form a chain. How many hands are held? ______

4 A birdcage was built in the shape of a rectangular prism using 3 vertical posts for each side (one at each end and one in the middle). How many posts were needed to build the cage? ______

5 A snail is climbing a bamboo pole which is 20 m high. It climbs up 4 m every day and slides down 3 m every night. How long does it take the snail to reach the top? ______

Examine each pattern below. From the three possibilities given, circle the one that correctly completes the pattern.

Look for patterns.

INVESTIGATION

6

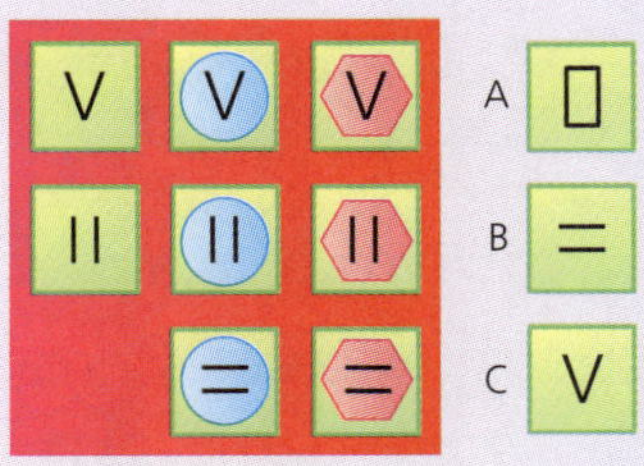

7

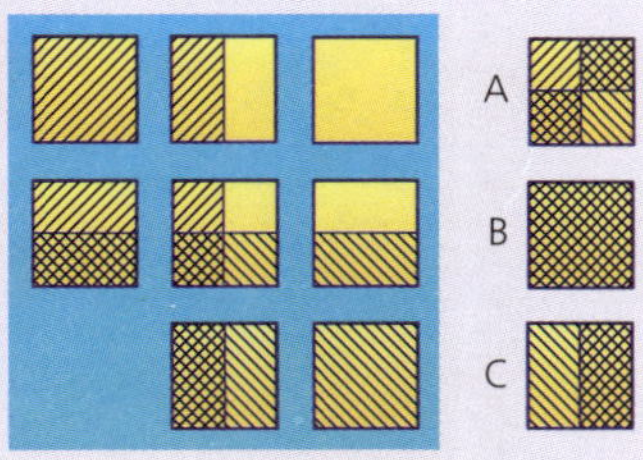

8

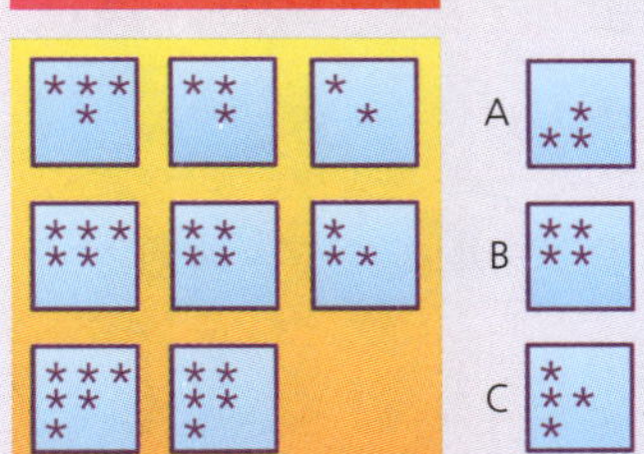

9

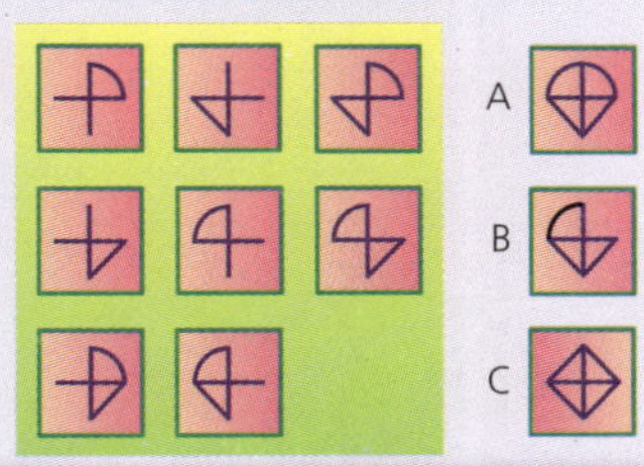

3:01 Centimetres and Millimetres

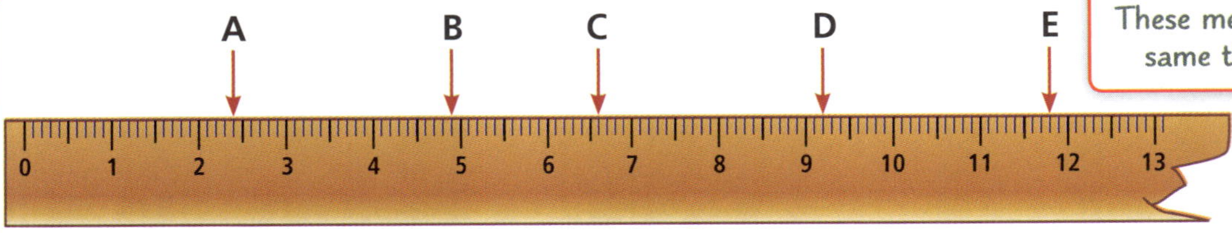

1 Record each length shown on the ruler.

A ☐ mm ☐ cm B ☐ mm ☐ cm C ☐ mm ☐ cm D ☐ mm ☐ cm E ☐ mm ☐ cm

2 Use the decimal form to write these as centimetres.

a 29 mm ☐	**b** 36 mm ☐	**c** 99 mm ☐	**d** 51 mm ☐
e 64 mm ☐	**f** 82 mm ☐	**g** 48 mm ☐	**h** 73 mm ☐

3 Write these as millimetres.

a 3·9 cm ☐	**b** 7·2 cm ☐	**c** 9·1 cm ☐	**d** 4·5 cm ☐
e 6·8 cm ☐	**f** 2·4 cm ☐	**g** 5·3 cm ☐	**h** 10·6 cm ☐

4 Find the perimeter of each shape (in **cm**).

a

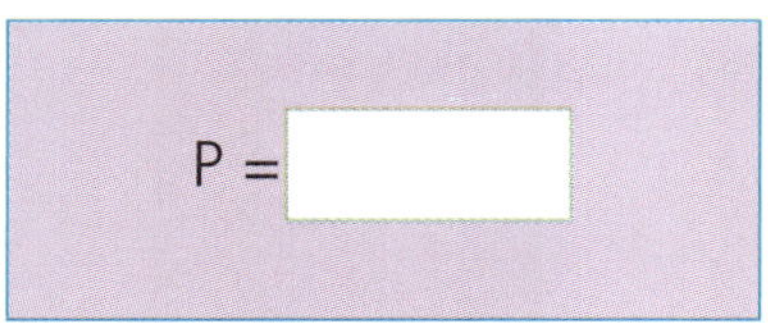

b

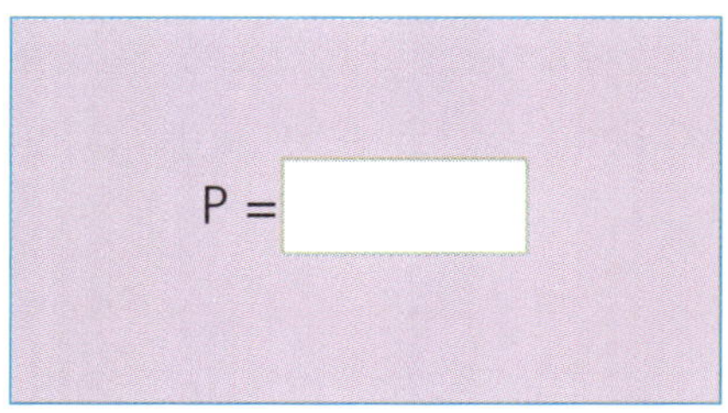

c

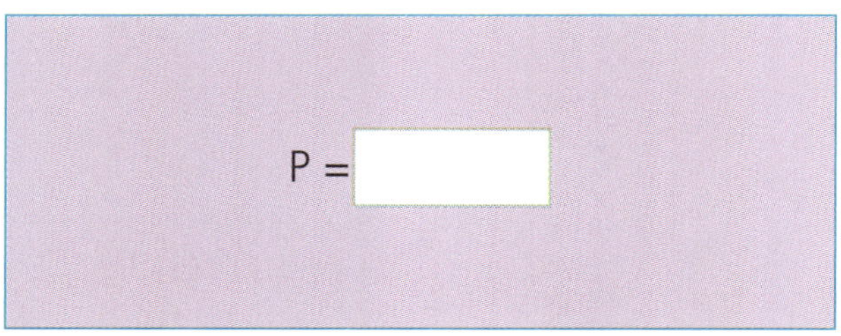

d

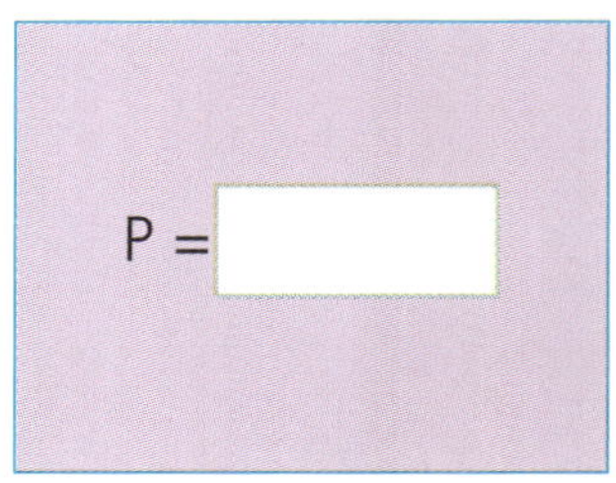

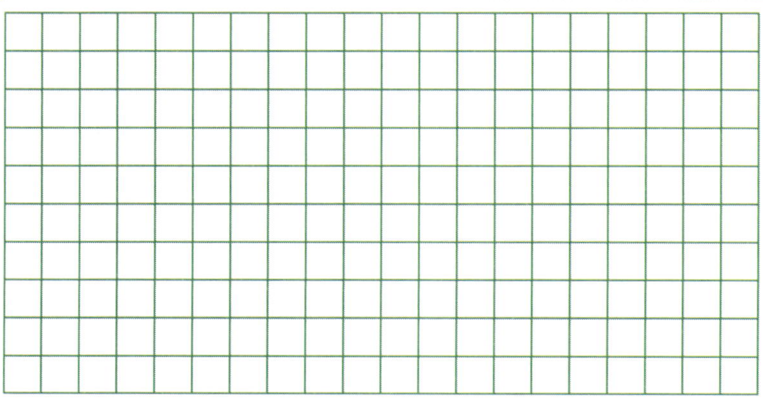

5 On the 5 mm grid paper, draw rectangles with the following measurements.

a 3·5 cm and 2·5 cm wide

Perimeter ☐

b 5·5 cm long and 4 cm wide

Perimeter ☐

 • *AUSTRALIAN SIGNPOST MATHS 6* • ISBN 978 1 4886 2193 2

3:02 Kilometres

1 Write the length in metres.

a 3 km ☐ m b 5 km ☐ m c 9 km ☐ m
d 7 km ☐ m e 10 km ☐ m f 8 km ☐ m
g 20 km ☐ m h 15 km ☐ m i 12 km ☐ m

2 Write the length in kilometres.

a 2000 m ☐ km b 6000 m ☐ km c 1000 m ☐ km
d 4000 m ☐ km e 10 000 m ☐ km f 11 000 m ☐ km

3 Choose the most suitable unit of length (**km**, **m**, **cm**, **mm**) to measure:

a the distance from Geelong to Bright ☐
b the length of a swimming pool ☐
c the length of your foot ☐
d the width of a match ☐
e the width of Tasmania ☐
f the length of a marathon race ☐
g the height of Mt Kosciuszko ☐
h the width of a yacht ☐

4

Distances in Kilometres

	Goulburn	Gundagai	Albury	Benalla	Melbourne
Sydney	196	386	572	684	877
Goulburn		190	376	488	681
Gundagai			186	298	491
Albury				112	305
Benalla					193

From Goulburn to Benalla is 488 km.

Use the table to find the distance between:

a Sydney and Gundagai ☐
b Goulburn and Albury ☐
c Gundagai and Melbourne ☐
d Melbourne and Goulburn ☐
e Sydney and Melbourne ☐

5 Convert these measurements to metres.

a $\frac{1}{2}$ km ☐ m b $\frac{1}{4}$ km ☐ m c $\frac{3}{4}$ km ☐ m
d $2\frac{1}{10}$ km ☐ m e $1\frac{1}{2}$ km ☐ m f $1\frac{7}{10}$ km ☐ m
g $1\frac{3}{4}$ km ☐ m h $\frac{4}{10}$ km ☐ m i $6\frac{1}{4}$ km ☐ m
j $7\frac{3}{10}$ km ☐ m k 7·3 km ☐ m l 2·5 km ☐ m
m $4\frac{9}{10}$ km ☐ m n 4·9 km ☐ m o 6·7 km ☐ m

6 Find the speed (**km/h**) for each of the following:

a 87 kilometres in one hour ☐
b 186 kilometres in two hours ☐
c 340 kilometres in four hours ☐
d 390 kilometres in five hours ☐

 • *AUSTRALIAN SIGNPOST MATHS 6* • ISBN 978 1 4886 2193 2

3:03 Converting Measurements

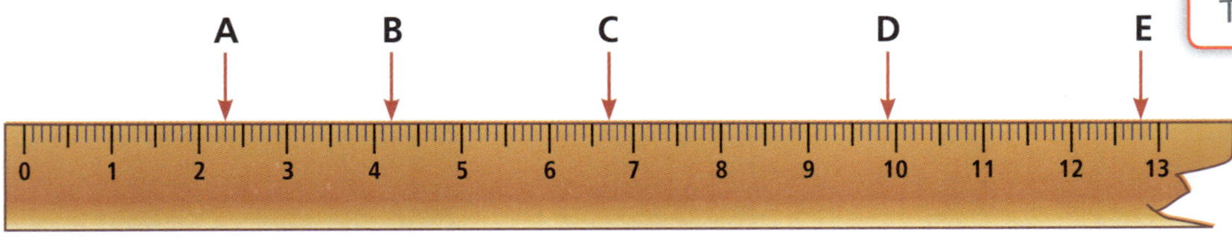

1 Record the lengths shown on the ruler.

A ___ mm ___ cm B ___ mm ___ cm C ___ mm ___ cm D ___ mm ___ cm E ___ mm ___ cm

2 Use the decimal form to write these as centimetres.

a 63 mm ___ b 95 mm ___ c 79 mm ___ d 36 mm ___

e 81 mm ___ f 103 mm ___ g 41 mm ___ h 162 mm ___

3 Use the decimal form to write these as metres.

a 135 cm ___ b 431 cm ___ c 565 cm ___ d 984 cm ___

e 729 cm ___ f 385 cm ___ g 895 cm ___ h 1075 cm ___

4 Choose the most suitable unit (**km**, **m**, **cm**, **mm**) to measure:

a the height of a wall ___ b the thickness of a door ___

c the distance to London ___ d the length of your foot ___

e the width of Australia ___ f the width of a match ___

ICT

To change metres to kilometres, divide by 1000.
e.g. 2364 m = 2364 ÷ 1000
= 2·364 km

When dividing, the decimal point moves to the left.

When multiplying, the decimal point moves to the right.

5 Use a calculator to complete the table.

Metres	1659	4752	9225	5076	3855	7125
Kilometres						

6 To change kilometres to metres, multiply by 1000. Use a calculator to complete the table.

Kilometres	3·575	9·635	2·749	7·36	6·4	8
Metres						

3:04 Converting Measurements

1 Use the decimal form to write these as centimetres.

a 49 mm ☐ b 67 mm ☐ c 96 mm ☐ d 32 mm ☐
e 83 mm ☐ f 54 mm ☐ g 145 mm ☐ h 463 mm ☐

2 Use the decimal form to write these as metres.

a 135 cm ☐ b 326 cm ☐ c 951 cm ☐ d 796 cm ☐
e 478 cm ☐ f 659 cm ☐ g 1 234 cm ☐ h 3 865 cm ☐

3 Write the following as metres.

a 1·657 km ☐ b 3·425 km ☐ c 2·976 km ☐
d 5·125 km ☐ e 9·816 km ☐ f 4·291 km ☐
g 6·95 km ☐ h 8·75 km ☐ i 7·1 km ☐

1 cm = 10 mm
1 m = 100 cm
1 km = 1 000 m

4 Choose the most suitable unit of length (**km**, **m**, **cm**, **mm**) to measure:

a the width of a tennis court ☐ b the thickness of a toothpick ☐
c the length of a highway ☐ d the width of this book ☐
e the length of a garden ☐ f the distance to New Zealand ☐

5 Complete the table.

	Object	Millimetres	Centimetres	Metres
a	Watchband Width	17 mm	1·7 cm	0·017 m
b	Pencil Length		9·4 cm	
c	Eraser Length	38 mm		
d	Felt Pen Length		14·5 cm	
e	Paper Width			0·21
f	Calculator Thickness	12 mm		

- In pairs, plan a fun run of 1 km within the school grounds.
- Use tapes or trundle wheels to check the length of the course.
- Draw the course on the plan of the school so that others can follow it.
- Discuss the scale used on the plan of the school.

3:05 Units of Length

1 Choose the most suitable unit of measurement (**km**, **m**, **cm**, **mm**) for:

- **a** the distance from Perth to Melbourne
- **b** the length of this page
- **c** the length of a basketball court
- **d** the width of a key
- **e** the thickness of this book
- **f** the length of a swimming pool
- **g** the distance you throw a ball
- **h** the height of a girl

2 Use the decimal form to write each in metres.

- **a** 1 m 56 cm
- **b** 3 m 39 cm
- **c** 2 m 94 cm
- **d** 5 m 67 cm
- **e** 4 m 86 cm
- **f** 7 m 24 cm
- **g** 9 m 45 cm
- **h** 8 m 17 cm

3 Use the decimal form to write each in metres.

- **a** 627 cm
- **b** 495 cm
- **c** 261 cm
- **d** 478 cm
- **e** 894 cm
- **f** 134 cm
- **g** 716 cm
- **h** 548 cm

4 On this map, 1 cm represents 1 km. Following the red path, measure distances correct to one decimal place (the nearest mm) to answer the following questions.

- **a** How much further is it in kilometres from the swamp to the treasure than from the mountain to the treasure?
- **b** Which distance is longer: from the swamp to the waterfall (**A**) or from the swamp to the treasure (**B**)?

  What is the difference in distance between these two trips?
- **c** How far is it from the ship to the treasure?

ACTIVITY

Estimate then measure each object in millimetres.

- length of a pen
- height of a desk
- width of a door
- thickness of a door

3:06 Variation in Measurement

When we measure, we may make mistakes or not be as accurate as we should be.

CONCEPT

1 John measured the length of a piece of wood 5 times.

Measurements were: 95 mm | 9 cm 5 mm | 96 mm | 19 cm 6 mm | 9·5 cm

a Which is most likely to be the real length? ____

Why is this so? ____

b Which is least likely to be the real length? ____

Why is this so? ____

ACTIVITY

2 Have five different people measure the length of a desk.

a Write the measurements below.

b Were all of the measurements the same? ____

c Why do you think these measurements might vary?

3 Have four different people measure the arm span of a volunteer.

a Write the measurements below.

b Were all of the measurements the same? ____

c Write possible reasons why these measurements could be different.

3:07 Perimeter

1 Find the perimeter of each shape.

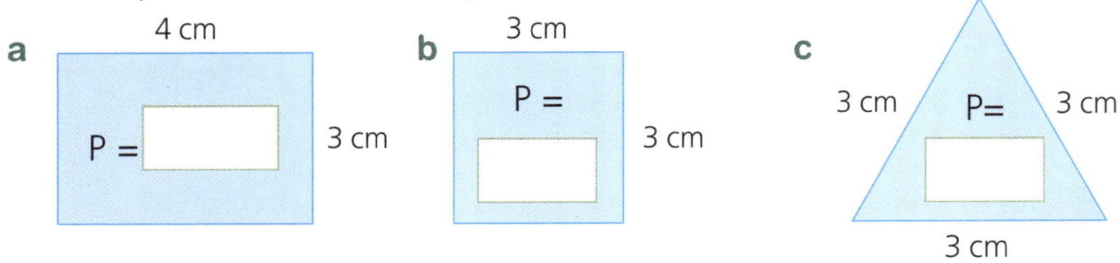

Discuss shortcut methods for calculating these perimeters.

2 Use shortcut methods to calculate the perimeter of each shape.

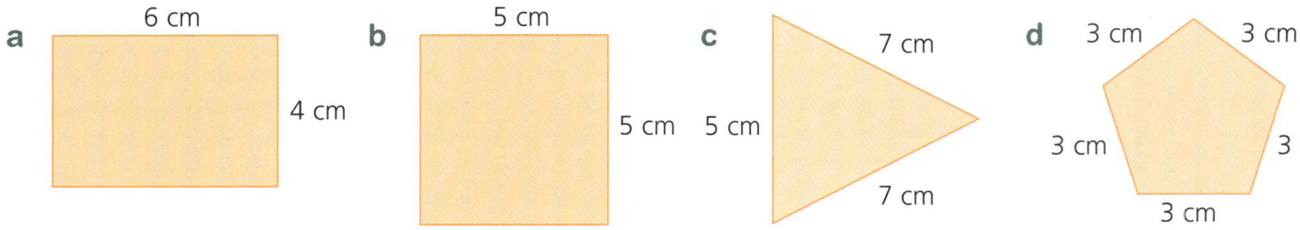

a P = b P = c P = d P =

3 On grid paper, draw rectangles with the given dimensions. Find their perimeters and complete the table.

Length	Width	Perimeter
6 cm	4 cm	
9 cm	7 cm	
8 cm	5 cm	
7·5 cm	4·5 cm	
6·5 cm	3·5 cm	

4 Circle the correct perimeter for these rectangles.

a length 10 cm, width 6 cm
perimeter: 16 cm, 32 cm, 45 cm

b length 12 m, width 3 m
perimeter: 15 m, 25 m, 30 m

c length 8 cm, width 6 cm
perimeter: 14 cm, 20 cm, 28 cm

d length 9 km, width 3 km
perimeter: 24 km, 27 km, 30 km

5 Find the perimeter of each shape.

a rectangle with sides 5 cm and 3 cm

b rectangle with sides 7 m and 6 m

c rectangle with sides 8 cm and 6 cm

d rectangle with sides 5 km and 4 km

e rectangle with sides 3·2 m and 6·3 m

f square with sides 9·5 cm

6 Which shape in Question 2 has the smallest perimeter?

 • *AUSTRALIAN SIGNPOST MATHS 6* • ISBN 978 1 4886 2193 2

3:08 Perimeter

1 Find the perimeter of each shape.

a

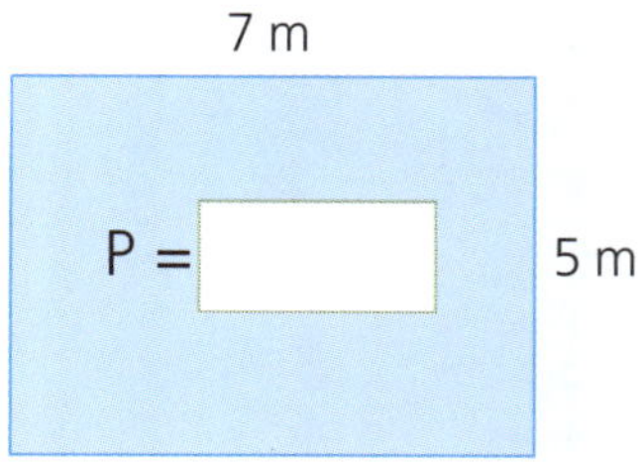

b

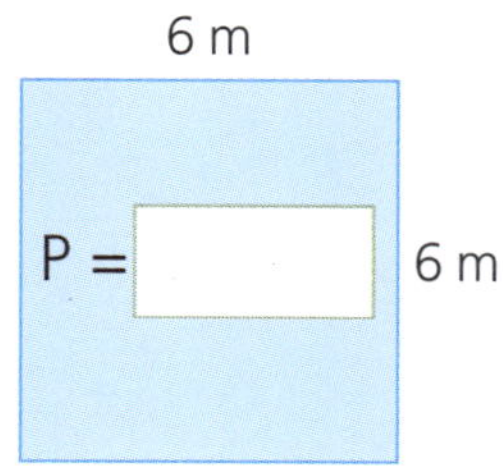

c

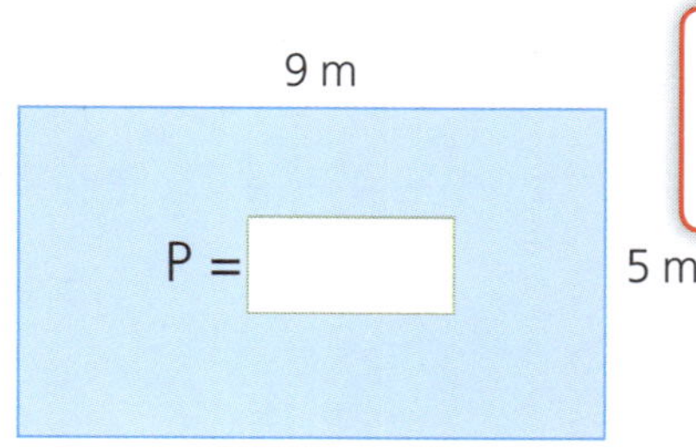

d

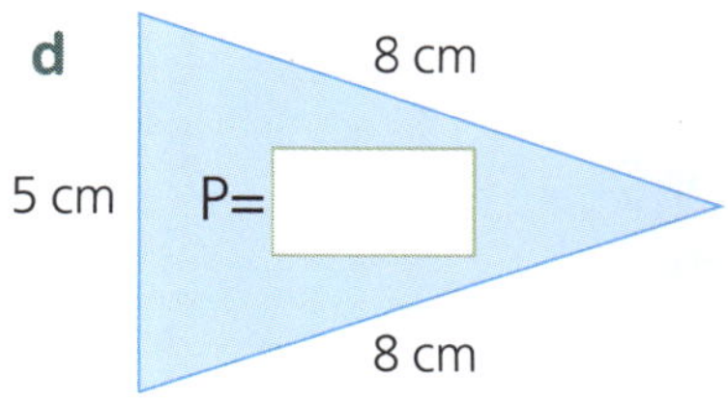

e

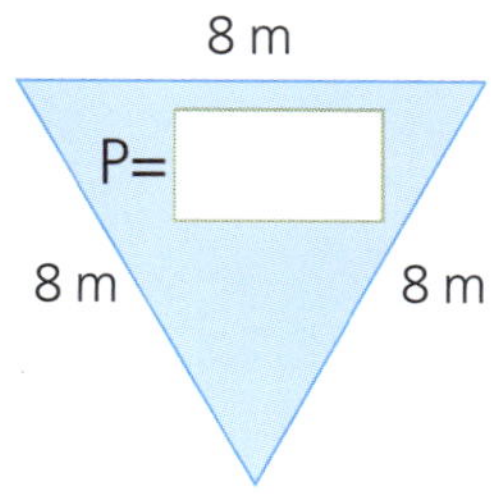

f

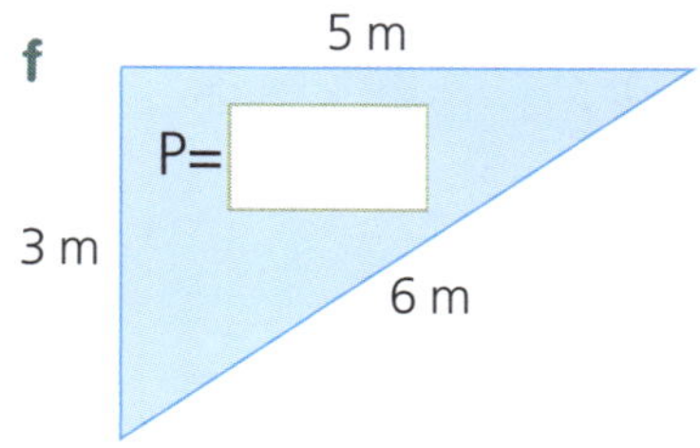

2 Circle the correct perimeter for these rectangles.

a length 9 cm, width 6 cm
perimeter: 15 cm, 54 cm, 30 cm

b length 15 m, width 9 m
perimeter: 135 m, 48 m, 24 m

c length 7·5 cm, width 3 cm
perimeter: 21 cm, 22·5 cm, 10·5 cm

d length 16 km, width 5 km
perimeter: 21 km, 42 km, 80 km

3 Complete the table.

	Shape	Length of Sides	Perimeter
a	Square	6·4 cm	
b	Regular Hexagon	9 cm	
c	Regular Pentagon	4·9 cm	
d	Regular Octagon	12 m	
e	Regular Heptagon	10·2 cm	

f Which of these shapes has the largest perimeter?

ACTIVITY

- In a group, measure and mark out a square with 25 m sides.
- Use a trundle wheel to measure the perimeter of the square.
- Measure and mark out squares of different sizes.
- Estimate and measure the perimeter of each square.

 • *AUSTRALIAN SIGNPOST MATHS 6* • ISBN 978 1 4886 2193 2

3:09 Elapsed Time

CONCEPT

Elapsed time is the time that has passed.
e.g. 9:45 pm to 4:20 am

Elapsed time 9:45 pm —15 mins→ 10 pm —2 hours→ 12 midnight —4 hours 20 mins→ 4:20 am

Time elapsed = 2 hours + 4 hours + 15 mins + 20 mins
= 6 h 35 min

1 What time has elapsed between:

- **a** 3 am and 4 pm?
- **b** 11 pm and 7 am?
- **c** 4:10 am and 11:00 am?
- **d** 10:30 pm and 4:10 am?
- **e** 3 pm Mon and 2 pm Tues?
- **f** 17:47 and 23:12?
- **g** 4:17 am Thurs to 2:15 am Fri?
- **h** 01:42 and 14:28?
- **i** 16:47 and 08:15 (next day)?
- **j** 2:22 am Mon and 2:22 am Tues?

2 Use **am** or **pm** to write the time that is one hour after:

- **a** 16:39
- **b** 07:28
- **c** 17:03
- **d** 23:35

3 Use **am** or **pm** to write the time the train arrives at:

- **a** Keeton
- **b** Musgrave
- **c** Price
- **d** Elmont
- **e** Westing
- **f** Tewalla
- **g** Flinders

Sometimes 24-hour time is written without the colon.

Train Timetable	
Westing	09:46
Keeton	10:07
Price	10:39
Moreton	11:16
Flinders	11:49
Musgrave	12:31
Elmont	12:59
Barren Point	13:26
Tewalla	13:48
Dutton	14:19
Town Centre	14:37

4 How long does it take to go from:

- **a** Keeton to Price?
- **b** Musgrave to Elmont?
- **c** Price to Moreton?
- **d** Tewalla to Dutton?

ACTIVITY

- Collect a number of different timetables, e.g. train, bus, ferry, plane. Practise reading these in 12-hour and 24-hour time.
- Use a square piece of paper for each day of the month to build a time line for a whole month.

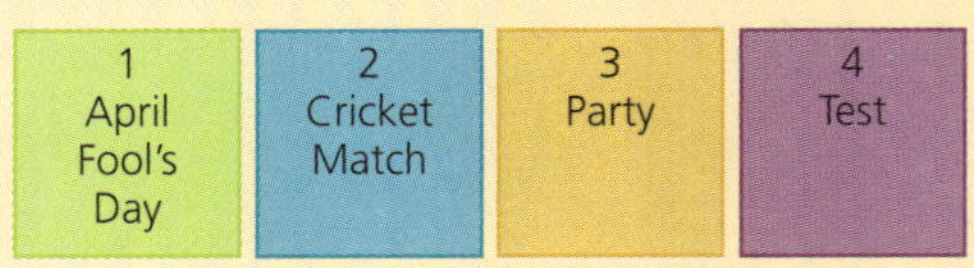

3:10 Timetables

1 Study this timetable carefully and answer the questions below.

Edmont to City Centre – Mondays to Fridays										
	am	am	am	am	am	am	am	pm	pm	pm
Edmont	11:03	11:10	11:19	11:27	11:36	11:45	11:55	12:05	12:16	12:28
Johnson	11:06	11:13		11:30	11:39		11:58	12:08		12:31
Mason	11:10	11:17		11:34	11:43		12:02	12:12		12:35
Broken Point	11:14	11:21		11:38	11:47		12:06	12:16		12:39
Cory Bay	11:17	11:24		11:41	11:50		12:09	12:19		12:42
Flinders	11:21	11:28		11:45	11:54		12:13	12:23		12:46
Parker Street	11:24	11:31	11:34	11:48	11:57	12:00	12:16	12:26	12:31	12:49
City Centre	11:28	11:35	11:38	11:52	12:01	12:04	12:20	12:30	12:35	12:53

a How long does it take an **all-stations** train to travel from Edmont to City Centre?

b How long does it take an **express** train to travel from Edmont to City Centre?

c What time does the 11:36 am train from Edmont arrive at Flinders?

d What time does the 11:55 am train from Edmont arrive at Cory Bay?

e What is the latest train that could be taken from Edmont to arrive at City Centre by 12 noon?

f If you missed the 11:27 am train from Edmont, how long would you have to wait for the next one?

2 Use 24-hour time to write:

a 7:15 am **b** 1:32 pm **c** 4:50 am **d** 11:13 am
e 5:21 pm **f** 6:48 pm **g** 2:19 pm **h** 8:03 pm

3 Use **am** or **pm** to write:

a 06:33 **b** 22:19 **c** 10:04 **d** 19:39
e 11:24 **f** 07:44 **g** 13:56 **h** 20:34

Planning a Trip

ACTIVITY

You live at Edmont. You wish to travel to City Centre for lunch and then take a 10-minute bus trip (buses leave on the $\frac{1}{2}$-hour) to see a 2-hour movie at 2:30 pm. Plan the trip.

When I leave

When I reach city

Lunch

Catch the bus

Movie

Catch bus back

Arrive City Centre

Start for home

Comments

 • *AUSTRALIAN SIGNPOST MATHS 6* • ISBN 978 1 4886 2193 2

3:11 Square Centimetres

1 Calculate the area of each shape.

a
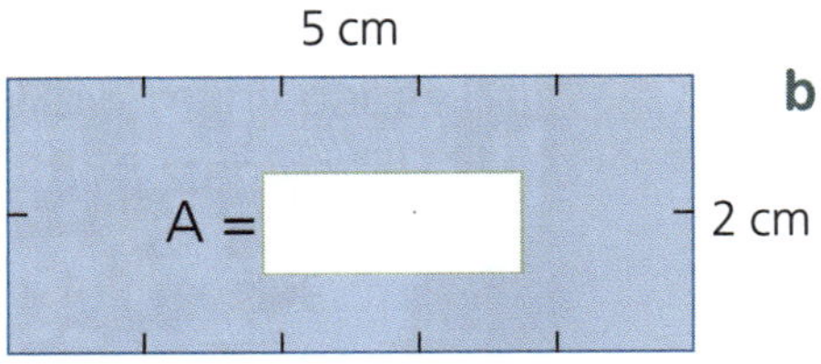

b
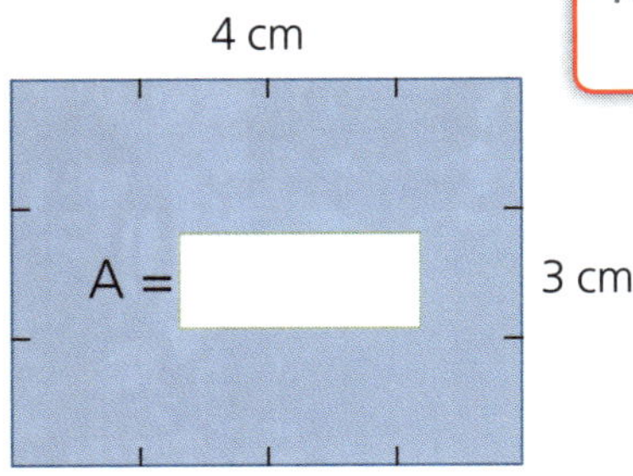

c
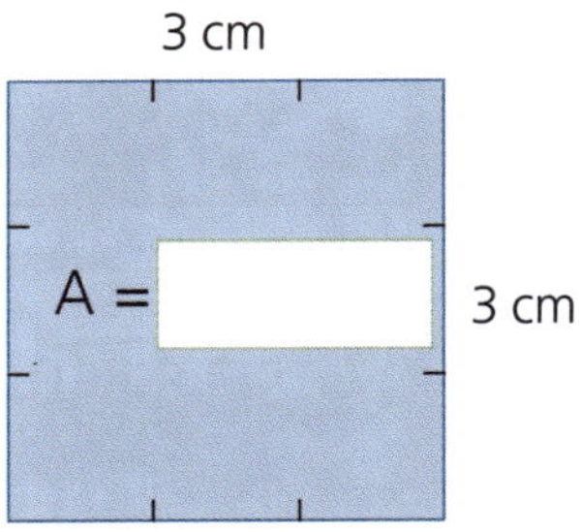

d
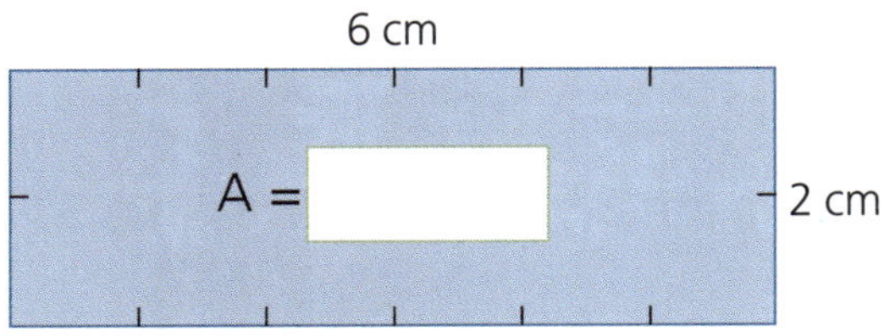

e
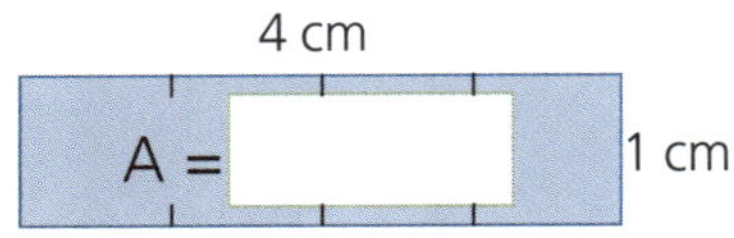

2 Use the measurements below to draw rectangles on 1 cm grid paper. Find the area of each and complete the tables.

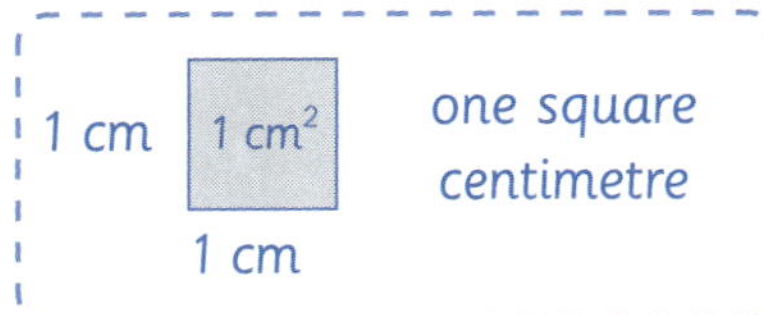

	Length in cm	Width in cm	Area in cm²
a	5	3	
b	4	4	
c	5	6	
d	6	5	

	Length in cm	Width in cm	Area in cm²
e	4	6	
f	7	2	
g	8	3	
h	5	4	

3 Calculate the area of each rectangle with the following measurements:

a length 10 cm, width 4 cm

b length 5 cm, width 6 cm

c length 8 cm, width 4 cm

d length 7 cm, width 3 cm

e length 8 cm, width 3 cm

f length 10 cm, width 6 cm

Making Plans of Areas

ACTIVITY

Sometimes we need to find the area of an irregular shape like the one shown.

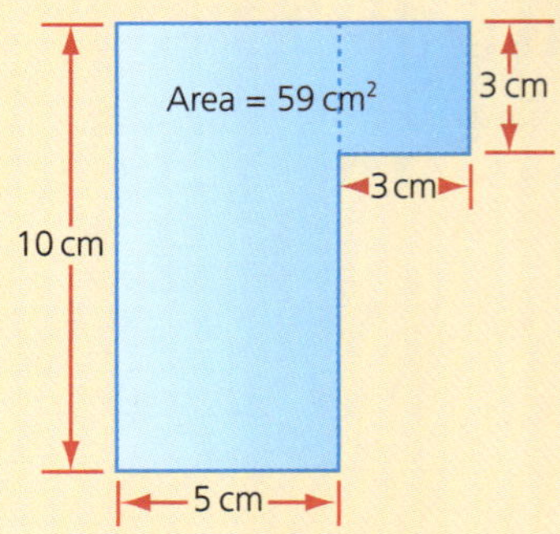

- Discuss ways to find its area.
- Write a number sentence to show how to calculate the area.
- Measure a section of your classroom, to the nearest metre. Draw a scale plan of it on grid paper and calculate the area.
- Repeat the activity for other areas.

3:12 Square Metres

1 Calculate the area of each shape.

a

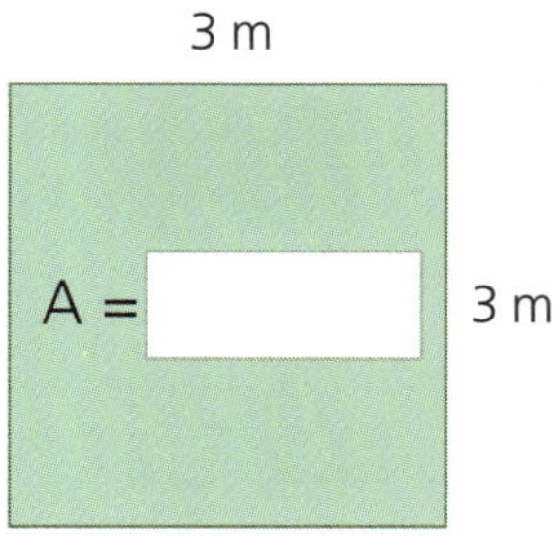

b

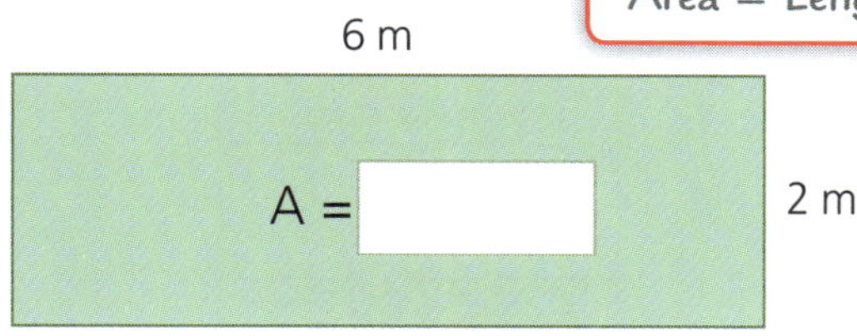

c

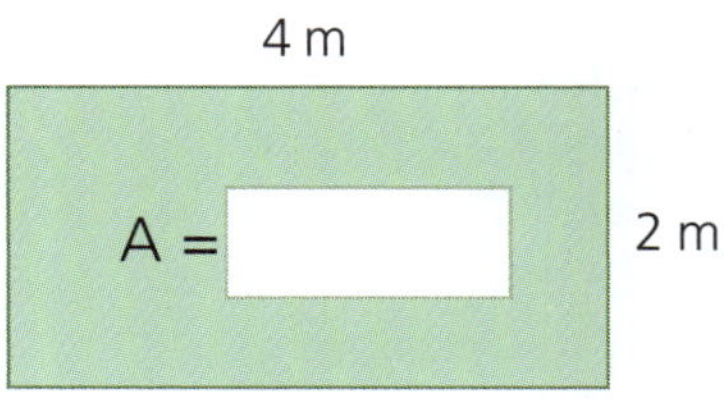

d

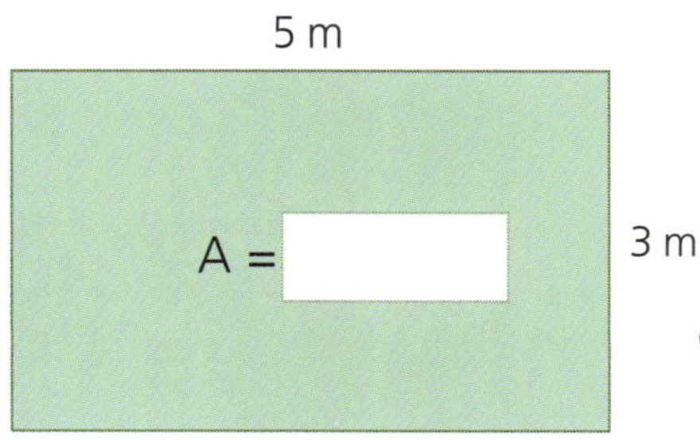

A = L × W

e

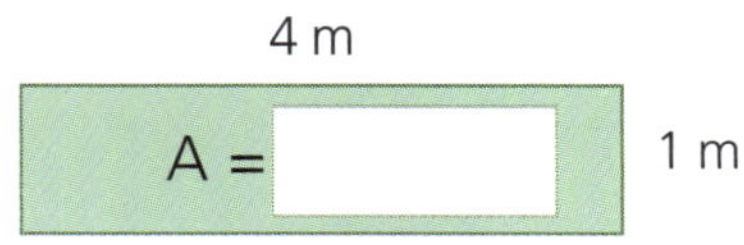

2 Circle the correct area for each rectangle with the following measurements.

a Length 6 m, Width 5 m
Area: 11 m^2, 22 m^2, 30 m^2

b Length 10 m, Width 7 m
Area: 17 m^2, 70 m^2, 35 m^2

c Length 8 m, Width 7 m
Area: 15 m^2, 30 m^2, 56 m^2

d Length 9 m, Width 8 m
Area: 72 m^2, 17 m^2, 34 m^2

3 The rectangles in each table have the same perimeter. Calculate the area of each rectangle.

	Length in m	Width in m	Area in m^2
a	1	7	
b	2	6	
c	3	5	
d	4	4	

	Length in m	Width in m	Area in m^2
e	6	6	
f	7	5	
g	8	4	
h	9	3	

i In each table, what is the type of rectangle that has the greatest area?

ACTIVITY

- On 1 cm grid paper, draw a rectangle that has an area of 12 m^2. (Use a scale of 1 cm = 1 m.)
- How many different rectangles can you draw that have the same area?

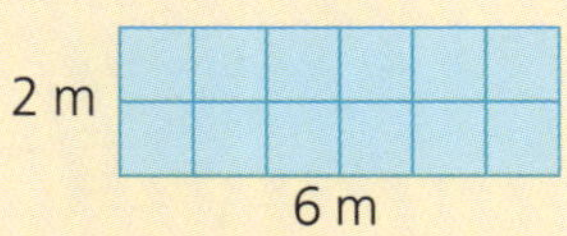

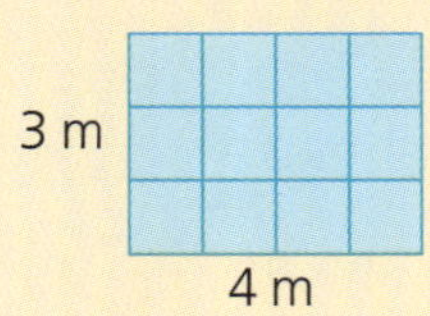

3:13 Comparing Perimeter and Area

Comparing Areas

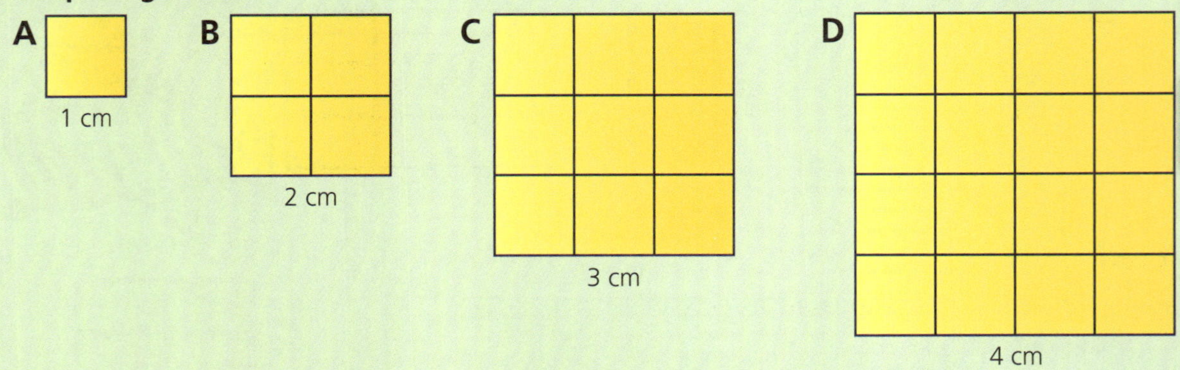

If we double the dimensions of a plane shape, we multiply the area by 4.

1. Use the diagrams above to find the perimeter of:
   a A ______ b B ______ c C ______ d D ______

2. a If we multiply the dimensions of a plane shape by 2, we multiply the perimeter by ______.
   b If we multiply the dimensions of a plane shape by 3, we multiply the perimeter by ______.
   c If we multiply the dimensions of a plane shape by 4, we multiply the perimeter by ______.

3. Use the diagrams above to find the area of:
   a A ______ b B ______ c C ______ d D ______

4. a If we multiply the dimensions of a plane shape by 2, we multiply the area of the shape by ______.
   b If we multiply the dimensions of a plane shape by 3, we multiply the area of the shape by ______.
   c If we multiply the dimensions of a plane shape by 4, we multiply the area of the shape by ______.

5. Find the area and perimeter of each shape if each square on the grid represents 1 m^2.

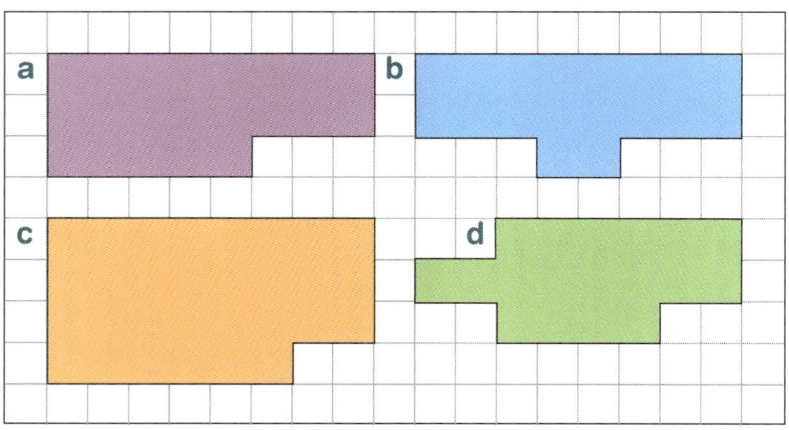

   a Area: ______ Perimeter: ______
   b Area: ______ Perimeter: ______
   c Area: ______ Perimeter: ______
   d Area: ______ Perimeter: ______

6. Investigate the change in volume of a cube if the side lengths are doubled. What change occurs?
   ______

 • *AUSTRALIAN SIGNPOST MATHS 6* • ISBN 978 1 4886 2193 2

3:14 Packing and Stacking

Cans in Carton

View from the Top

View from the Side

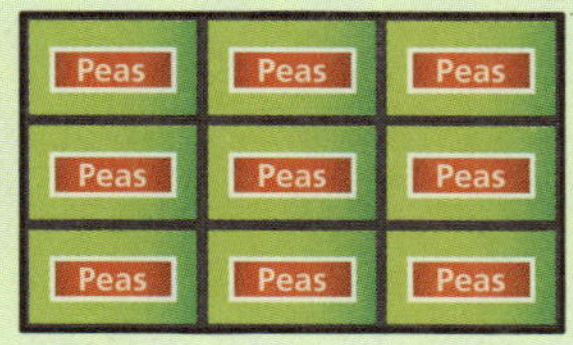

CONCEPT

An **efficient** carton or box:

- would have as much of its space used as possible
- would be a rectangular prism
- would be easy to carry
- could be easy to stack.

View from the End

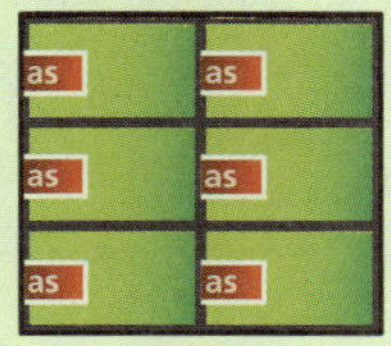

1 Use the pictures above to answer these questions.

a How many cans are in one level of the carton? ______

b How many levels of cans are in the carton? ______

c How many cans are in the carton? ______

d Why is this box an efficient carton for these cans?

______

______

______

______

2 The carton above has 3 levels. It is 3 cans long and 2 cans wide.

a Use the table to design 4 other cartons that would hold these cans so that they are held firm.

b Which of your cartons do you think is the most efficient? ______

Number of Levels	Cans in the Length	Cans in the Width	Number of Cans
			18
			18
			18
			18

3 Design a carton to hold 24 cans that are 7 cm wide and 4 cm high.

ACTIVITY

Number of Levels	Cans in the Length	Cans in the Width

Length of Carton	Width of Carton	Height of Carton

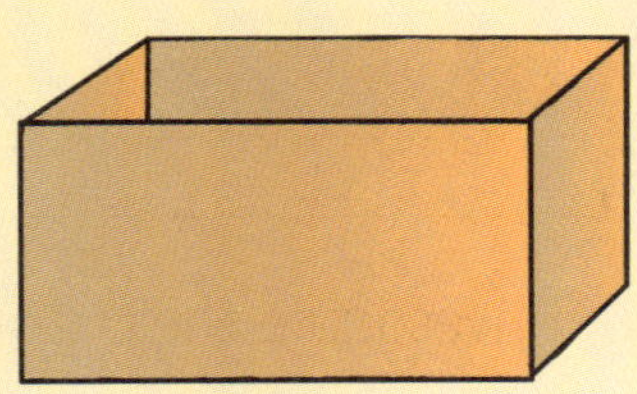

4 Use 12 objects of the same shape and size. Design a carton to hold the objects efficiently.

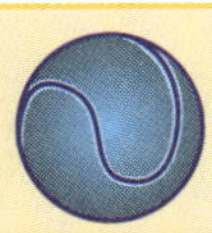

 • • ISBN 978 1 4886 2193 2

3:15 Volume of Prisms

CONCEPT

Volume of a Rectangular Prism

To calculate the volume of a rectangular prism, we work out how many cubes are in one layer, and multiply by the number of layers.

For this prism we can see there are 2 layers of 5 × 2 cubes.

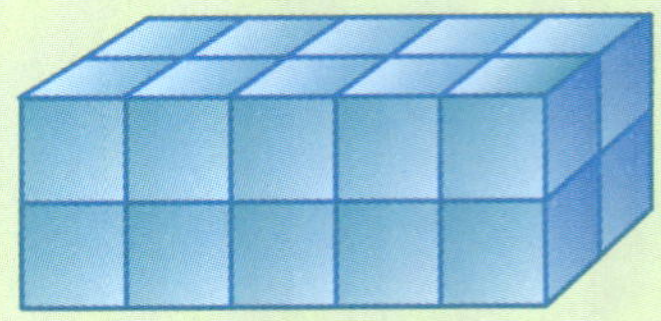

Volume of a prism = Length × Width × Height

V = L × W × H

---

1 Would you use cubic centimetres (**cm³**) or cubic metres (**m³**) to find the volume of each?

a a matchbox ☐ **b** a school hall ☐ **c** a shipping container ☐

d a briefcase ☐ **e** a lunch box ☐ **f** a shoe box ☐

---

2 Calculate the volume of each rectangular prism (made of cubic centimetres).

a

V = ☐

b

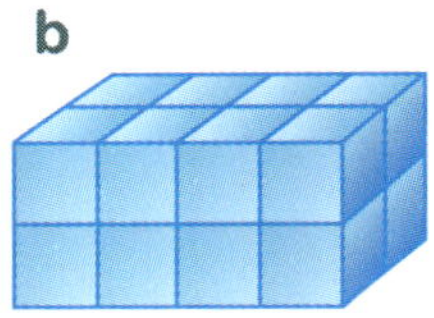

V = ☐

c

V = ☐

d

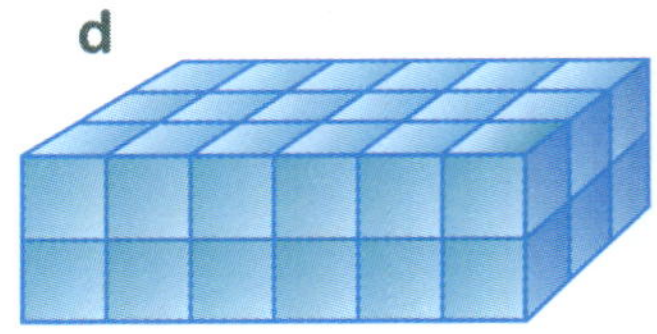

V = ☐

---

3 Calculate the volume of each rectangular prism, using Volume = Length × Width × Height.

a

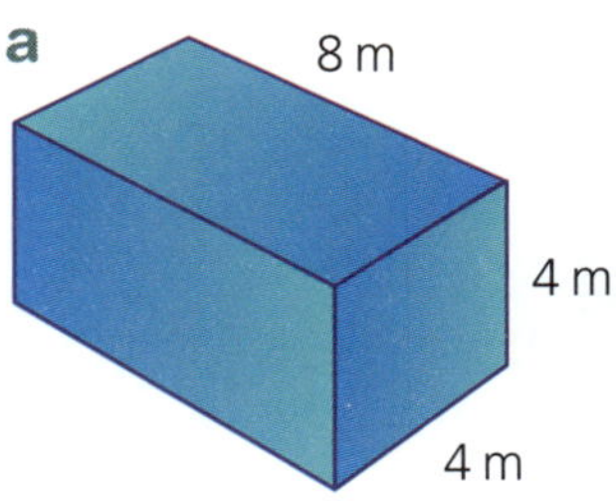

V = ☐

b

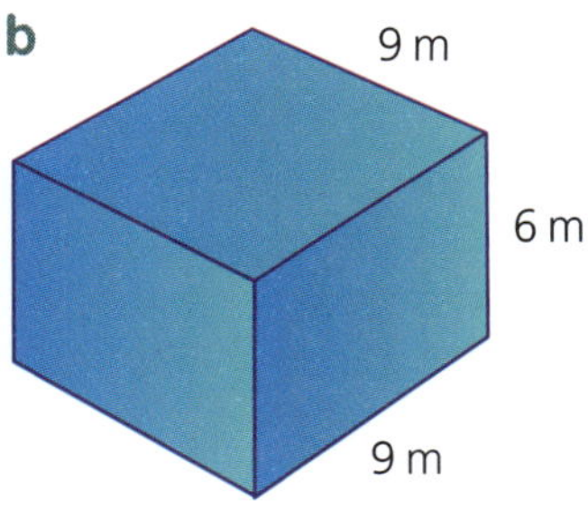

V = ☐

c

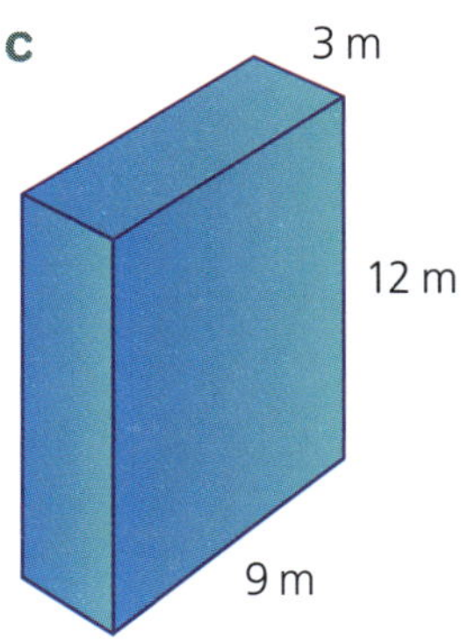

V = ☐

d

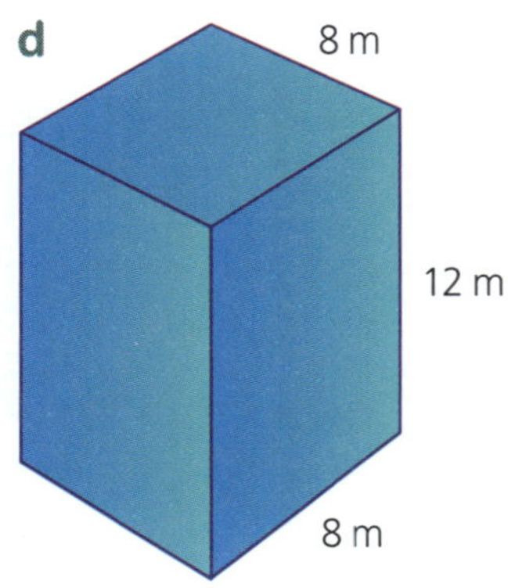

V = ☐

3:16 Volume of Rectangular Prisms

1 Calculate the volume of each rectangular prism.

a

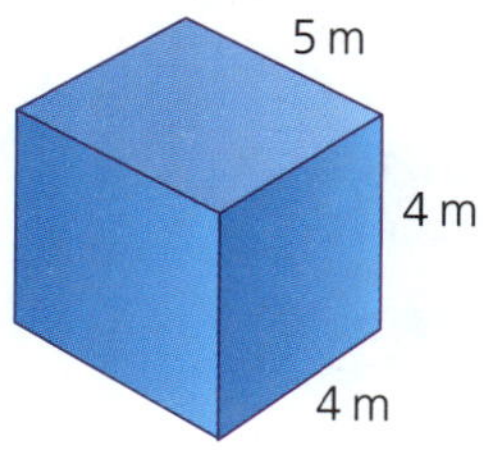

V =

b

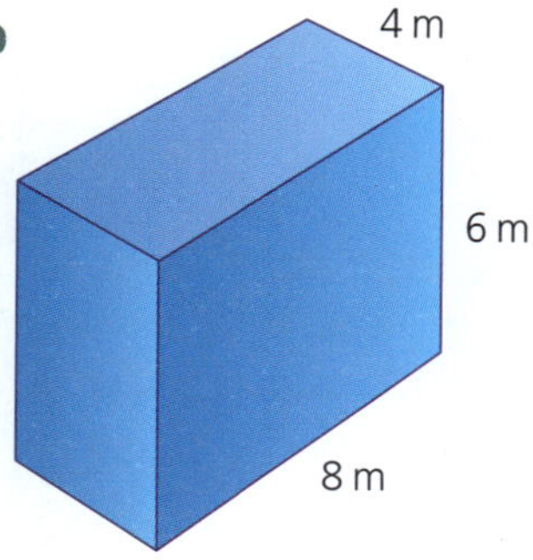

V =

c

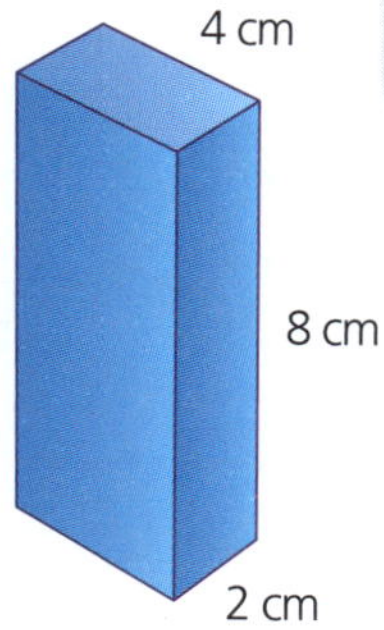

V =

d

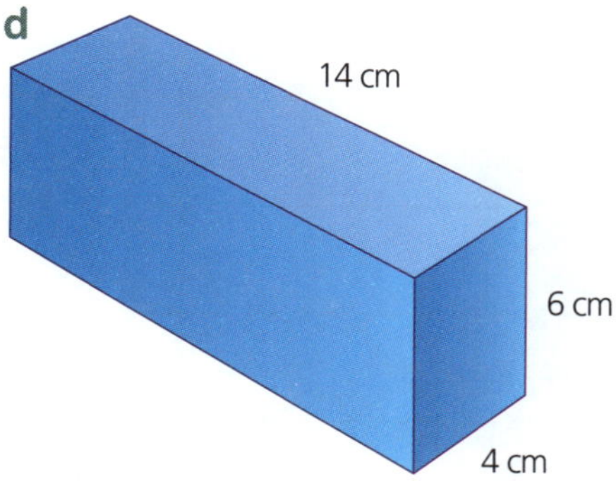

V =

e

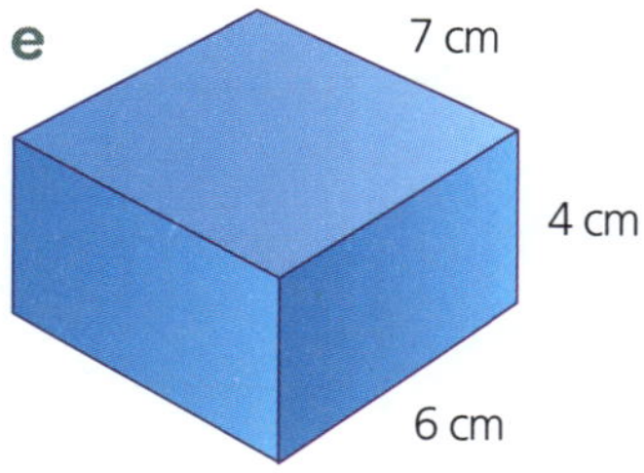

V =

f

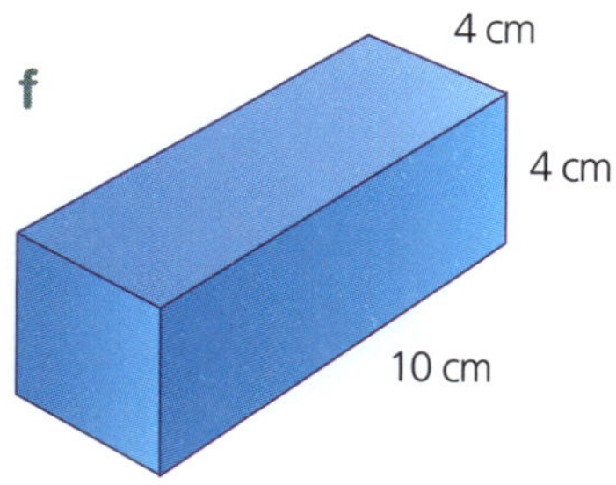

V =

2 Complete the tables.

	Length	Width	Height	Volume
a	6 cm	5 cm	4 cm	
b	8 cm	3 cm	2 cm	
c	10 cm	4 cm	5 cm	
d	9 cm	3 cm	2 cm	

	Length	Width	Height	Volume
e	6 m	3 m	2 m	
f	5 m	5 m	2 m	
g	8 m	4 m	5 m	
h	10 m	6 m	3 m	

3 **a** For these two prisms to have the same volume, how many blocks must be added to prism **B**?

b How many times as much is the volume of **A** greater than the volume of **B**?

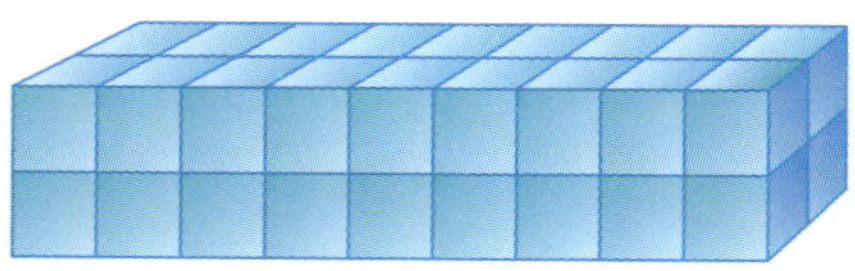

A

B

INVESTIGATION

Find examples of objects that have a volume between 500 and 1 000 cm^3.

3:17 Cubic Centimetres

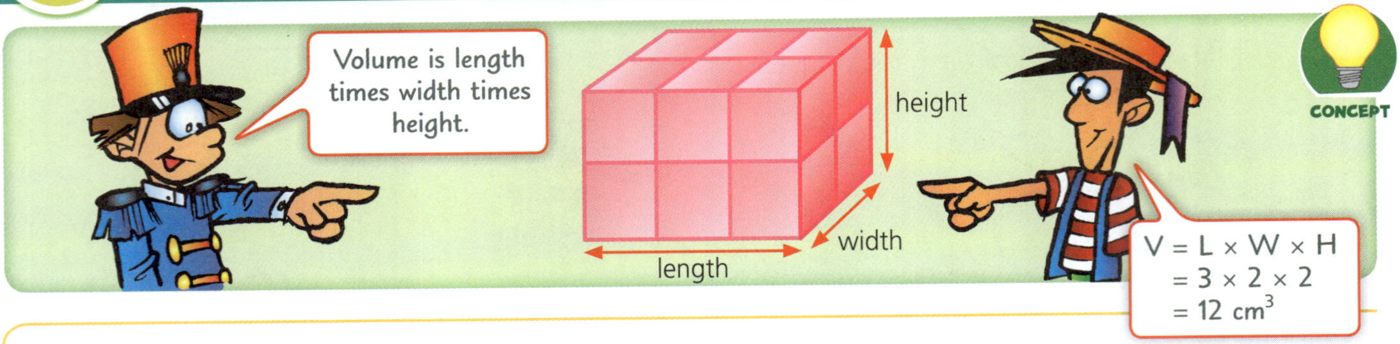

1 Calculate the volume of each rectangular prism.

a
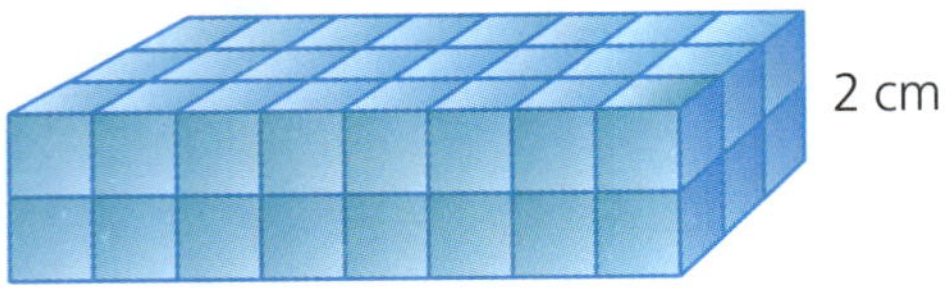

b
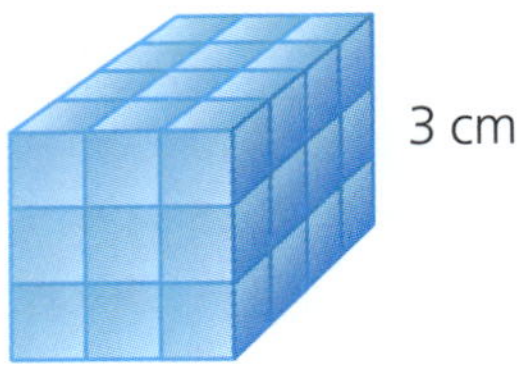

c
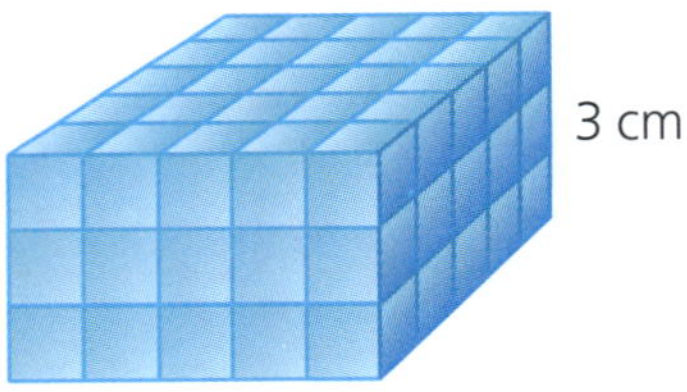

	Length	Width	Height	Volume
a	cm	cm	cm	cm^3
b	cm	cm	cm	cm^3
c	cm	cm	cm	cm^3
d	cm	cm	cm	cm^3
e	cm	cm	cm	cm^3
f	cm	cm	cm	cm^3
g	cm	cm	cm	cm^3

d

e
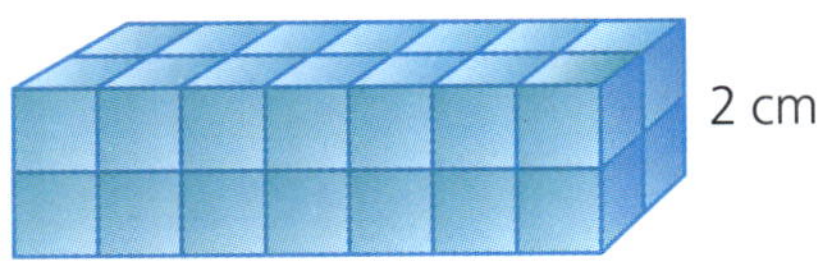

f
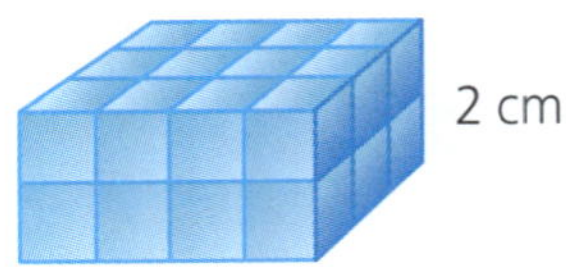

g
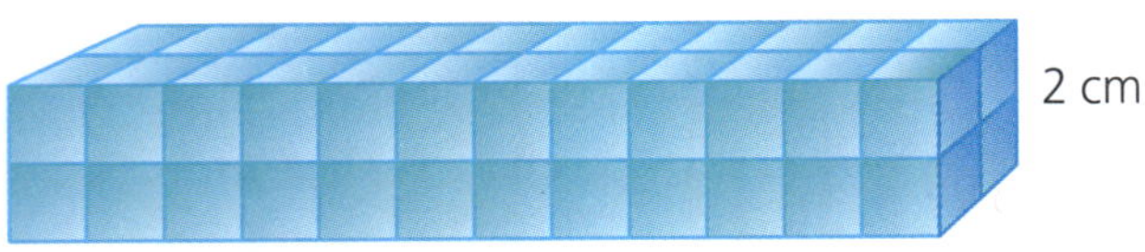

This is one cubic centimetre.

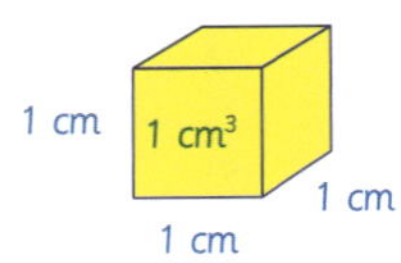

2 Calculate the volume of each rectangular prism.

a
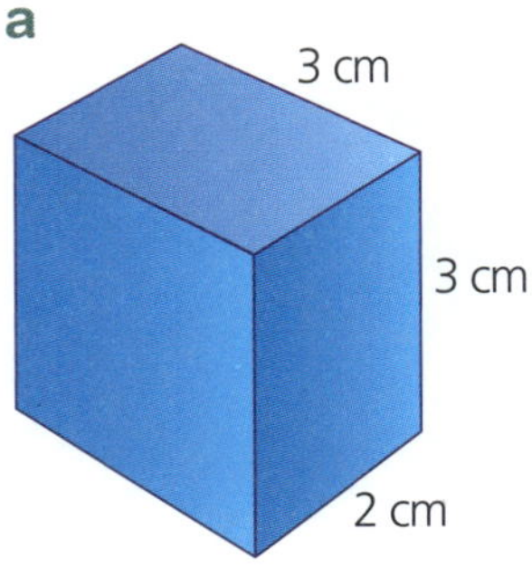

V = ☐ cm^3

b
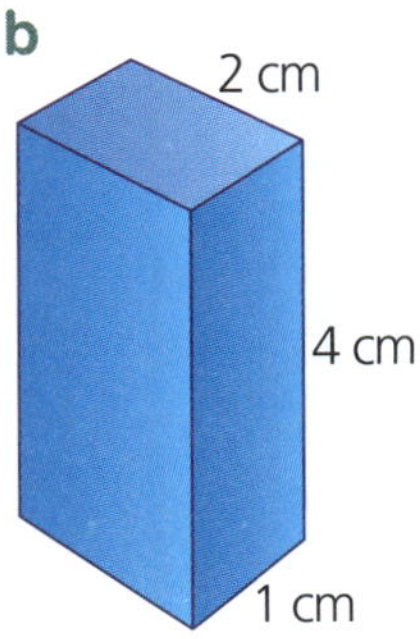

V = ☐ cm^3

c
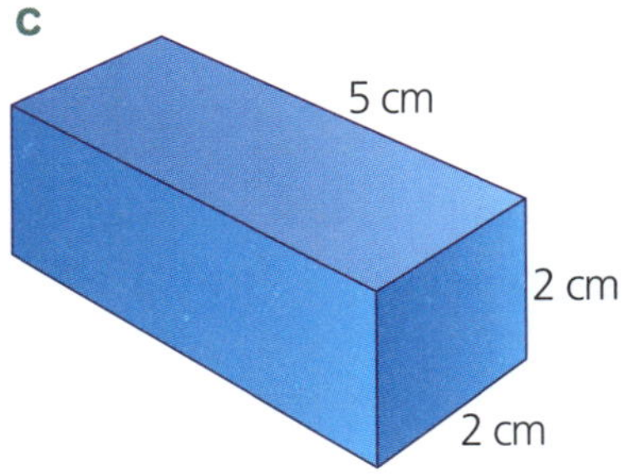

V = ☐ cm^3

d
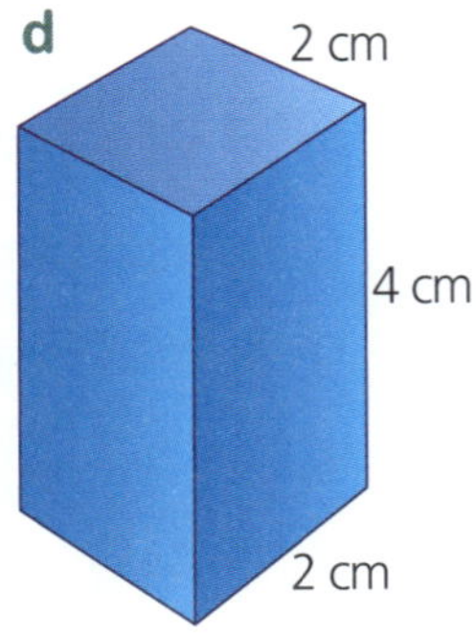

V = ☐ cm^3

 • *AUSTRALIAN SIGNPOST MATHS 6* • ISBN 978 1 4886 2193 2

3:18 Cubic Metres

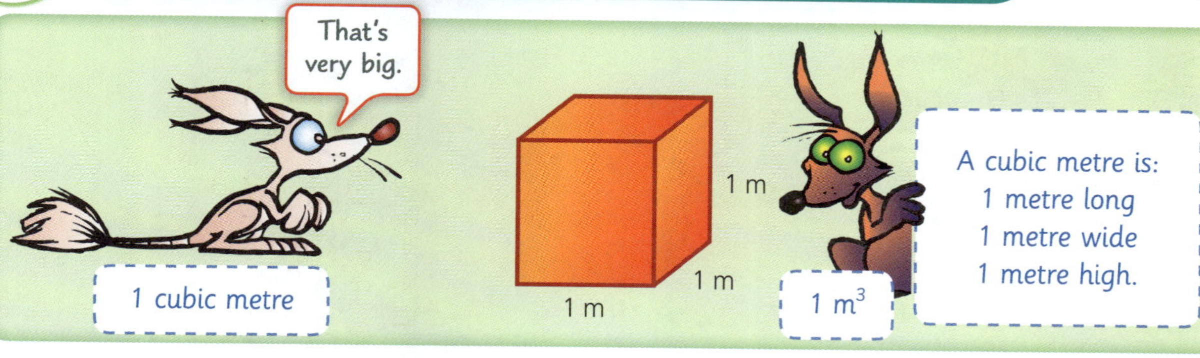

CONCEPT

1 Use the abbreviated form to write:

a 6 cubic metres ☐ **b** 30 cubic metres ☐ **c** 15 cubic metres ☐

d 20 cubic metres ☐ **e** 75 cubic metres ☐ **f** 34 cubic metres ☐

g 300 cubic centimetres ☐ **h** 76 cubic centimetres ☐

2 Would you use **cm³** or **m³** for the capacity of:

a a garage? ☐ **b** a frying pan? ☐ **c** a wardrobe? ☐

d a service station? ☐ **e** a pencil case? ☐ **f** a train carriage? ☐

3 Choose the most suitable unit (**cm³** or **m³**) for each of these measurements.

a brick: 1 540 ☐ **b** tool box: 12 000 ☐ **c** garden shed: 40 ☐

d grain silo: 9 000 ☐ **e** office: 60 ☐ **f** cereal box: 3 600 ☐

g matchbox: 30 ☐ **h** hall: 1 800 ☐ **i** trailer: 2 ☐

4 Find the volume of each rectangular prism.

a

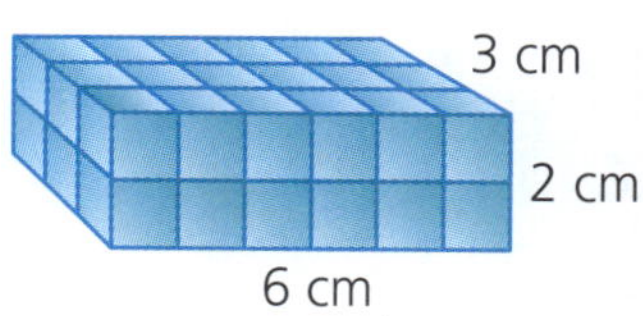

V = ☐ m³

b

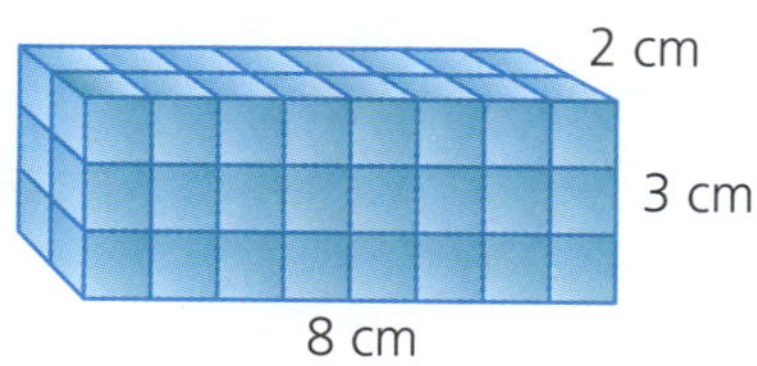

V = ☐ m³

c

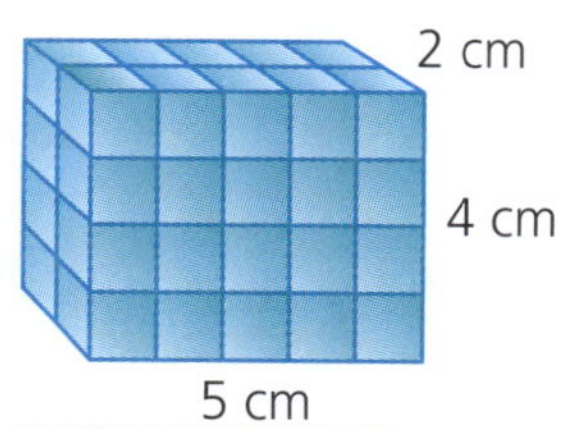

V = ☐ m³

d

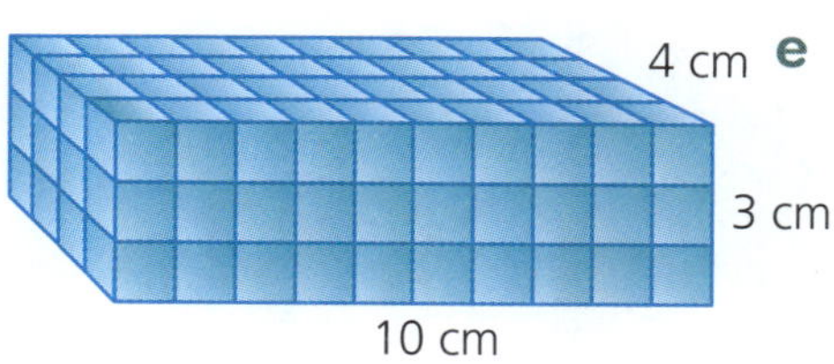

V = ☐ m³

e

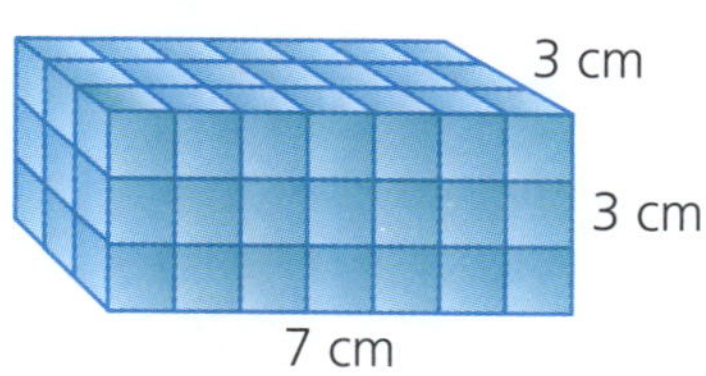

V = ☐ m³

f

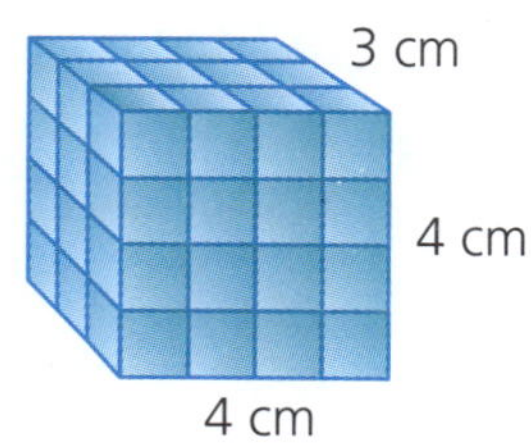

V = ☐ m³

 • *AUSTRALIAN SIGNPOST MATHS 6* • ISBN 978 1 4886 2193 2

3:19 Millilitres and Litres

1 Use decimal notation to write these as litres.

- **a** 439 mL
- **b** 835 mL
- **c** 146 mL
- **d** 340 mL
- **e** 207 mL
- **f** 67 mL

Remember, litres and millilitres are related by 1000.

2 Rewrite the following as millilitres.

- **a** 0·641 L
- **b** 0·793 L
- **c** 0·145 L
- **d** 0·839 L
- **e** 0·095 L
- **f** 0·026 L

3 Use decimal notation to write the following as litres.

- **a** 2 L 346 mL
- **b** 4 L 475 mL
- **c** 3 L 149 mL
- **d** 4 L 207 mL
- **e** 1 L 947 mL
- **f** 6 L 503 mL

4 Rewrite the following as millilitres.

- **a** 3·104 L
- **b** 5·624 L
- **c** 2·891 L
- **d** 6·516 L
- **e** 1·487 L
- **f** 7·235 L
- **g** 8·705 L
- **h** 3·065 L
- **i** 9·094 L

5 Match one illustration to each liquid measure.

600 mL		955 mL	
375 mL		0·4 L	
0·55 L		635 mL	
0·25 L		0·175 L	

Containers with different shapes may have the same capacity.

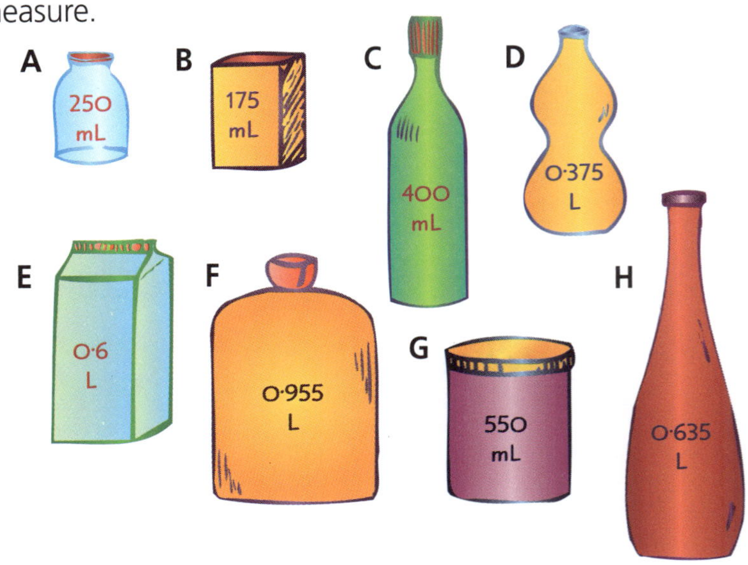

3:20 Kilolitres and Megalitres

CONCEPT

4 bathtubs hold about 1 kilolitre.

4 × 250 L = 1 kL

An Olympic swimming pool holds about 3 megalitres.

kL stands for kilolitres.
1 kL = 1 000 L
1 kL is equal to 1 cubic metre.

ML stands for megalitres.
1 ML = 1 000 000 L or 1 000 kL
1 ML is equal to 1 000 cubic metres.

kilo = 1 000
mega = 1 000 000

1 Write these measurements as kilolitres.

a 1 000 L ____ **b** 9 000 L ____ **c** 4 000 L ____
d 15 000 L ____ **e** 36 000 L ____ **f** 42 000 L ____

2 Use the decimal form to write these as kilolitres.

a 3 500 L ____ **b** 13 570 L ____ **c** 29 591 L ____
d 76 208 L ____ **e** 64 250 L ____ **f** 812 750 L ____

3 Write these measurements as megalitres.

a 4 000 kL ____ **b** 14 000 kL ____ **c** 53 000 kL ____
d 50 000 kL ____ **e** 725 000 kL ____ **f** 491 000 kL ____

4 Use the decimal form to write these measurements as megalitres.

a 6 500 kL ____ **b** 39 800 kL ____ **c** 32 629 kL ____
d 94 250 kL ____ **e** 640 750 kL ____ **f** 371 423 kL ____

5 Write these measurements as litres.

a 1 kL ____ **b** 3 kL ____ **c** 6·3 kL ____
d 29·45 kL ____ **e** 32·513 kL ____ **f** 4·662 kL ____

6 Write these measurements as kilolitres.

a 3 ML ____ **b** 17 ML ____ **c** 8·3 ML ____
d 28·24 ML ____ **e** 87·56 ML ____ **f** 97·465 ML ____

7 **a** How many kilolitres of water would fit into seven 320 L bathtubs? ____

b 32 000 ML of water was added to the Murray River in January and 36 000 ML was added in March. How much was added altogether? ____

3:21 Cubic Centimetres and Millilitres

$1 \text{ cm}^3 = 1 \text{ mL}$

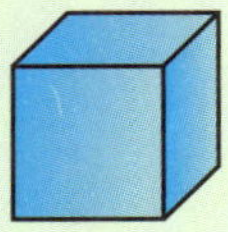

$1\,000 \text{ cm}^3 = 1 \text{ L}$

CONCEPT

1 Write as cubic centimetres (**cm³**).

a 20 mL		**b** 95 mL		**c** 50 mL		**d** 65 mL	
e 140 mL		**f** 650 mL		**g** 825 mL		**h** 975 mL	

2 Write as millilitres (**mL**).

a 45 cm³		**b** 75 cm³		**c** 90 cm³		**d** 60 cm³	
e 350 cm³		**f** 590 cm³		**g** 750 cm³		**h** 980 cm³	

3 Write as cubic centimetres (**cm³**).

a 3 L		**b** 5 L		**c** 4 L		**d** 9 L	
e 10 L		**f** 12 L		**g** 15 L		**h** 20 L	

4 Write as litres (**L**).

a 6 000 cm³		**b** 8 000 cm³		**c** 2 000 cm³		**d** 7 000 cm³	
e 9 000 cm³		**f** 10 000 cm³		**g** 11 000 cm³		**h** 16 000 cm³	

5 Calculate the answers and complete the table.

	Container	Problem	Answer
a		How much juice would there be in eight 250 mL bottles?	L
b		9 L of milk was equally shared by twenty chefs. How much did each receive?	mL
c		Twenty-four children each drank 125 mL of juice. How much juice was that in total?	L

3:22 Millilitres and Litres

1 Use decimal notation to write the following as litres.

a 457 mL ☐ b 379 mL ☐ c 955 mL ☐ d 157 mL ☐
e 58 mL ☐ f 734 mL ☐ g 29 mL ☐ h 5 mL ☐

2 Rewrite these as millilitres (**mL**).

a 0·216 L ☐ b 0·351 L ☐ c 0·625 L ☐ d 0·475 L ☐
e 0·75 L ☐ f 0·5 L ☐ g 0·8 L ☐ h 0·15 L ☐

3 Use decimal notation to write the following as litres (**L**).

a 1 L 255 mL ☐ b 3 L 5 mL ☐ c 2 L 25 mL ☐
d 4 675 mL ☐ e 247 mL ☐ f 75 mL ☐

4 Rewrite these as millilitres (**mL**).

a 2·104 L ☐ b 1·768 L ☐ c 4·5 L ☐
d 3·43 L ☐ e 7·23 L ☐ f 5·8 L ☐

5 How many cubic centimetres (**cm³**) are there in:

a 0·355 L? ☐ b 0·985 L? ☐ c 0·6 L? ☐
d 1·256 L? ☐ e 1·47 L? ☐ f 2·825 L? ☐

ACTIVITY

- Pour 500 mL of water into a calibrated measuring jug.
- Immerse an object such as a rock into the water.
- Use the scale to calculate the volume of the object.
- Repeat with other suitable objects and complete the table.

Object	Estimated Volume	Change in Jug
	mL	mL
	mL	mL
	mL	mL
	mL	mL

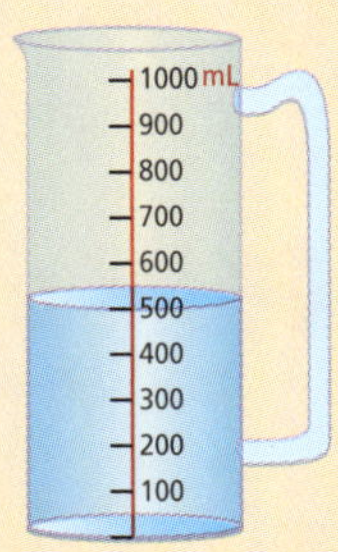

$1\ mL = 1\ cm^3$

3:23 Cubic Centimetres and Litres

ACTIVITY

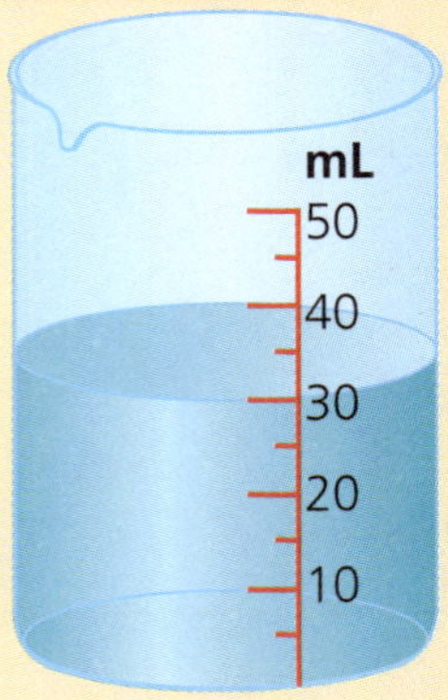

1 cm³ = 1 mL

- Pour 30 mL of water into a measuring cylinder.
- One by one, place ten Centicubes into the cylinder and note the change in water level. Make sure the Centicubes are fully immersed.

1000 cm³ = 1 L

- Fill an ice-cream container to the top with water.
- Immerse a thousand block in the water and collect the water that overflows into the tray.
- Measure the volume of the water collected.

1 Write the capacity (**mL**) of a container that can hold:

a 35 cm^3 ☐ **b** 40 cm^3 ☐ **c** 95 cm^3 ☐
d 65 cm^3 ☐ **e** 100 cm^3 ☐ **f** 125 cm^3 ☐
g 500 cm^3 ☐ **h** 750 cm^3 ☐

2 Write as cubic centimetres (**cm³**).

a 25 mL ☐ **b** 80 mL ☐ **c** 45 mL ☐ **d** 75 mL ☐
e 60 mL ☐ **f** 100 mL ☐ **g** 30 mL ☐ **h** 90 mL ☐
i 250 mL ☐ **j** 850 mL ☐ **k** 425 mL ☐ **l** 950 mL ☐

3 Write the capacity (**L**) of a container that can hold:

a 1000 cm^3 ☐ **b** 3000 cm^3 ☐ **c** 5000 cm^3 ☐ **d** 8000 cm^3 ☐
e 4000 cm^3 ☐ **f** 6000 cm^3 ☐ **g** 9000 cm^3 ☐ **h** 7000 cm^3 ☐

4 Calculate as cubic centimetres (**cm³**).

a 1 L ☐ **b** 6 L ☐ **c** 9 L ☐ **d** 10 L ☐
e 8 L ☐ **f** 5 L ☐ **g** 12 L ☐ **h** 15 L ☐

5 Three bottles each held 300 mL of water. If all this water was poured into a square container with a capacity of 1 L, what volume remained unfilled? ☐

6 A 2 L container was filled with water. From this container, 200 mL of water was poured into each of seven bottles. What volume of water remained in the container? ☐

7 If 3 L of juice was shared equally among six children, how much did each get? ☐

ISBN 978 1 4886 2193 2

3:24 Time Lines

1 How many years were there between:

a the Federation of Australia and the Federal Parliament in Canberra? ☐

b The starts of World War I and World War II? ☐

c the Melbourne Olympics and the Sydney Olympics? ☐

d the Sydney Harbour Bridge and the Sydney Harbour Tunnel? ☐

e the Federation of Australia and the New Parliament House? ☐

Australian History Time Line

2000	Sydney Olympics
1992	Sydney Harbour Tunnel
1988	New Parliament House
1973	Sydney Opera House
1956	Melbourne Olympics
1939	World War II starts
1932	Sydney Harbour Bridge
1927	Federal Parliament in Canberra
1914	World War I starts
1901	Federation of Australia

2 Major earthquakes have occurred in many places around the world. Use the letters below to show the listed earthquakes on the time line.

A Mexico (1985)
B India (1993)
C Turkey (1999)
D Japan (2011)
E Iran (1990)
F China (2008)
G USA (1994)
H Indian Ocean (2004)

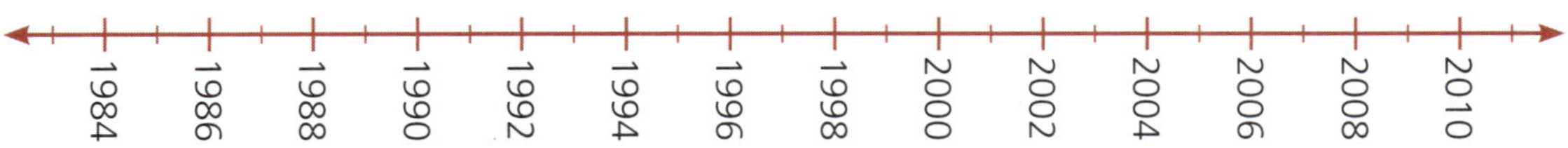

ACTIVITY

- Design a time line to show these important events in Mia's life.

had heart surgery in 2002

started school in 2005

moved to Melbourne in 2004

immigrated to Australia in 2001

born in 2000

had heart surgery in 2008

- Construct a personal time line that records important events in your life. Use an appropriate scale.
- Construct a time line for the first 12 months of a baby's life.
- Compare the scale you used for each time line. What did you notice?

 • *AUSTRALIAN SIGNPOST MATHS 6* • ISBN 978 1 4886 2193 2

3:25 Tonnes

1 000 grams = 1 kilogram
1 000 kilograms = 1 tonne

1 Calculate as kilograms.

a 3 t ____ b 6 t ____ c 10 t ____ d 7 t ____
e 5 t ____ f 9 t ____ g 15 t ____ h 20 t ____

2 Calculate as tonnes.

a 4 000 kg ____ b 6 000 kg ____ c 2 000 kg ____ d 8 000 kg ____
e 11 000 kg ____ f 16 000 kg ____ g 25 000 kg ____ h 40 000 kg ____

3 Calculate as kilograms.

a 2 t 750 kg ____ b 5 t 375 kg ____ c 3 t 495 kg ____
d 6 t 550 kg ____ e 9 t 625 kg ____ f 4 t 125 kg ____
g 1·5 t ____ h 3·5 t ____ i 7·5 t ____

4 Complete the table.

Kilograms	4 kg	7 kg		3·5 kg		8·5 kg
Grams	4 000 g		9 000 g		6 500 g	

5 Choose the most suitable measuring device to find the mass of:

	Measuring Device
A	standard balance scales
B	kitchen scales
C	spring balance
D	bathroom scales
E	weighbridge
F	displacement tank

a yacht ____

b flour ____

c glue ____

d grapes ____

e Con ____

f truck ____

Use the internet to investigate any measuring device that you do not know.

 • *AUSTRALIAN SIGNPOST MATHS 6* • ISBN 978 1 4886 2193 2

3:26 Units of Mass

1 Find how many kilograms in each.

a 6 t ☐ b 3·5 t ☐ c 9·5 t ☐ d 23 t ☐

e 7·25 t ☐ f 5·75 t ☐ g 11·5 t ☐ h 16 t ☐

2 Find how many tonnes in each.

a 3000 kg ☐ b 6500 kg ☐ c 8750 kg ☐ d 10 500 kg ☐

e 4500 kg ☐ f 7250 kg ☐ g 9500 kg ☐ h 5750 kg ☐

3 Complete the tables to show equivalent masses.

a	1 t 965 kg	1·965 t
b	2 t 645 kg	
c		1·375 t
d		5·697 t
e	3 t 125 kg	

f	7 t 850 kg	
g	1 kg 375 g	1·375 kg
h		1·962 kg
i	2 kg 395 g	
j		3·625 kg

k	4 kg 625 g	
l		5·835 kg

4 How much water would have a mass of:

1 mL of water has a mass of 1 g.

a 6 kg? ☐ b 9 kg? ☐ c 5 kg? ☐ d 10 kg? ☐

e 350 g? ☐ f 750 g? ☐ g 825 g? ☐ h 125 g? ☐

5 Write the mass equivalents for the following quantities of water.

a 3 L ☐

b 7 L ☐

c 12 L ☐

d 8 L ☐

e 650 mL ☐

f 375 mL ☐

g 925 mL ☐

h 635 mL ☐

6 Tick the most suitable unit of mass for the following items.

	Mass	g	kg	t
a	a frog			
b	a bus			
c	a boy			
d	a ferry			
e	a woman			
f	a plane			

 • *AUSTRALIAN SIGNPOST MATHS 6* • ISBN 978 1 4886 2193 2

3:27 Units of Mass

CONCEPT

1000 grams = 1 kilogram
1000 kilograms = 1 tonne

1 Calculate how many grams there are in:

a 0·357 kg ______ **b** 0·624 kg ______ **c** 0·506 kg ______
d 0·975 kg ______ **e** 0·167 kg ______ **f** 0·719 kg ______
g 0·412 kg ______ **h** 0·832 kg ______ **i** 0·235 kg ______

2 Use decimal notation to write:

a 1 kg 359 g ______ **b** 1 kg 725 g ______ **c** 3 kg 403 g ______
d 2 kg 635 g ______ **e** 5 kg 943 g ______ **f** 7 kg 875 g ______
g 4 kg 264 g ______ **h** 9 kg 517 g ______ **i** 6 kg 125 g ______

3 Calculate how many grams there are in:

a 2·124 kg ______ **b** 5·275 kg ______ **c** 3·506 kg ______
d 8·407 kg ______ **e** 1·327 kg ______ **f** 6·712 kg ______
g 4·851 kg ______ **h** 7·963 kg ______ **i** 9·615 kg ______

4 Complete the tables to show equivalent masses.

a	1 t 357 kg	1·357 t
b	1 t 945 kg	
c		1·625 t
d		2·375 t
e	3 t 285 kg	
f	5 t 649 kg	

g		4·125 t
h	6 t 456 kg	
i		5·364 t
j		7·056 t
k	9 t 835 kg	
l	8 t 224 kg	

5 Find examples of objects that have a mass between 1 kg and 2 kg.

______

3:28 Introducing the Milligram

CONCEPT

1 tonne = 1 000 kilograms

1 t = 1000 kg

1 small car weighs about 1 t.

1 kilogram = 1 000 grams

1 kg = 1000 g

A 1 L carton of milk weighs about 1 kg.

1 gram = 1 000 milligrams

1 g = 1000 mg

A small paperclip weighs about 1 g.

1 Choose the best unit to measure the mass of each item (**mg**, **g**, **kg** or **t**).

a a grain of sand ☐ **b** a butterfly ☐ **c** a cat ☐ **d** a raindrop ☐

e a bus ☐ **f** a flea ☐ **g** a pencil ☐ **h** an envelope ☐

2 Convert between grams and milligrams.

a 2 g = ☐ mg **b** 23 g = ☐ mg **c** 9000 mg = ☐ g

d 10 g = ☐ mg **e** 1 800 mg = ☐ g **f** 5 600 mg = ☐ g

g 5 g = ☐ mg **h** 8 923 mg = ☐ g **i** 3 402 mg = ☐ g

j 4 g = ☐ mg **k** 9 000 mg = ☐ g **l** 97 g = ☐ mg

3 Convert these measurements.

a 7 kg = ☐ g **b** 2 370 g = ☐ kg **c** 2 t = ☐ kg

d 5 000 kg = ☐ t **e** 5 419 mg = ☐ g **f** 65 g = ☐ mg

g 12·5 kg = ☐ g **h** 4 823 g = ☐ kg **i** 45 t = ☐ kg

j 1 446 kg = ☐ t **k** 2 406 mg = ☐ g **l** 10 g = ☐ mg

4 **a** Flynn caught three insects. Their total mass was 99 mg. What is the average mass? ☐

b Hudson's lizard loses 2 mg of water per gram of body weight each hour. If the lizard weighs 15 g, how much water does it lose each hour? ☐

c William used 1 mg of pencil lead for every 10 minutes of drawing with a pencil. How much did he use in an hour? ☐

d Laura found that 10 sheets of paper weighed 50 g. What would each sheet weigh? ☐

INVESTIGATION

Use the internet to investigate the mass of very light objects.

Object	Mass	Object	Mass
a flea			

3:29 Hectares

CONCEPT

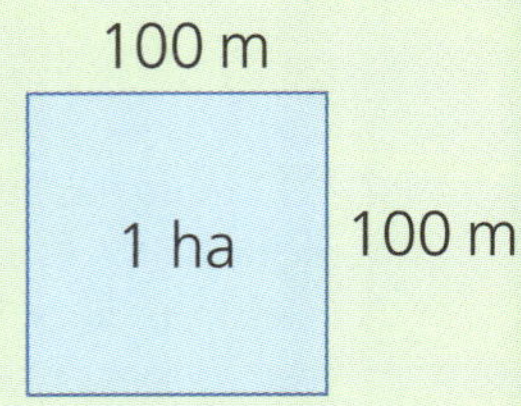

1 Use the short form to write:

a 5 hectares ______ b 9 hectares ______ c 23 hectares ______

d 48 hectares ______ e 3·6 hectares ______ f 9·5 hectares ______

2 Write the most suitable unit of measure (**cm²**, **m²**, **ha**) to record the area of:

a a tennis court ______ b a wheat paddock ______ c this page ______

d a footpath ______ e a city block ______ f a national park ______

g a dairy farm ______ h a beach ______ i a phone screen ______

3 Convert the following to hectares (**ha**).

a 10 000 m^2 ______ b 20 000 m^2 ______ c 60 000 m^2 ______

d 90 000 m^2 ______ e 70 000 m^2 ______ f 50 000 m^2 ______

g 110 000 m^2 ______ h 130 000 m^2 ______ i 180 000 m^2 ______

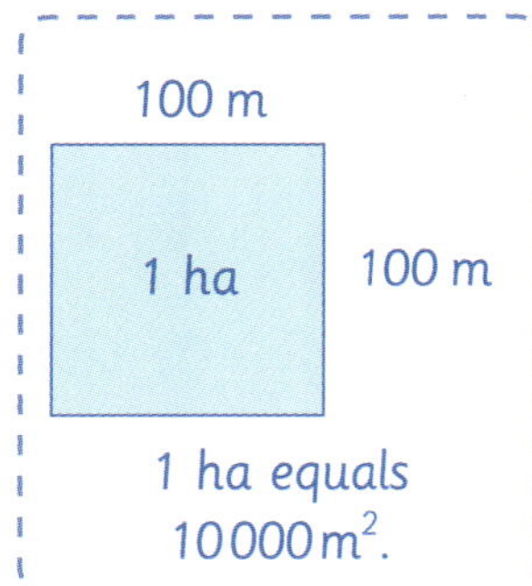

4 Convert the following to square metres (**m²**).

a 4 ha ______ b 8 ha ______ c 3 ha ______

d 10 ha ______ e 7 ha ______ f 9 ha ______

g 12 ha ______ h 16 ha ______ i 24 ha ______

ACTIVITY

- From 1 mm grid paper, cut a number of squares with 10 cm sides.
- Cut each square so that it can be re-formed to make a rectangle.
- How many different rectangles can you make? ______
- If each mm stands for 1 m, what is the area of each shape? ______

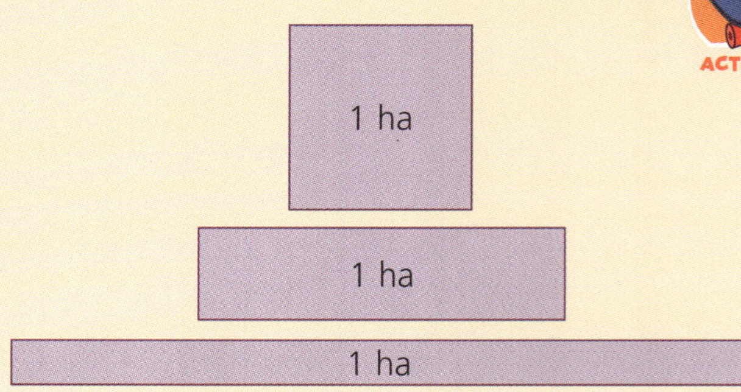

 • *AUSTRALIAN SIGNPOST MATHS 6* • ISBN 978 1 4886 2193 2

3:30 Square Kilometres

1 Convert the following to hectares (**ha**).

a 1 km^2 ☐ b 3 km^2 ☐ c 4 km^2 ☐

d 9 km^2 ☐ e 6 km^2 ☐ f 5 km^2 ☐

g 11 km^2 ☐ h 20 km^2 ☐ i 12 km^2 ☐

2 Convert the following to square kilometres (**km²**).

a 200 ha ☐ b 600 ha ☐ c 1000 ha ☐ d 1200 ha ☐

e 800 ha ☐ f 1700 ha ☐ g 3000 ha ☐ h 2500 ha ☐

3 Write the most suitable unit of measure (**m²**, **ha**, **km²**) to record the area of:

a the classroom ☐ b a local park ☐ c a tennis court ☐

d France ☐ e the playground ☐ f a shopping centre ☐

g a dining table ☐ h a golf course ☐ i Tasmania ☐

4

Country/State	Area
Austria	83 854 km^2
France	551 000 km^2
Germany	357 868 km^2
Hungary	93 033 km^2
Italy	301 252 km^2
Netherlands	37 330 km^2
Portugal	91 630 km^2
Spain	492 431 km^2
Tasmania	67 800 km^2

Use the information in the table to find:

a a country smaller than Tasmania ☐

b a country bigger than Spain ☐

c how much bigger Portugal is than Tasmania ☐

d how much bigger Germany is than Italy ☐

e the total area of Portugal and Spain combined ☐

f the total area of France and Germany combined ☐

g the country that is about eight times bigger than Tasmania ☐

ACTIVITY

- Locate the scale used in a street directory (usually 1 cm : 200 m).
- Use the scale to draw a 1 km^2 square on a page of the directory.
- Estimate the areas of various places inside the square, e.g. schools, parks, ovals etc.
- Discuss your results.

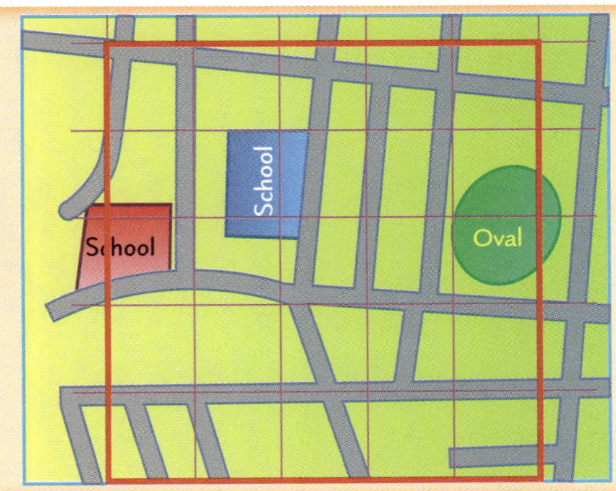

3:31 Time Zones

Eastern Standard Time (**EST**)

Central Standard Time (**CST**) ½ hour behind EST

Western Standard Time (**WST**) 2 hours behind EST

1 Complete the tables to show the time in each time zone.

	WST	CST	EST
a	06:30		
b			08:45
c		13:15	
d			19:20

	WST	CST	EST
e			12:25
f		09:50	
g	14:40		
h		16:35	

	WST	CST	EST
i			01:10
j	18:13		
k		10:47	
l	23:55		

2 Record the times that match each given time.

Darwin	**9:37 am**	**a**	Perth		**b**	Sydney		**c**	Adelaide	
Melbourne	**12:03 pm**	**d**	Darwin		**e**	Perth		**f**	Brisbane	
Brisbane	**1:54 am**	**g**	Hobart		**h**	Adelaide		**i**	Perth	

INVESTIGATION

Daylight Saving Time is used in many Australian states and territories during summer so that people have more daylight hours after work. To begin Daylight Saving Time, clocks are put forward one hour; to finish, the clocks are put back one hour. In Australia, only NSW, ACT, Victoria, Tasmania and South Australia use Daylight Saving Time.

3 Show the time on each clock to match the Daylight Saving Time in Melbourne.

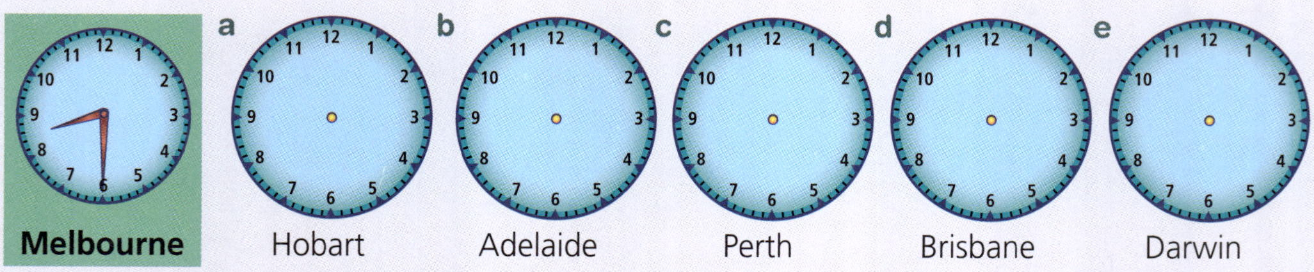

 • *AUSTRALIAN SIGNPOST MATHS 6* • ISBN 978 1 4886 2193 2

4:01 Space Review

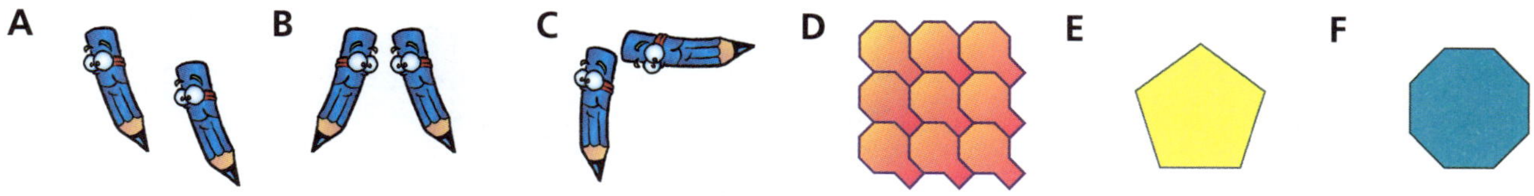

1 Which of pictures **A**, **B**, **C** or **D** is an example of a:

a flip? ☐ b slide? ☐ c turn? ☐ d tessellation? ☐

2 How many axes of symmetry has picture:

a **A**? ☐ b **B**? ☐ c **C**? ☐ d **E**? ☐ e **F**? ☐

3 Some cubes have been stacked in this box.

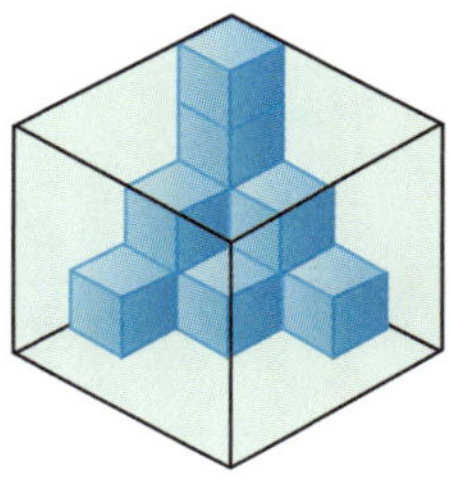

a How many cubes are in the stack? ☐

b How many cubes would fill the box? ☐

4 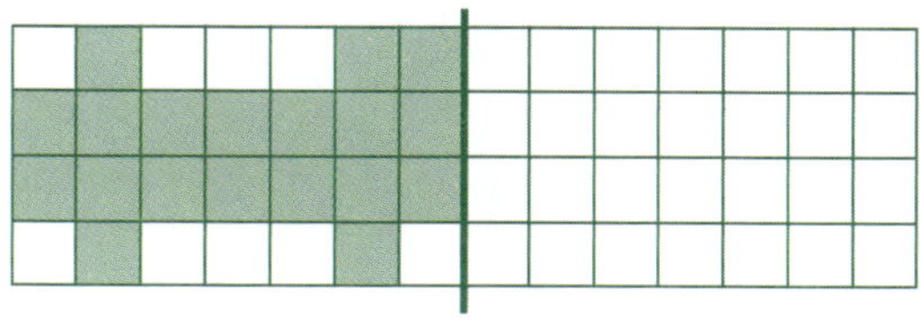

Shade squares to show the mirror image (the reflection) of the coloured shape.

5 Continue each pattern for two more figures, using the instructions given.

a Translate 2 cm right and reflect vertically.

b Translate 3 cm right and reflect horizontally.

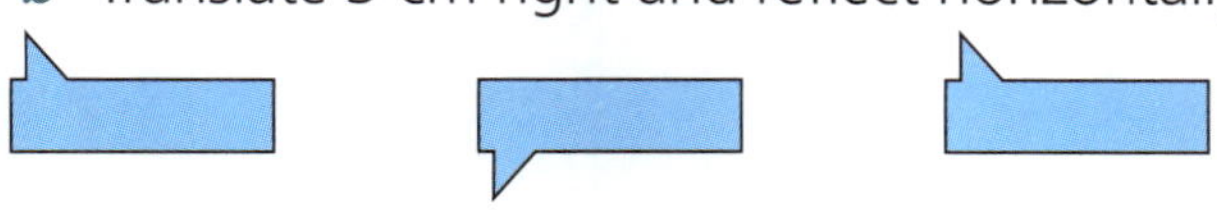

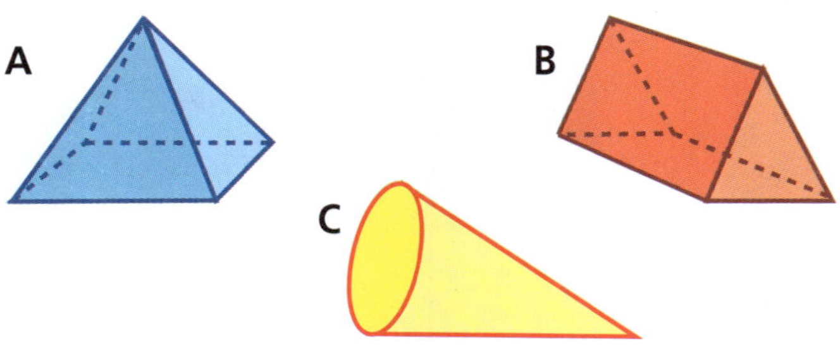

6 Complete the table below.

	Number of Surfaces	Number of Vertices	Number of Edges
A			
B			
C			

7 From Question 6, give the full name of solid:

a **A** ☐

b **B** ☐

c **C** ☐

d Which of these solids would have the net on the right? ☐

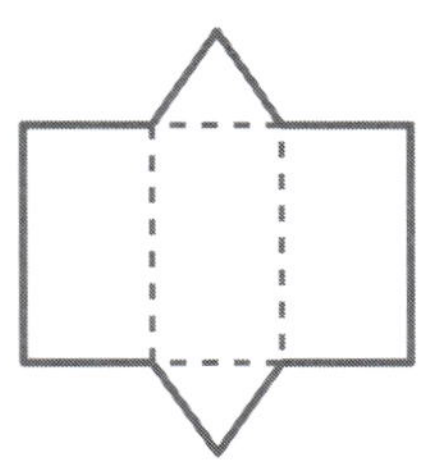

 • *AUSTRALIAN SIGNPOST MATHS 6* • ISBN 978 1 4886 2193 2

4:02 Angles and Informal Units

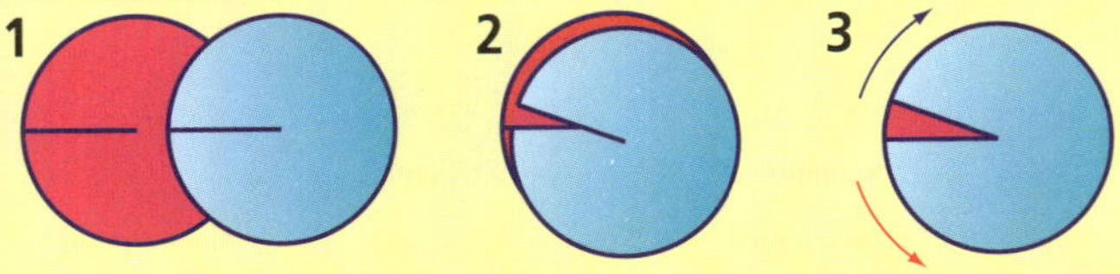

Step 1 Cut two circles the same size but of a different colour.

Step 2 Cut a slit in each one and fit them together.

Step 3 Move one circle on the other to make angles.

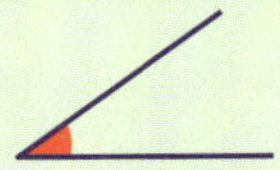

acute angle
smaller than a right angle

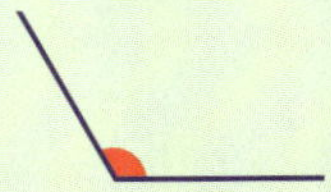

obtuse angle
greater than a right angle but less than a straight angle

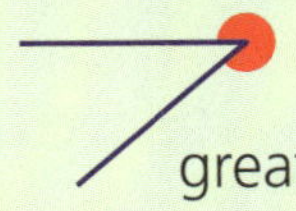

reflex angle
greater than a straight angle

FUN SPOT

1 Use your angle wheel to make:

a an acute angle **b** a right angle **c** an obtuse angle
d a straight angle **e** a reflex angle **f** a revolution

The size of an angle is the amount of turn.

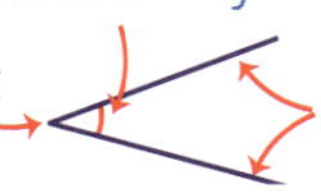

2 Which angle in this figure is:

a an acute angle? ☐ **b** a right angle? ☐
c an obtuse angle? ☐ **d** a straight angle? ☐
e a reflex angle? ☐ **f** a revolution? ☐

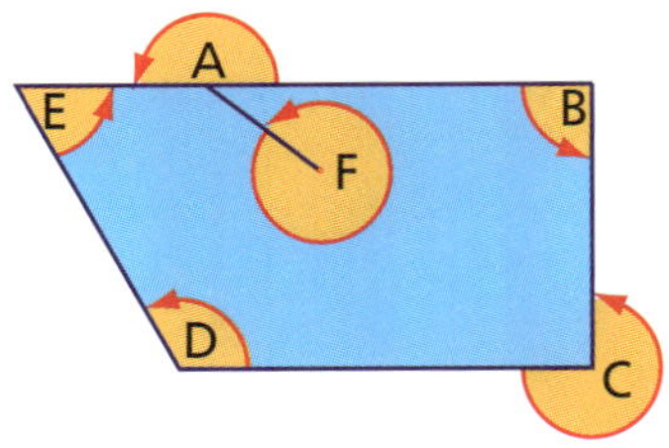

3 Use the letters to put these angles in order from smallest to largest. ☐

4 How many of the angles you are asked to make below are needed to cover:

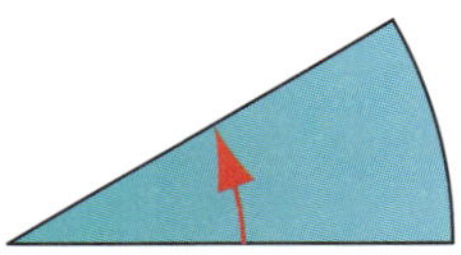

Trace and cut this angle 10 times.

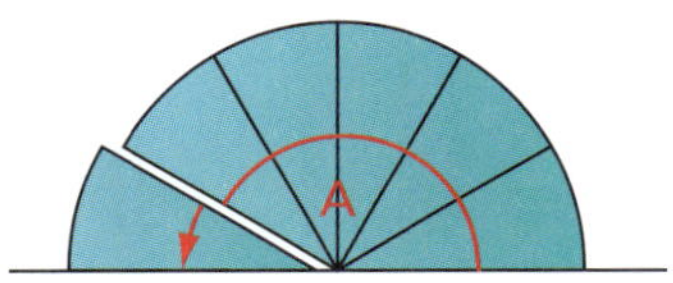

Six of these angles are needed to cover angle **A**.

a angle **E**? ☐
b angle **B**? ☐
c angle **C**? ☐
d angle **D**? ☐

5 The copies you made of the angle **A** can be used to measure other angles. Use terms like 'nearly 4' or 'a little more than 2' or 'about $1\frac{1}{2}$' to measure these angles.

a 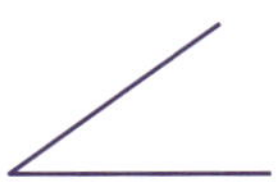 ☐

b  ☐

c 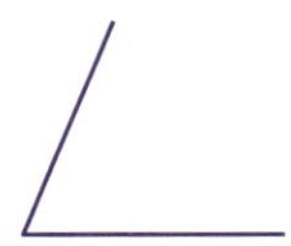 ☐

Make a chart with these headings.
Draw pictures which show these angles in real life.

Acute Angle	Obtuse Angle	Reflex Angle

Example:

ACTIVITY

 • *AUSTRALIAN SIGNPOST MATHS 6* • ISBN 978 1 4886 2193 2

4:03 The Protractor

CONCEPT

A straight angle

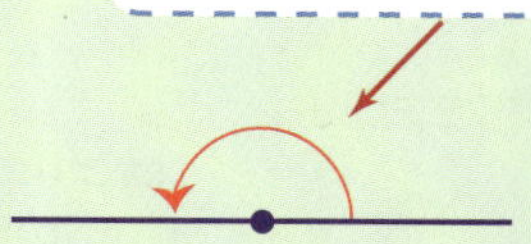

If we divide a straight angle into 180 parts, each part is one degree.

30 degrees is written as 30°.

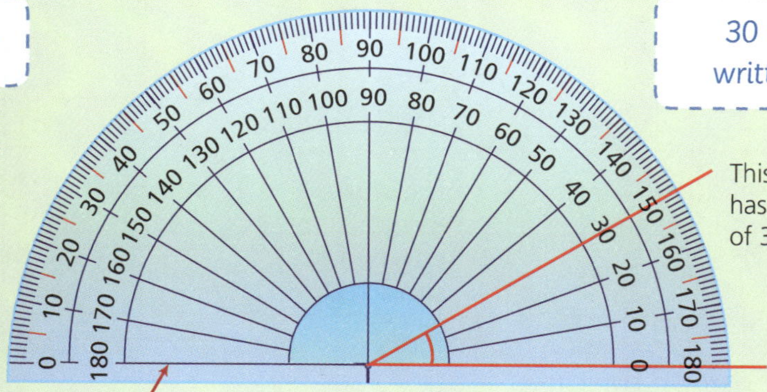

A protractor measures the amount of turn. It has:

1. a baseline
2. a centre point where the 90° line meets the baseline.
3. two scales, one on the inside and one on the outside. Each scale goes from 0° to 180°.

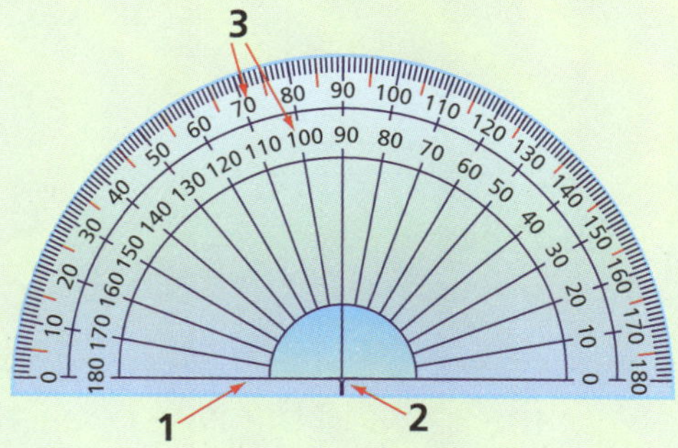

- Place the lower arm of the angle along the baseline.
- Slide the protractor until the vertex is at the centre point.
- Find 0° on the lower arm and use that scale to read off the size of the angle.

1 Write the size of each angle.

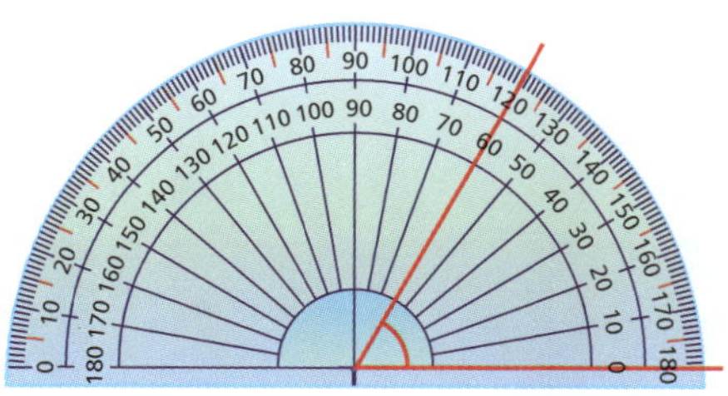

a ☐

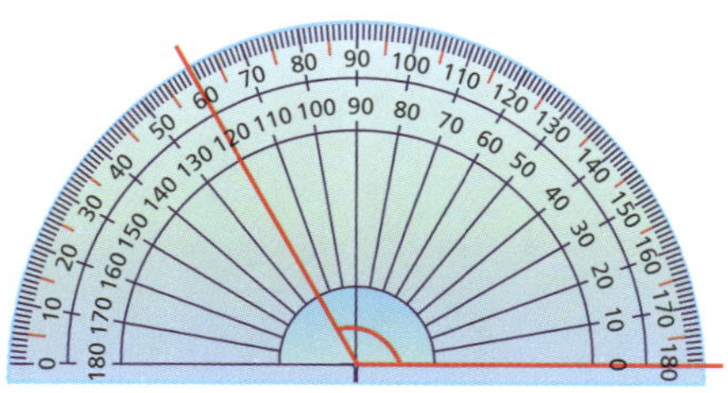

b ☐

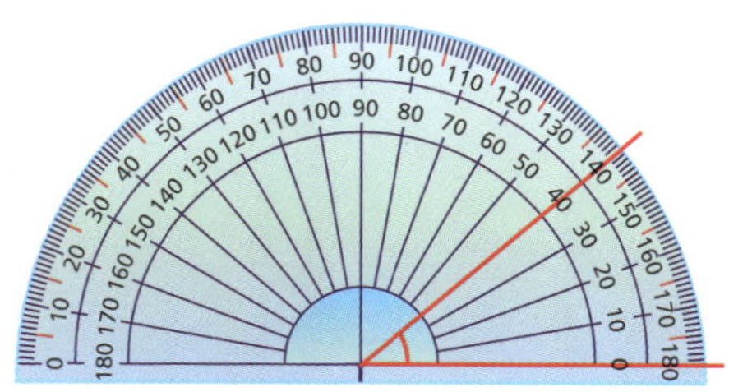

c ☐

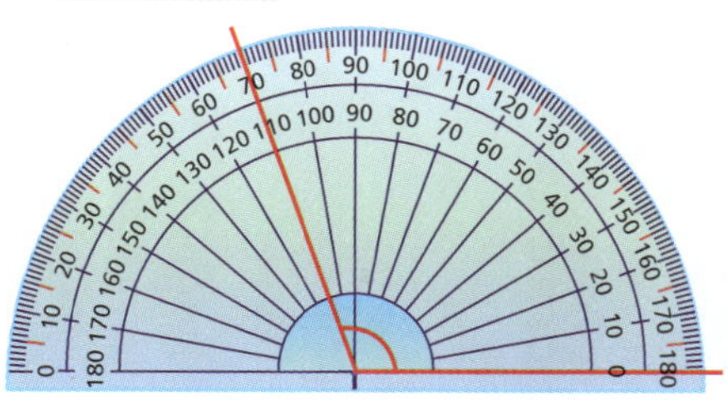

d ☐

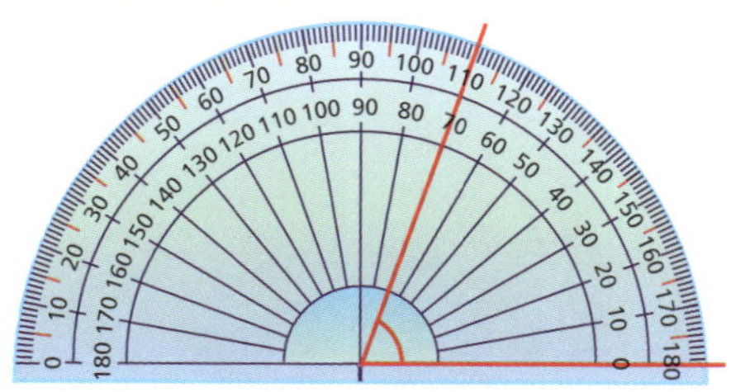

e ☐

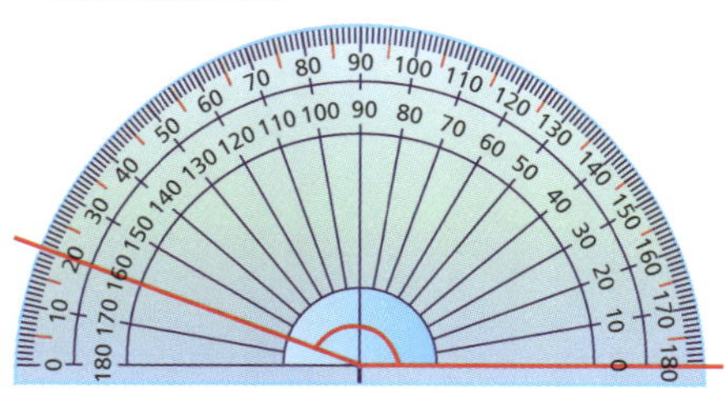

f ☐

4:04 Angle Types

Angle	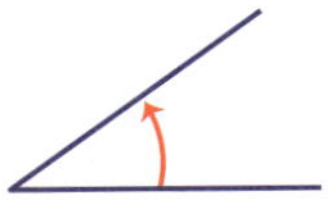	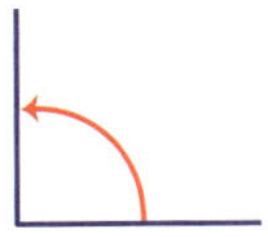	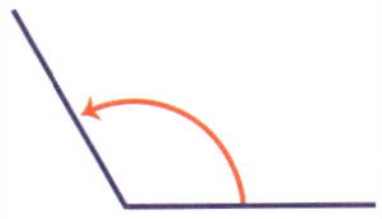	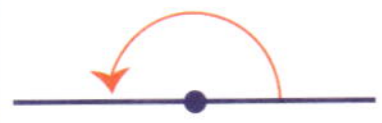	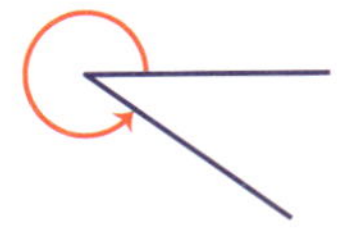	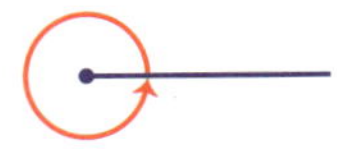
Type	acute	right	obtuse	straight	reflex	revolution (full turn)
Size	between 0° and 90°	90°	between 90° and 180°	180°	between 180° and 360°	360°

1 Give the type of angle if its size is:

a 45° ______	**b** 180° ______	**c** 360° ______	**d** 305° ______
e 71° ______	**f** 190° ______	**g** 90° ______	**h** 172° ______
i 92° ______	**j** 200° ______	**k** 310° ______	**l** 180° ______
m 9° ______	**n** 60° ______	**o** 112° ______	**p** 360° ______

2 Write the size and the type of each angle.

a

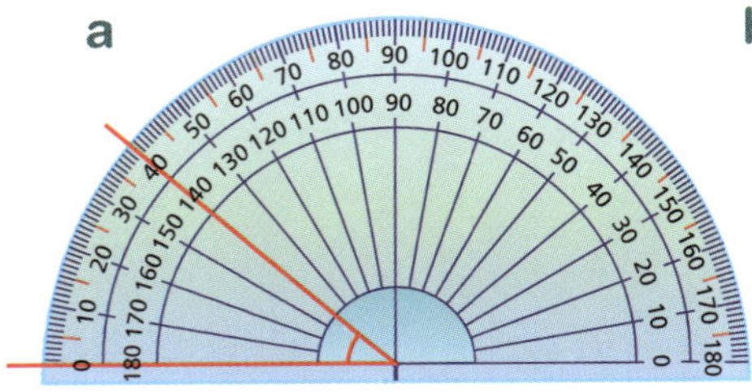

______ ° ______

b

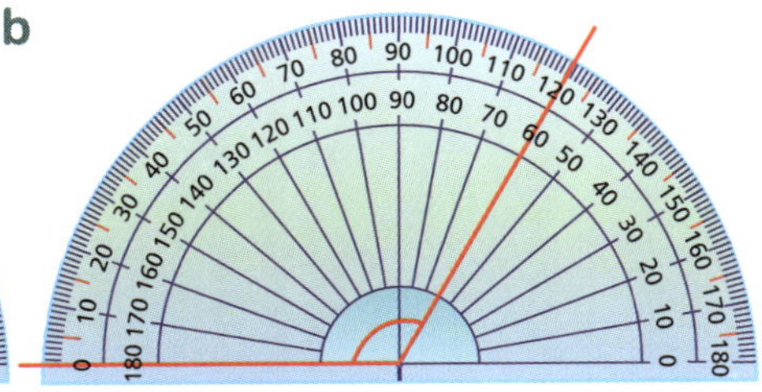

______ ° ______

Protractors can measure the amount of turning from either side.

c

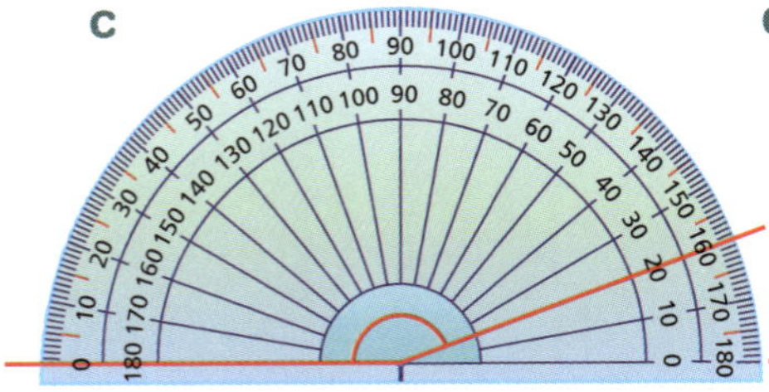

______ ° ______

d

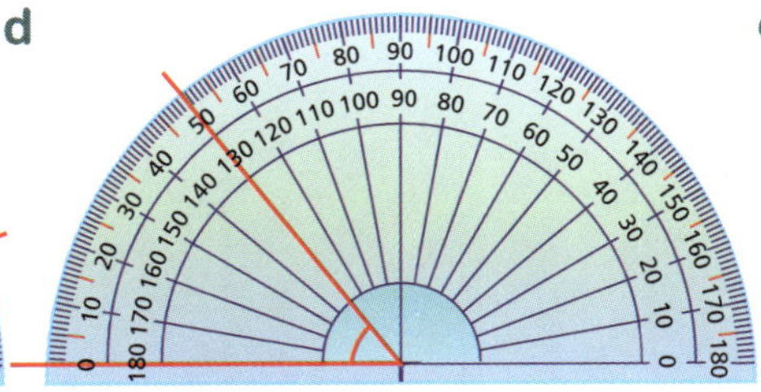

______ ° ______

e

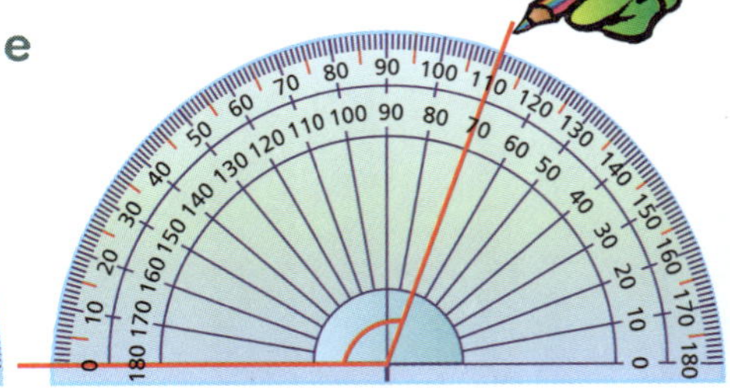

______ ° ______

f

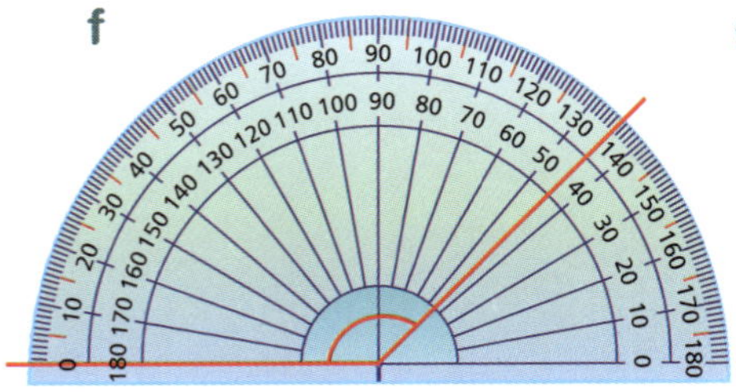

______ ° ______

g

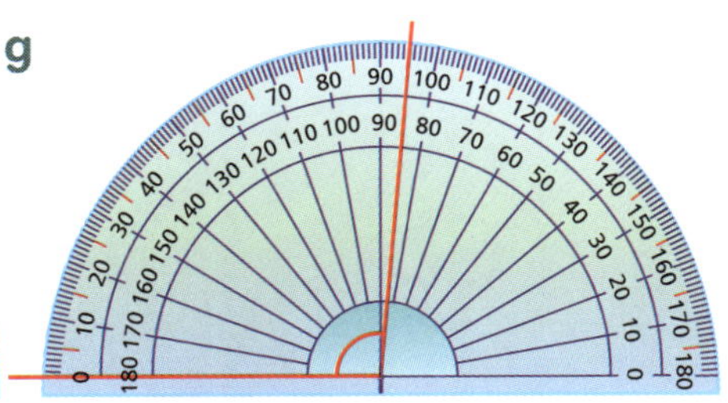

______ ° ______

h

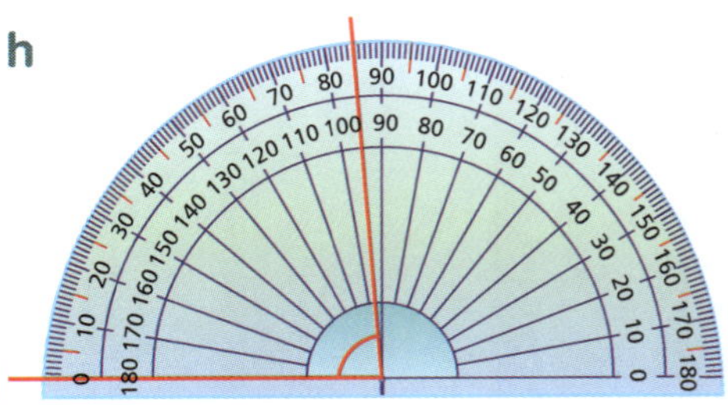

______ ° ______

4:05 Angles

CONCEPT

Learn to recognise these common angles.

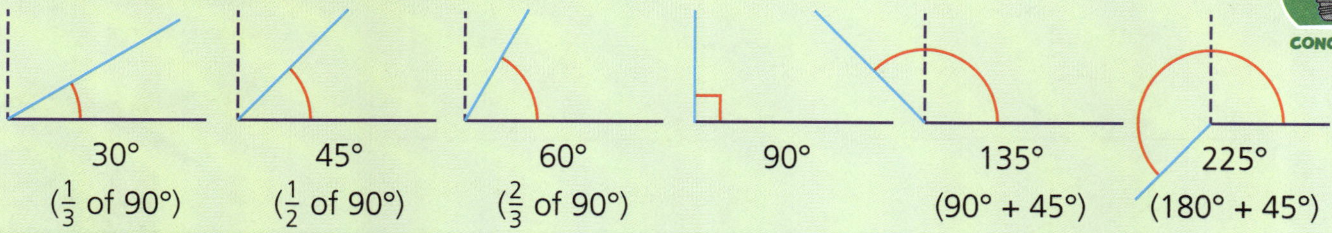

1 Estimate the size of each angle and then match it to a label.

45° | 30° | 315° | 240° | 120° | 210° | 330° | 270° | 60° | 90°

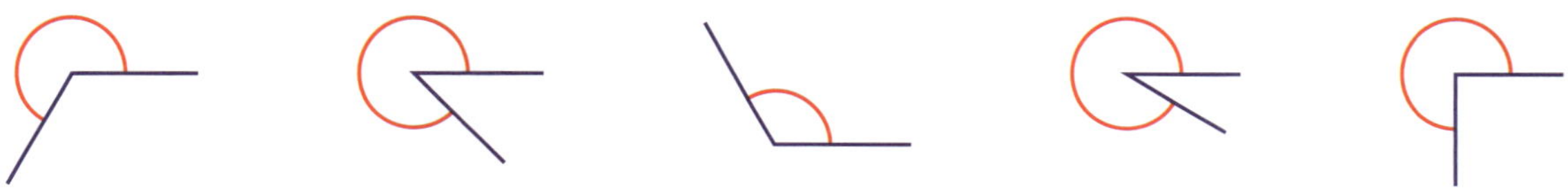

2 Use a protractor to draw these angles.

a

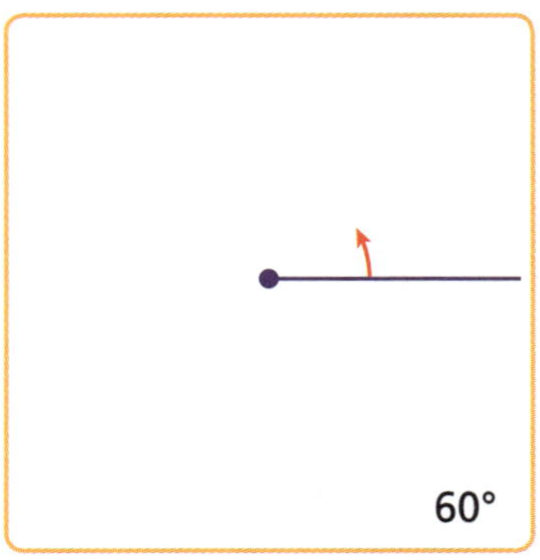

b

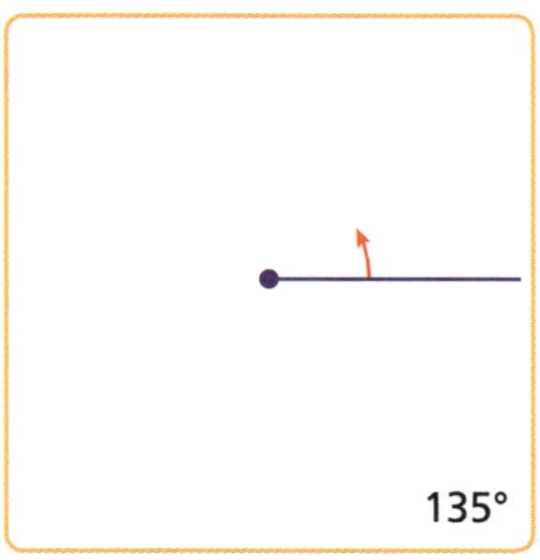

c

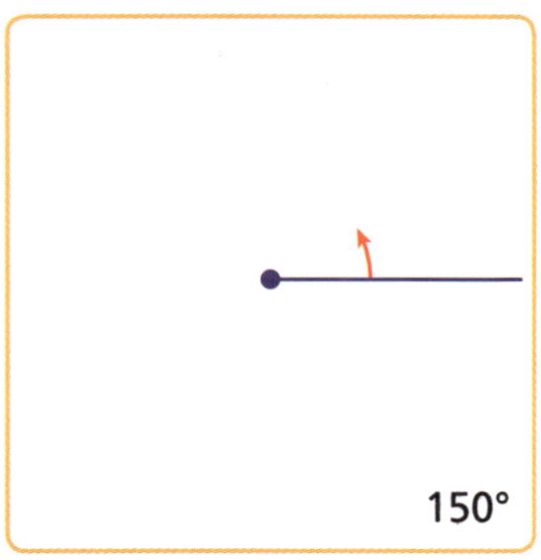

d

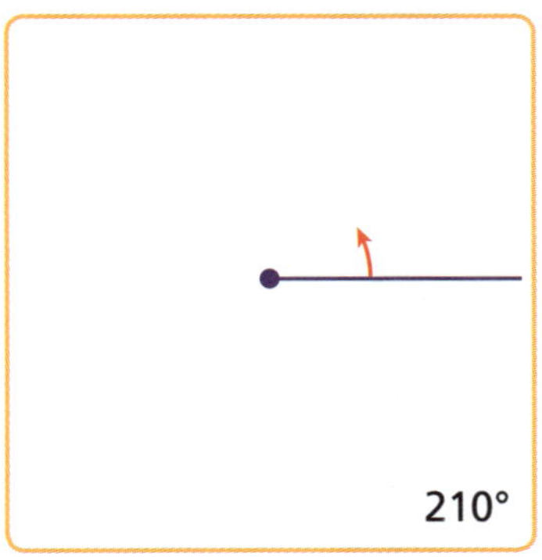

e

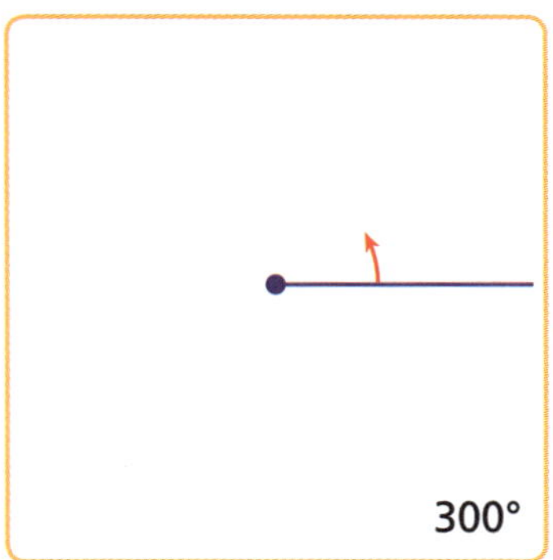

f

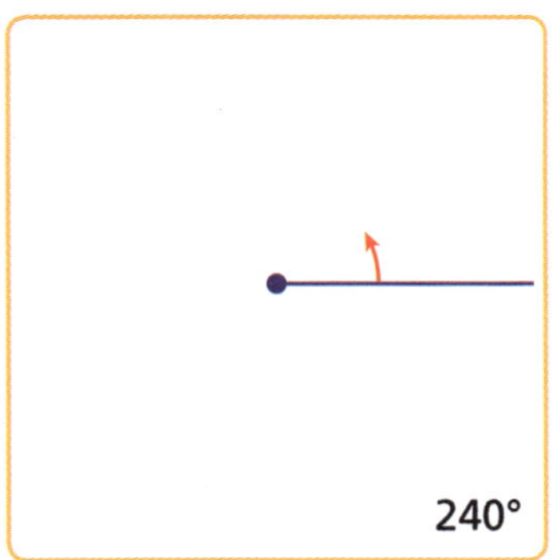

4:06 Angles on a Straight Line

1 Use a protractor to measure each angle.

a

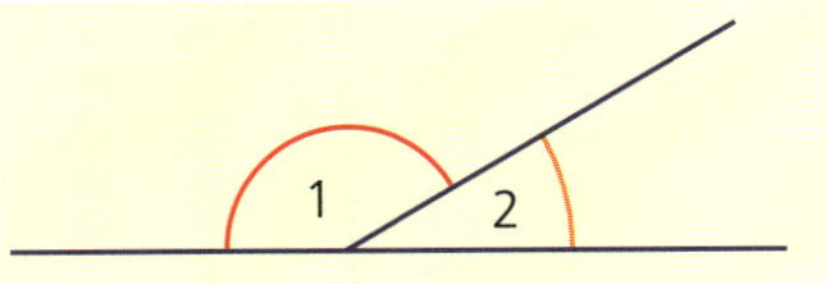

angle 1 = ☐°

angle 2 = ☐°

angle 1 + angle 2 = ☐°

b

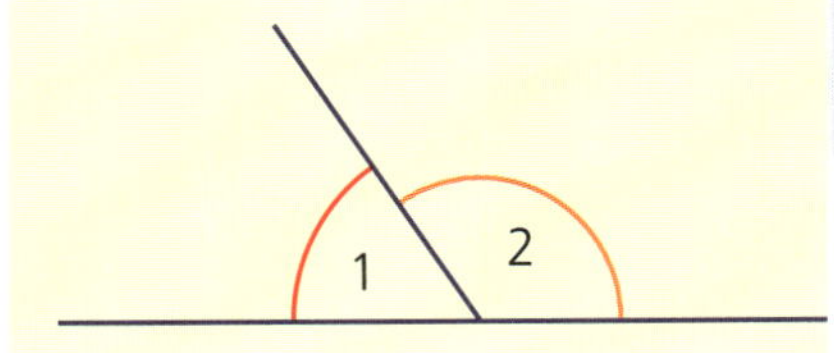

angle 1 = ☐°

angle 2 = ☐°

angle 1 + angle 2 = ☐°

Angles that make a straight angle add up to 180°.

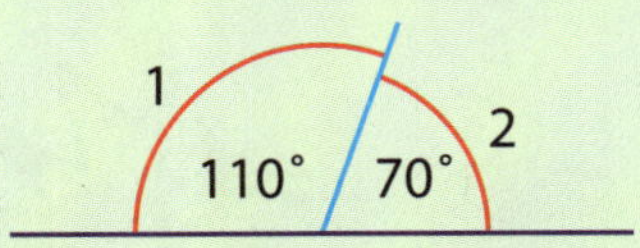

angle 1 = 110°
angle 2 = 70°
110° + 70° = 180°

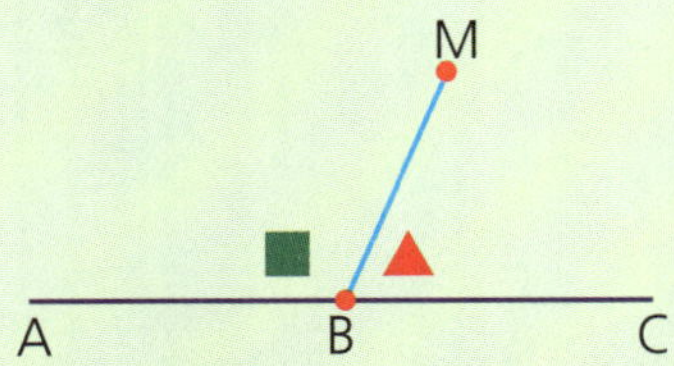

■ = 114.5°
▲ = 65.5°
■ + ▲ = 180°

General Case
M is a movable point.
As **M** is dragged to new positions, the measurements for ■ and ▲ change.
However, the sum of the two angles is always 180°.

2 Find the value of the unknown angle.

a

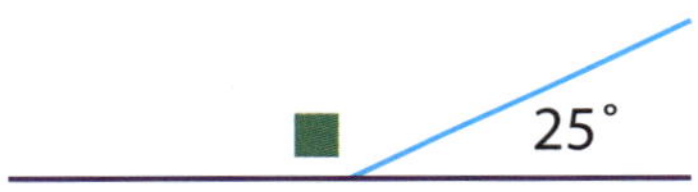

180° – 25° = ☐

■ = ☐

b

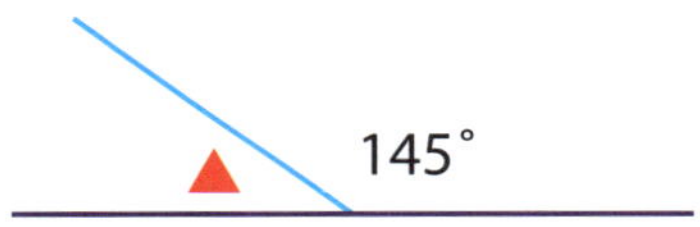

180° – 145° = ☐

▲ = ☐

c

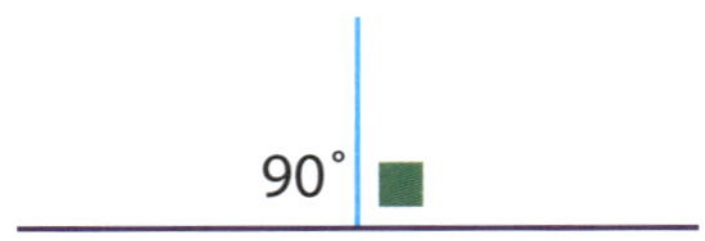

180° – 90° = ☐

■ = ☐

d

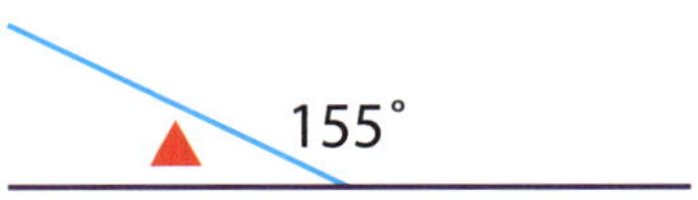

180° – 155° = ☐

▲ = ☐

e

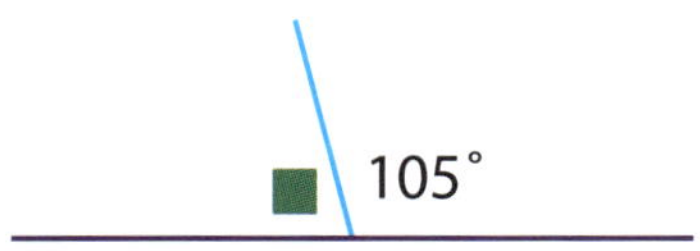

180° – 105° = ☐

■ = ☐

f

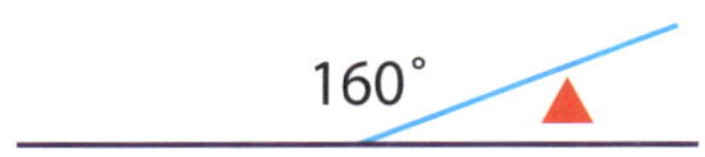

180° – 160° = ☐

▲ = ☐

g

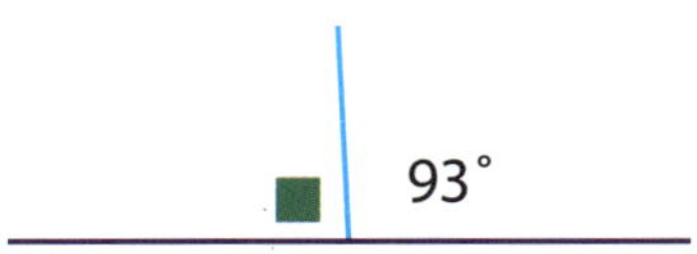

180° – 93° = ☐

■ = ☐

h

180° – 84° = ☐

▲ = ☐

i

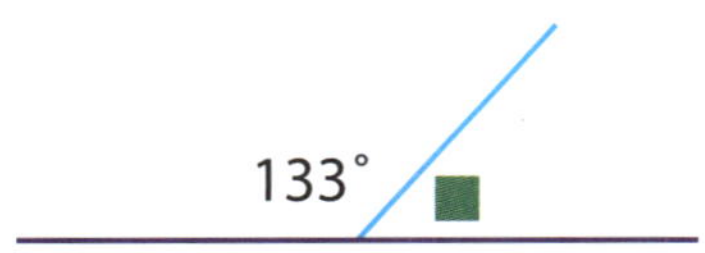

180° – 133° = ☐

■ = ☐

 • *AUSTRALIAN SIGNPOST MATHS 6* • ISBN 978 1 4886 2193 2

4:07 Angles at a Point

1 Use a protractor to measure each angle.

a

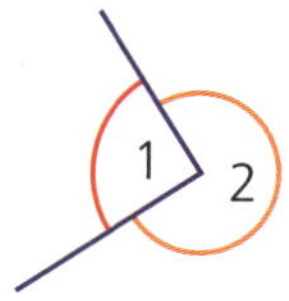

angle 1 = ____°

angle 2 = ____°

angle 1 + angle 2 = ____°

b

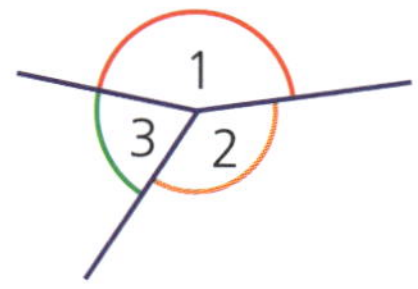

angle 1 = ____°

angle 2 = ____°

angle 3 = ____°

angle 1 + angle 2 + angle 3 = ____°

CONCEPT

Angles that meet at a point add up to 360°.

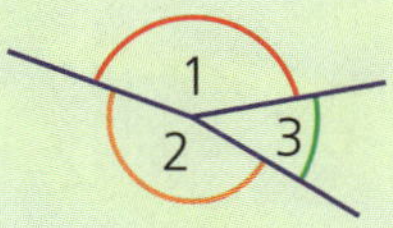

angle 1 = 150°
angle 2 = 170°
angle 3 = 40°
150° + 170° + 40° = 360°

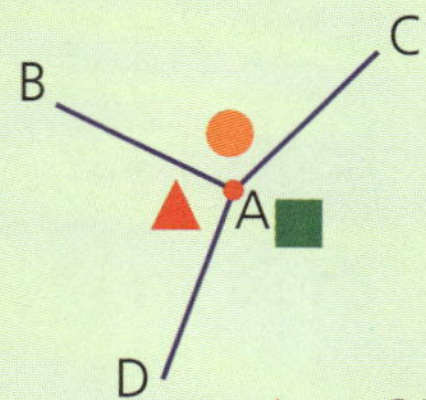

▲ = 95°
● = 110°
■ = 155°
▲ + ● + ■ = 360°

General Case
A cannot move (it is fixed).
B, C and D can be dragged to new positions so that the measurements for ■, ● and ▲ change.
However, the sum of the three angles is always 360°.

2 Find the value of the unknown angle.

a

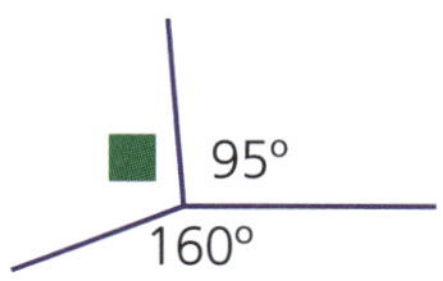

360° – 95° – 160° = ____

■ = ____

b

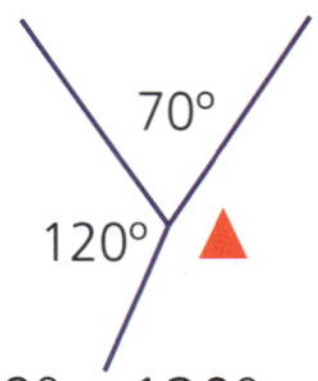

360° – 120° – 70° = ____

▲ = ____

c

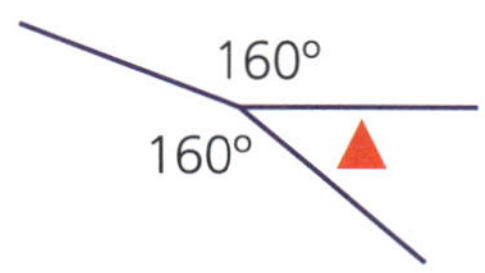

360° – 160° – 160° = ____

▲ = ____

d

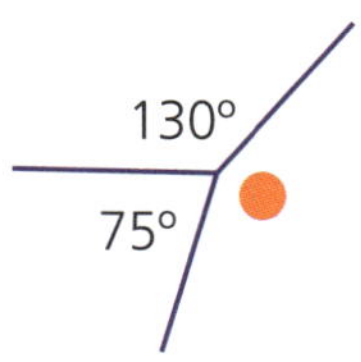

● = ____

e

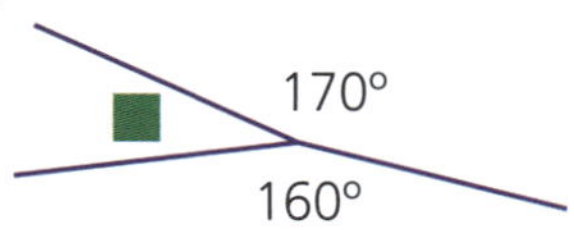

■ = ____

f

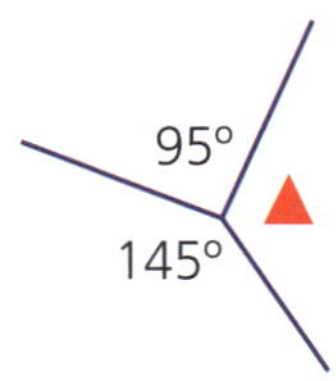

▲ = ____

g

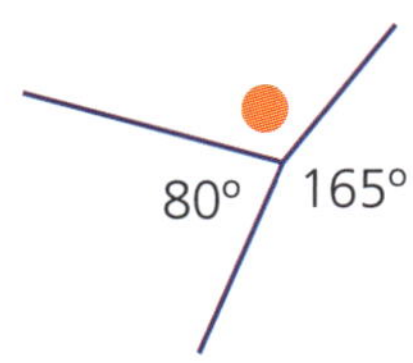

● = ____

h

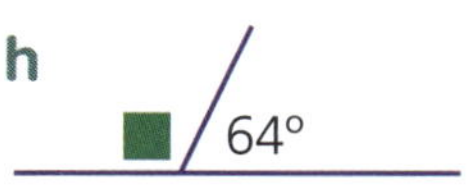

■ = ____

i

▲ = ____

j

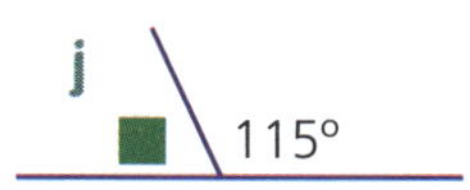

■ = ____

4:08 Vertically Opposite Angles

Vertically opposite angles are the opposite angles formed when two straight lines cross.

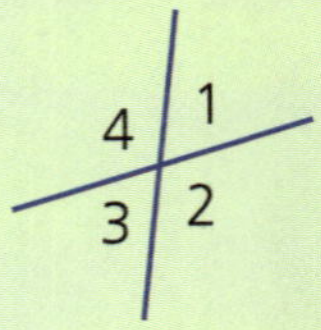

angle 1 = 67° angle 3 = 67°

angle 2 = 113° angle 4 = 113°

angle 1 = angle 3, angle 2 = angle 4

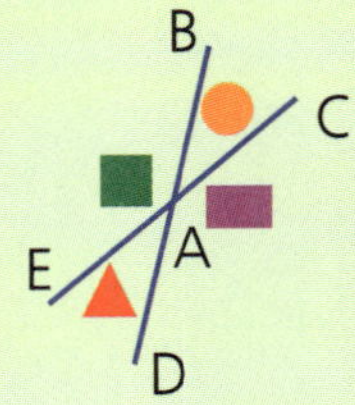

General Case

A cannot move.
The lines **EC** and **BD** can be turned to vary the angles.

▲ = 37° ● = 37°
■ = 143° ■ = 143°

CONCEPT

Rule: Vertically opposite angles are equal.

1 Use a protractor to measure each angle.

a

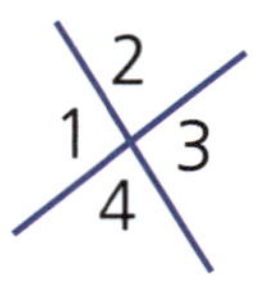

angle 1 = °
angle 2 = °
angle 3 = °
angle 4 = °
angle 1 = angle
angle 2 = angle

b

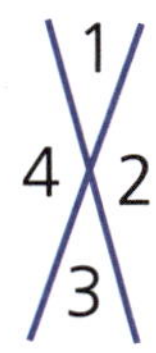

angle 1 = °
angle 2 = °
angle 3 = °
angle 4 = °
angle 1 = angle
angle 2 = angle

Estimate first.

2 Find the value of the unknown angle.

a

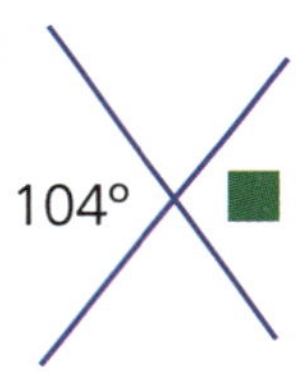

■ =

b

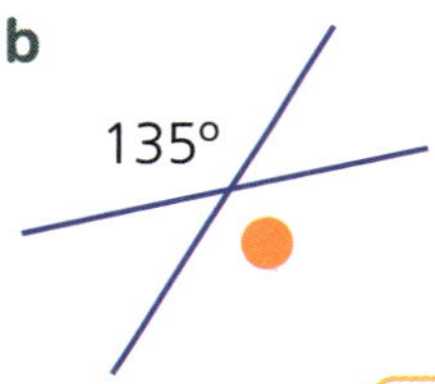

● =

c

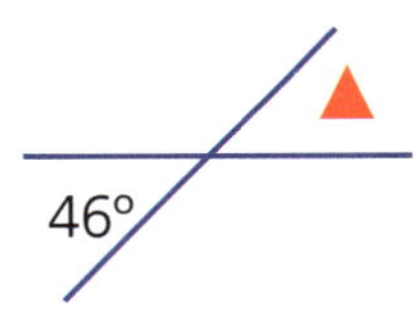

▲ =

d

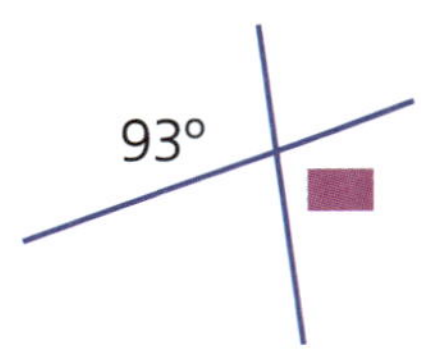

■ =

3 Find the missing angle.

a

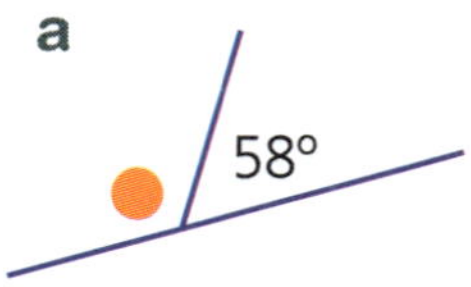

● =

b

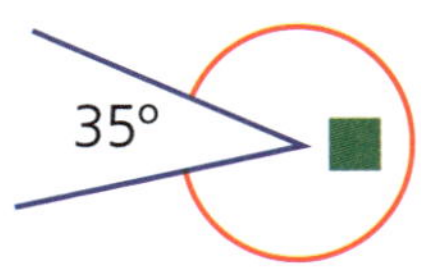

■ =

c

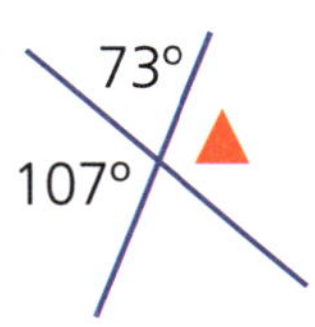

▲ =

d

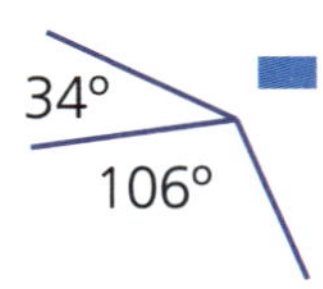

■ =

e

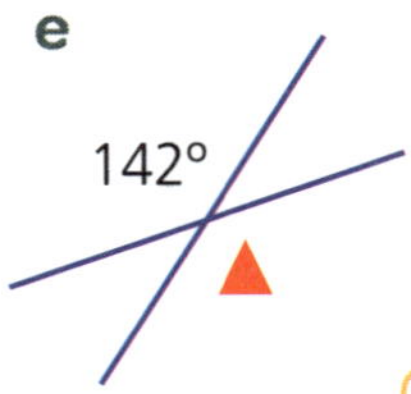

▲ =

f

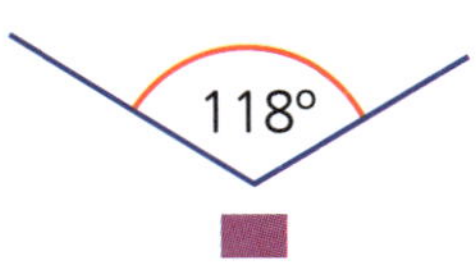

■ =

g

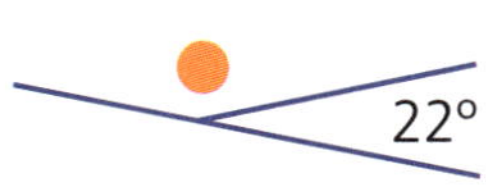

● =

h

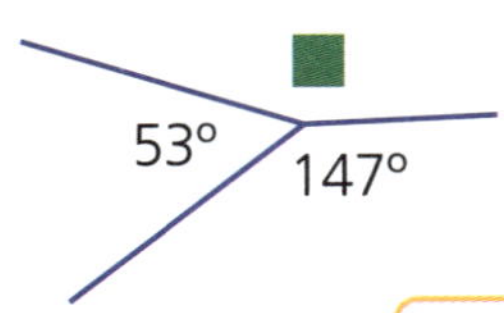

■ =

 • *AUSTRALIAN SIGNPOST MATHS 6* • ISBN 978 1 4886 2193 2

4:09 2D Shapes

CONCEPT

A parallelogram is a quadrilateral that has opposite sides parallel.

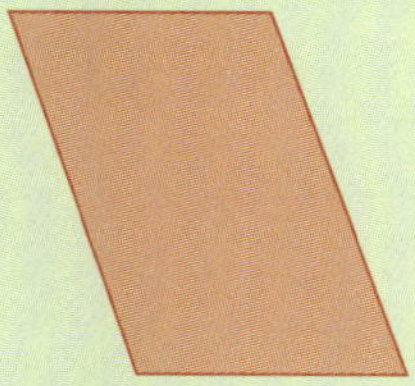

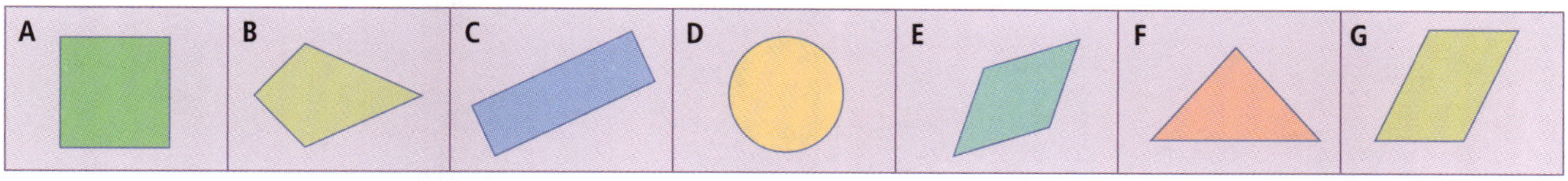

1. Which of the shapes above is a:
   a triangle? ☐ b square? ☐ c rhombus? ☐ d kite? ☐

2. a Which of the shapes above have opposite sides parallel? ☐
   b Which of the shapes above have four right angles? ☐
   c Which of the shapes above have all sides equal? ☐

A pentagon has 5 sides.
A hexagon has 6 sides.
An octagon has 8 sides.

3. Give the letter of a shape below that is a:
   a pentagon ☐ b rectangle ☐ c rhombus ☐
   d octagon ☐ e trapezium ☐ f hexagon ☐

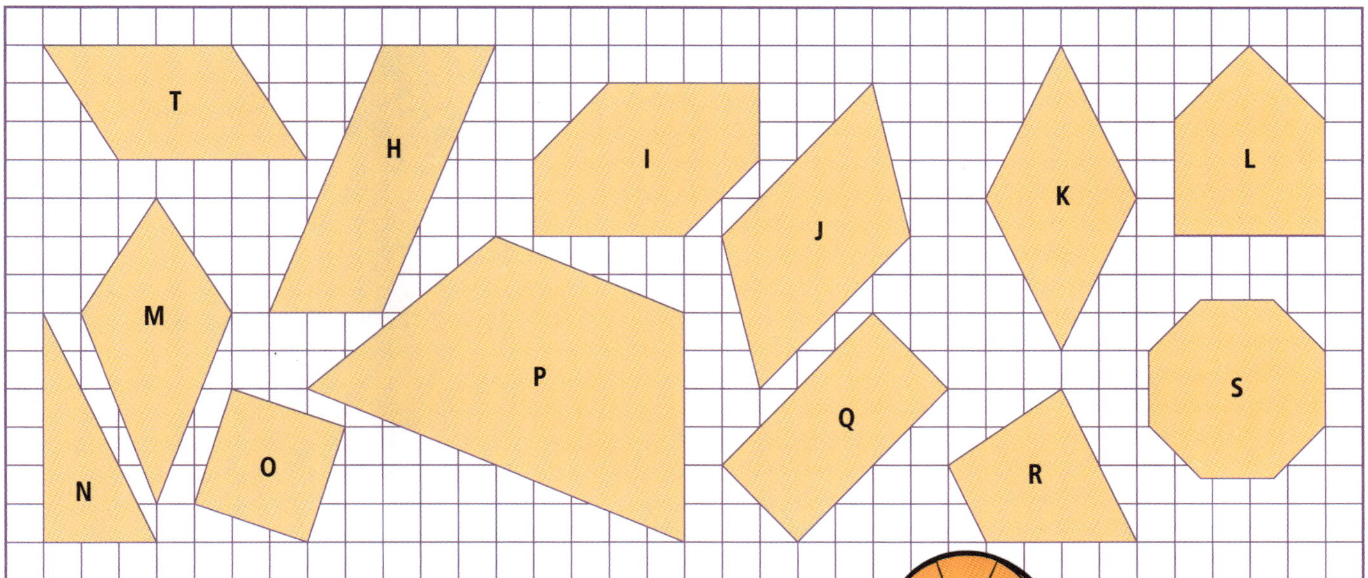

4. Measure the side lengths of parallelograms **T** and **J**. Are opposite sides equal? ☐

5. Measure the angles of parallelograms **T** and **J**. Are opposite angles equal? ☐

4:10 Finding Your Way

1 This is a map of my town.

a What place is at G4?

b What place is found at A5 or B5?

c Give two grid references for:

i school

ii hospital

d Colour the shortest path from:

i home to hospital in red

ii home to school in blue

e What is the shortest distance from:

i home to school?

ii home to hospital?

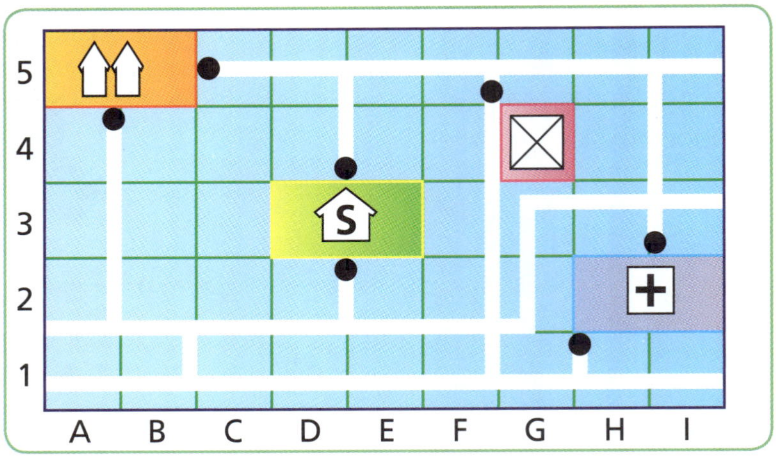

Scale: 40 m

Legend

school · hospital · gate · shops · home

2 A map of our park is on the right. The letters show positions in the park. What object is:

a north of **D** and west of **J**?

b east of **D** and south of **H**?

c west of **C** and north of **G**?

To what letter does the dog move if it goes:

d 2 boxes north then 1 box east?

e 1 box north then 4 boxes west?

f 1 box south then 1 box east?

g 2 boxes south then 4 boxes west?

h 2 boxes west?

i 1 box east?

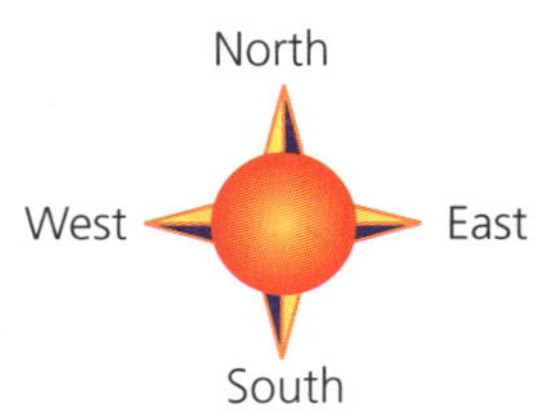

Remember NEWS: North, East, West, South.

3 On this map, what is:

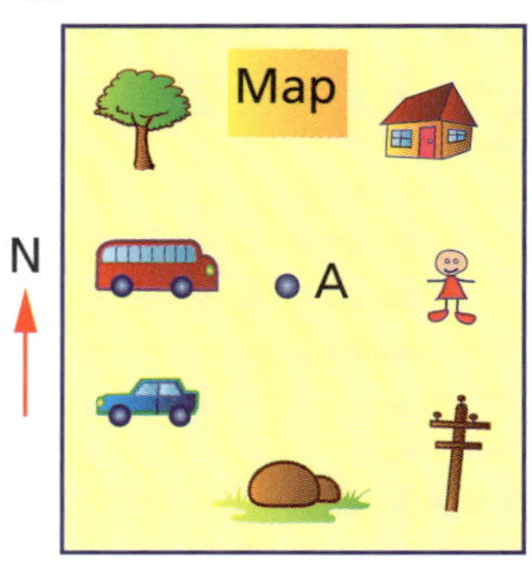

a west of A?

b south of A?

c south of the tree and west of the girl?

d north of the car and west of the house?

e What things are north of the power pole?

f What things are south of the tree?

 • *AUSTRALIAN SIGNPOST MATHS 6* • ISBN 978 1 4886 2193 2

4:11 Following Directions – The Compass

Chinese sailors were the first to direct their ships using the compass.

ACTIVITY

- Stroke a sewing needle about twenty times with the end of a magnet. Always stroke in the same direction with the one end of the magnet.
- Suspend the magnetic needle by cotton. It will point north and south.

1 A wind coming from the north is called a northerly wind. What would be the name of a wind coming from the:

a south? ______ **b** west? ______ **c** east? ______

2 The direction halfway between north and east is called north-east. What is the name of the direction that is halfway between:

a north and west? ______ **b** south and east? ______ **c** south and west? ______

3 If you are facing north, what direction is:

a to your left? ______ **b** to your right? ______ **c** behind you? ______

4 The side edges of a map run north to south. In what direction would the top and bottom edges of the map run? ______

5 What angle lies between the compass directions:

a north and east? ______ **b** south and west? ______ **c** north and south? ______

6 At midday, in which direction is the sun? Why is this? ______

7 Starting from the point ▪ above **X** (on the right of this question) and using the direction **N** for north, follow the instructions below. Where do you end up?

a Go 20 cm north, then 16 cm west, then 3 cm south. ______

b Go 22 cm north, then 14 cm west, then 5 cm south, then 3 cm east. ______

c Go 4 cm west, then 2 cm south, then 2·5 cm west, then 19 cm north. ______

d Go 2 cm south, then 16 cm west, then 18 cm north, then 15 cm east. ______

4:12 Compass Directions

A compass needle always points north.

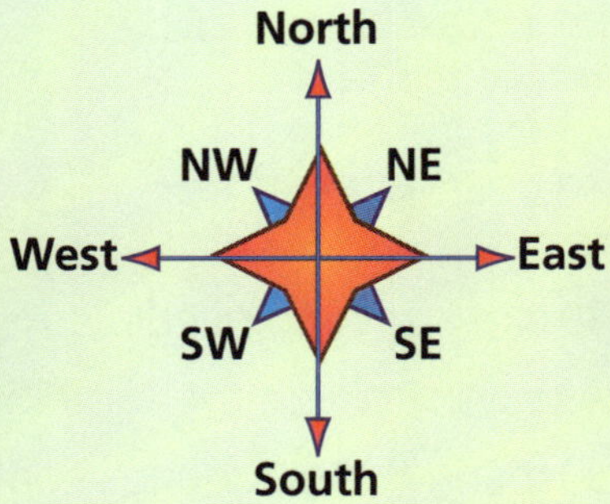

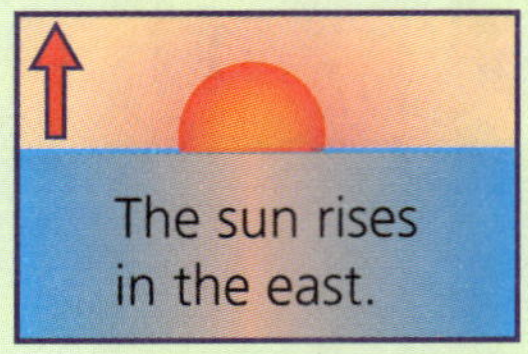

CONCEPT

- The direction **halfway** between north and east is called **north-east**.
- The direction **halfway** between south and west is called **south-west**.

1 a Which direction is halfway between south and east? ______

b Which direction is halfway between north and west? ______

2

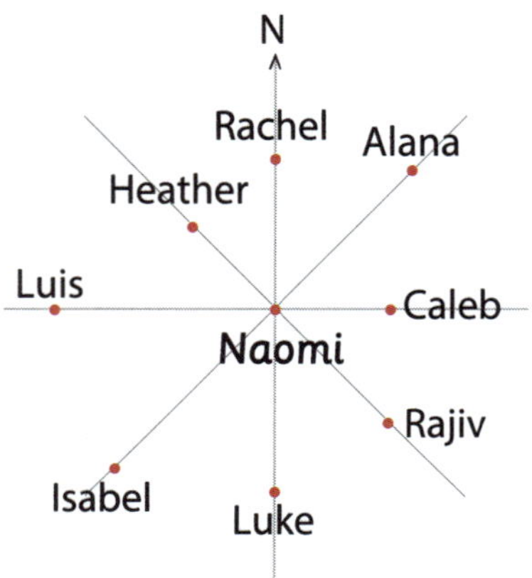

Rachel is north of Naomi. Who is to the:

a east? ______ b south? ______

c north-east? ______ d south-west? ______

From where Naomi stands, in what direction is:

e Luis? ______ f Rachel? ______

g Heather? ______ h Rajiv? ______

i Isabel? ______ j Alana? ______

3

From the centre of Australia, what is to the:

a north? ______ b east? ______

c west? ______ d south? ______

e south-east? ______

f north-east? ______

g north-west? ______

Remember your directions: NEWS is the collection of stories from North, East, West and South.

4

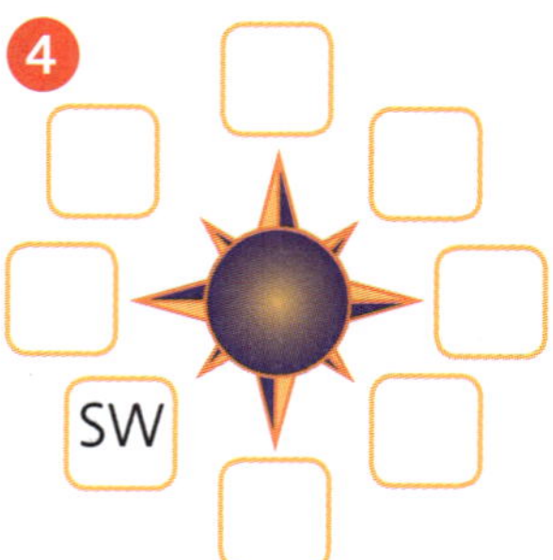

On this compass rose, write N, E, W, S, NE, NW, and SE to show the directions of the compass.

Use computer software to create simple shapes involving direction and angles.

ICT

 ISBN 978 1 4886 2193 2

4:13 Compass Directions

CONCEPT

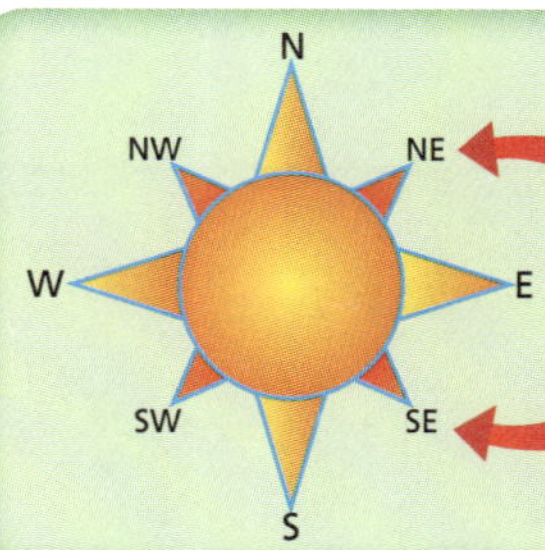

Halfway between north and east is called north-east.

Halfway between south and east is called south-east.

1 From this map, name the island that is:

- **a** east of Shell Island
- **b** south of Shell Island
- **c** north of Palm Island
- **d** west of Cup Island
- **e** north-east of Shell Island
- **f** north-west of the wharf
- **g** south-east of Shell Island

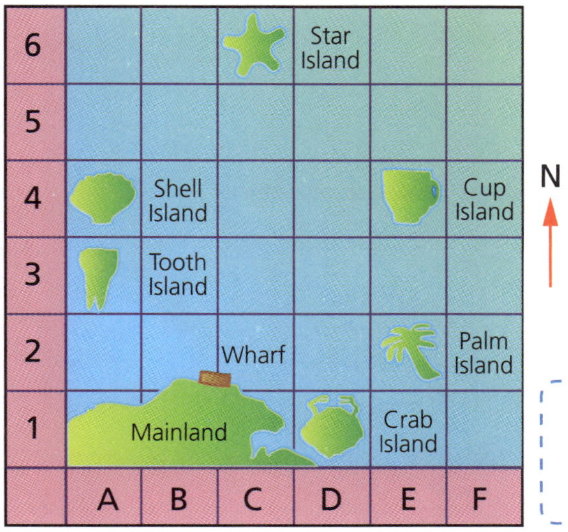

Scale
1 cm = 1 km
0 1

2 Name the island that has coordinates:

a E4 **b** A3 **c** E2

3 Give, to the nearest kilometre, the distance from the wharf to:

a Star Island **b** Palm Island **c** Shell Island **d** Cup Island

4 A ship sailed 2 km east from the wharf, then 3 km north and then 2 km west.

- **a** What is the direction of the ship from the wharf now?
- **b** What is the direction of the wharf from the ship now?
- **c** Which island is now south-east of the ship?

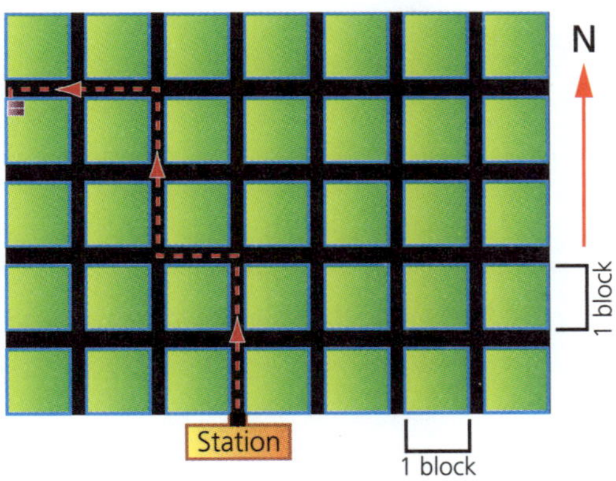

5 This is a map of some city streets.

- **a** Luke walked home from the station along the dotted path. How many blocks did he walk?
- **b** Luke arrived at the station at 5:52 pm. He walked the length of one block every 3 minutes. When did he reach home?

4:14 Using Maps

Only use the roads shown here.

1 Estimate the distance by air between:

a Mildura and Melbourne

b Portland and Echuca

c Ouyen and Ballarat

d Geelong and Echuca

Scale: 1cm = 100 km

0 100 200 300 400 kilometres

Scale: 1:10 000 000

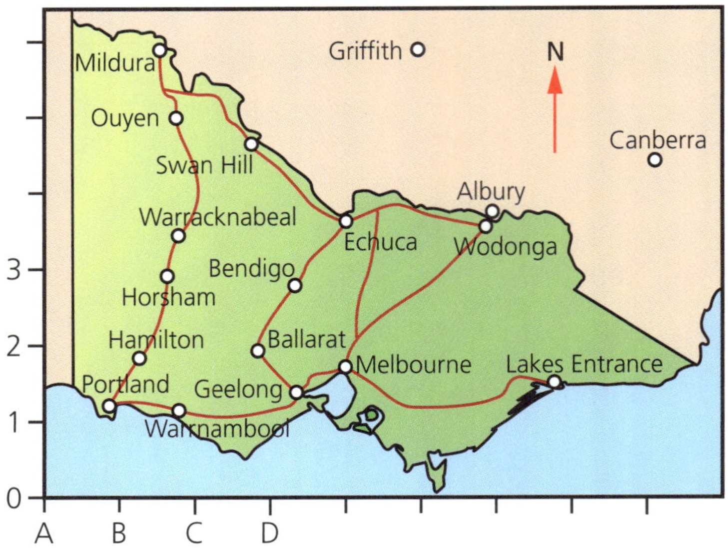

2 Draw a centimetre grid on this map using **A** to **I** along the horizontal axis and **0** to **6** on the vertical axis. Use your grid to give the nearest coordinates to:

a Melbourne

b Canberra

c Echuca

d Mildura

e Ouyen

f Bendigo

3 Use the map above to give the towns through which you would pass to go from:

a Horsham to Mildura

b Geelong to Echuca

c Melbourne to Hamilton

d Portland to Bendigo

e Ballarat to Mildura

4 Estimate the shortest distance by road from:

a Geelong to Echuca

b Melbourne to Hamilton

c Portland to Bendigo

d Ballarat to Mildura

5 Give the direction N, E, W, S, NE, NW, SE or SW from:

a Horsham to Bendigo

b Melbourne to Bendigo

c Ouyen to Swan Hill

d Ballarat to Geelong

e Echuca to Bendigo

f Warrnambool to Ballarat

6 Order the trips in Question 3 from shortest trip to longest.

Label the trips 3a, 3b, 3c, 3d and 3e:

4:15 Using Coordinates

Legend: • capital city

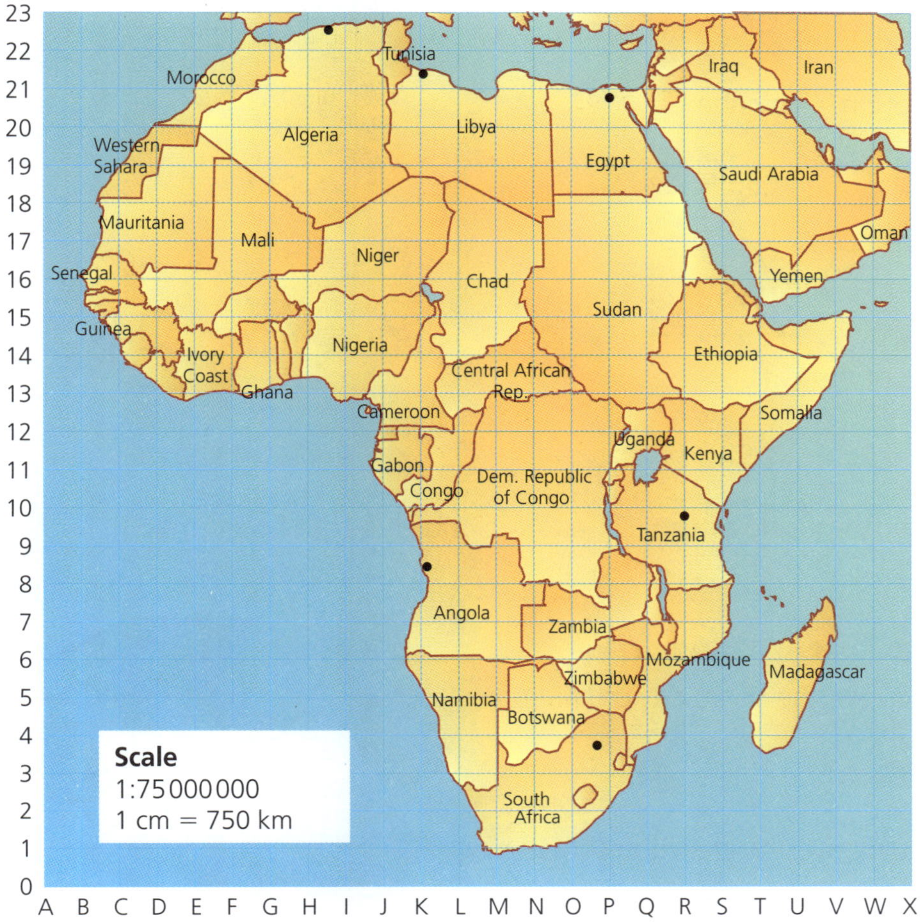

1 Write down the country that has coordinates:

a V22

b Q9

c R14

d M2

e F20

f J17

g U6

h S19

2 Which coordinates would be used for:

a Gabon? **b** Ghana? **c** Ivory Coast? **d** Western Sahara?

3 Give three sets of coordinates that could be used for the Central African Republic.

4 Use the scale to give the approximate distance across these countries, at the widest point (west to east). Give your answers correct to the nearest 100 km.

a Dem. Rep. of Congo **b** Namibia **c** Sudan **d** Zimbabwe

5 Give the approximate distance from north to south of these countries at their longest point. Give your answers correct to the nearest 100 km.

a Chad **b** Tanzania **c** Ethiopia **d** Ivory Coast

6 On the map, measure distances correct to the nearest millimetre and then use the scale. Use a calculator to find the real distance between the **capital cities** of:

a Angola and Egypt

b Angola and South Africa

c Algeria and Egypt

d Tanzania and South Africa

e Egypt and Tanzania

f Egypt and South Africa

4:16 Sections and Elevations of 3D Objects

ACTIVITY

- Make three models of a cone using clay or Plasticine.
- Use wire to cut sections through each cone as shown.
- Colour the cut sections with paint or ink and print the shapes on paper.
- Make three models of a cylinder and use wire to cut the sections as shown.
- Colour and print the shapes of the cut sections you have made.

CONCEPT

An **elevation** is a view from the front, side or top. Different 3D objects could have the same elevation.

A

B

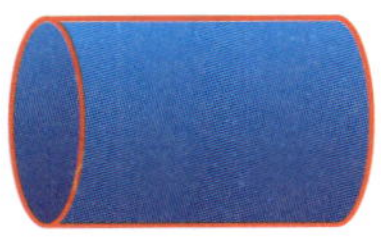

C

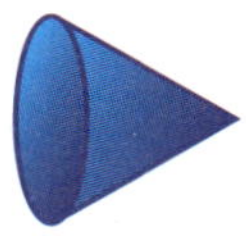

D

1 Which 3D objects above could have the elevation shown below?

a

b

c

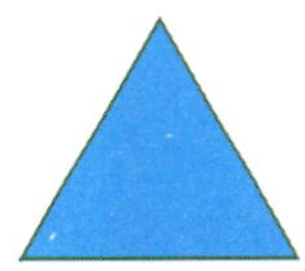

2 For each 3D object, draw the front, side and top elevations, as shown.

Example

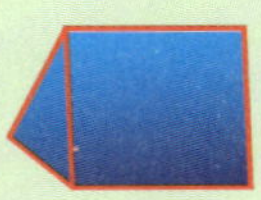

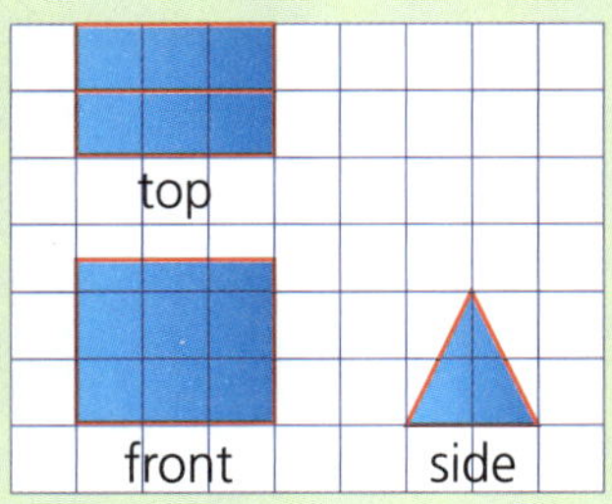

a

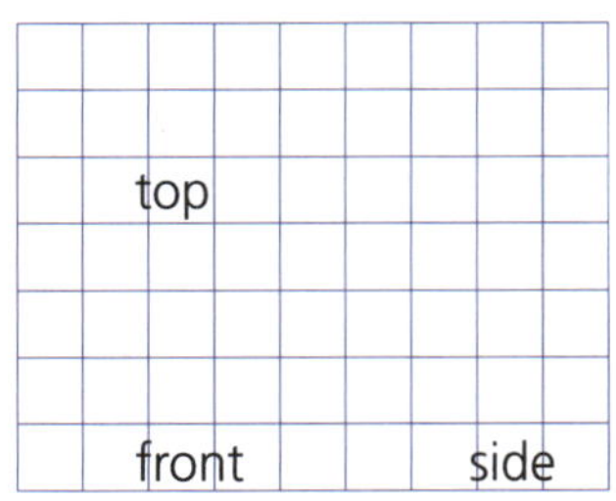

b

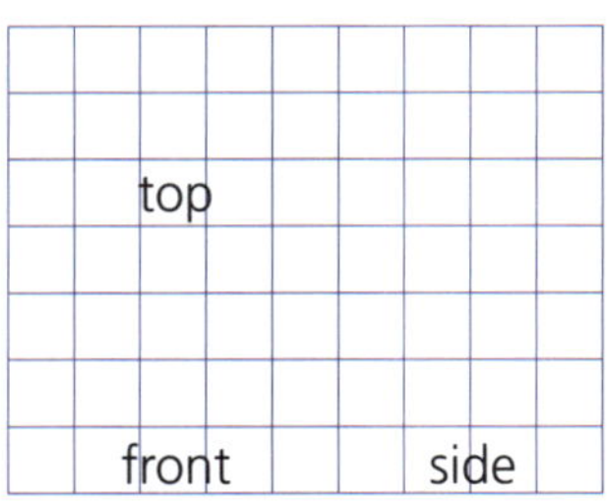

c

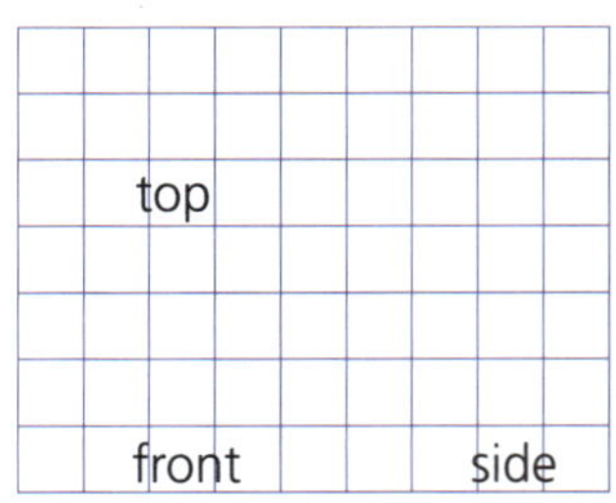

4:17 Naming 3D Objects

A prism or a pyramid is named according to the shape of its base.
These bases are octagons.

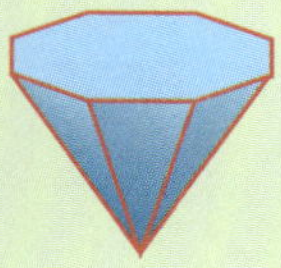

octagonal pyramid

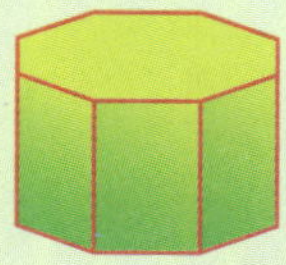

octagonal prism

CONCEPT

1 Match each drawing to its name.

A 

B 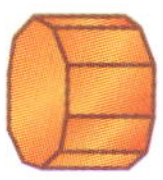

C 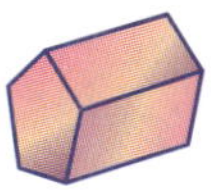

D 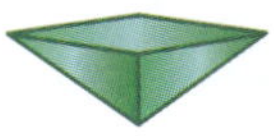

E 

F 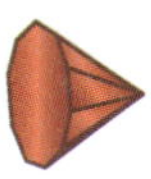

- rectangular pyramid
- rectangular prism
- decagonal prism
- decagonal pyramid
- pentagonal pyramid
- pentagonal prism
- hexagonal pyramid
- triangular prism
- octagonal pyramid
- octagonal prism
- hexagonal prism
- triangular pyramid

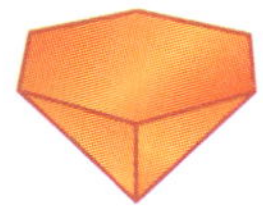 G

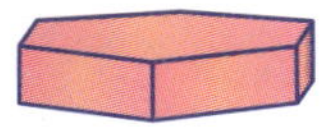 H

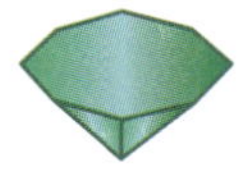 I

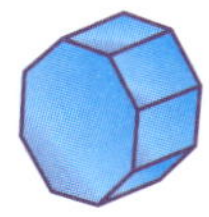 J

 K

 L

2 Which of the 3D objects in Question 1 are:

a prisms? ______

b pyramids? ______

Which of the 3D objects have:

c fewer than 7 faces? ______

d more than 8 faces? ______

e an even number of vertices? ______

f an odd number of edges? ______

3 Draw a:

a triangular prism

b hexagonal pyramid

c pentagonal prism

4:18 Nets of Prisms

1 Match each name to the net of the prism.

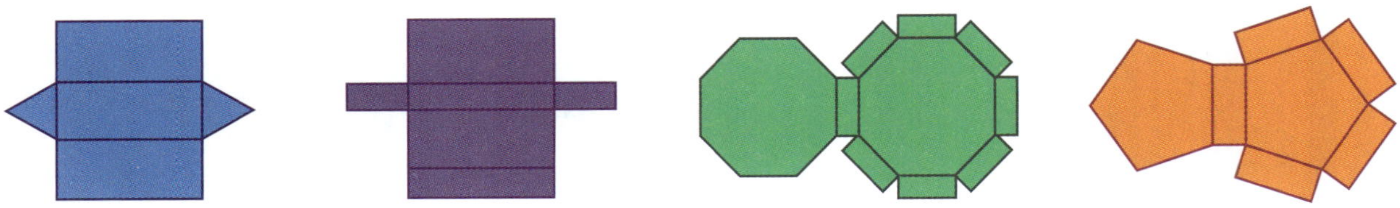

rectangular prism | triangular prism | pentagonal prism | octagonal prism

2 Copy this net and construct the prism.

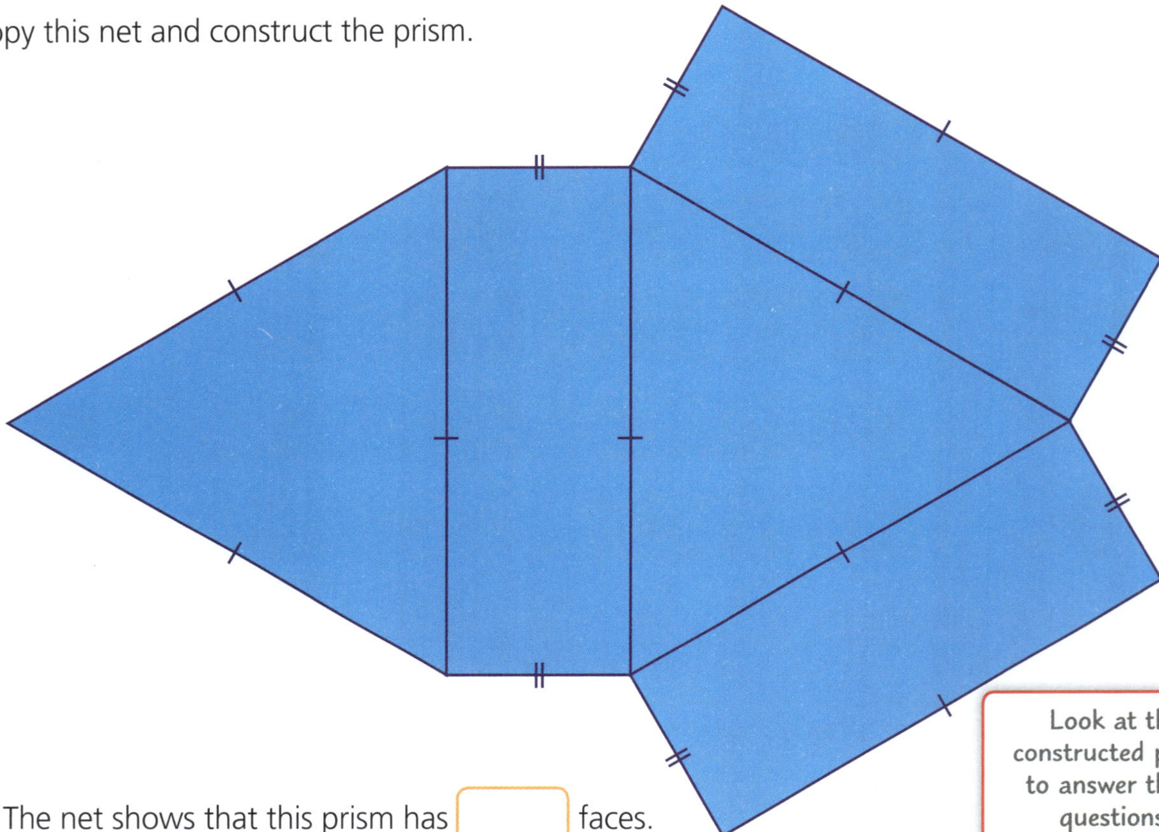

a The net shows that this prism has ______ faces.

b How many vertices (corners) does this prism have? ______

c How many edges does this prism have? ______

d This is the net of a ______.

e Construct prisms using other materials.

INVESTIGATION

- Use the internet to investigate and list the names of buildings that have prisms as part of their structure. (You could search for 'famous buildings'.)

______

______

4:19 Nets of Pyramids

1 Match each name to the net of the pyramid.

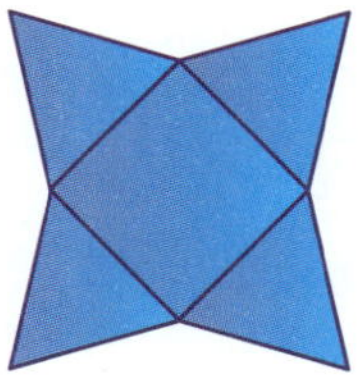

triangular pyramid | square pyramid | hexagonal pyramid | pentagonal pyramid

2 Copy this net and construct the pyramid.

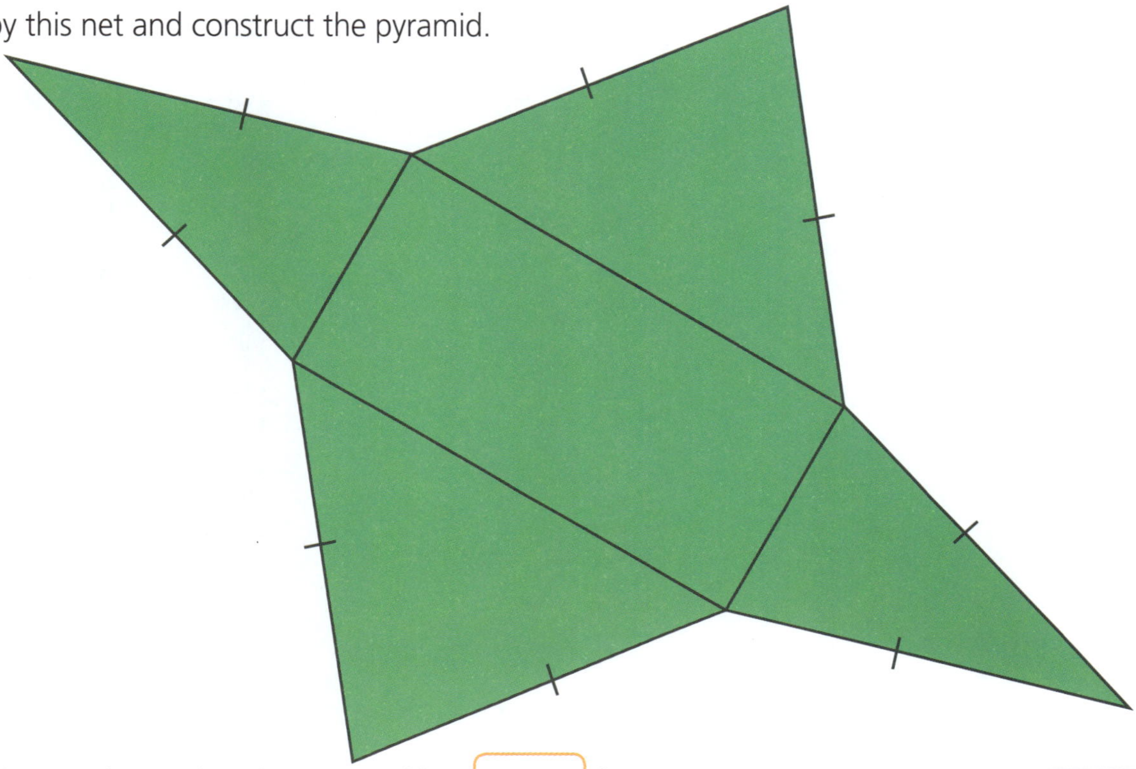

a The net shows that this pyramid has ______ faces.

b How many vertices (corners) does this pyramid have? ______

c How many edges does this pyramid have? ______

d This is the net of a ______.

e Construct pyramids using other materials.

INVESTIGATION

- Use the internet to investigate and list the names of buildings that have pyramids as part of their structure. (You could search for 'famous buildings'.) Search for architecture around the world, including Egypt, China, Korea and Indonesia.

______

______

4:20 The Number Plane

CONCEPT

- In a street directory, a letter and a number are used to name a position.
- On a number plane, two numbers are used.
  (2, 5) represents the point 2 to the right and 5 up from the **origin** which is (0, 0).
  The horizontal axis is called the *x*-axis. The vertical axis is called the *y*-axis.

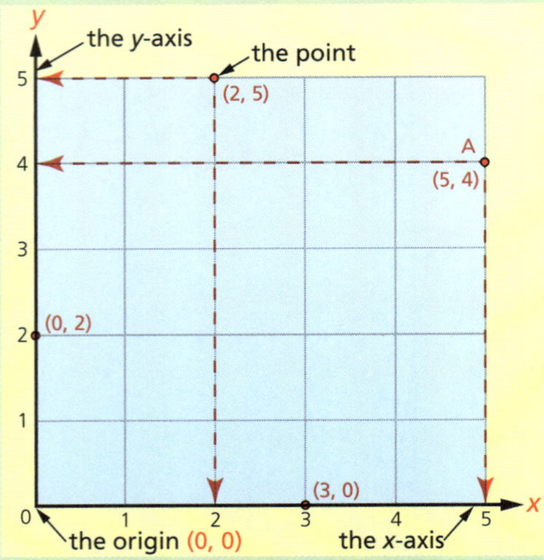

(5, 4) is called the **coordinates** of point A.
(5, 4) is 5 to the right of the origin and 4 up.
(5, 4) has a reading of 5 on the *x*-axis and a reading of 4 on the *y*-axis.

1 Write the letter used to name each point.

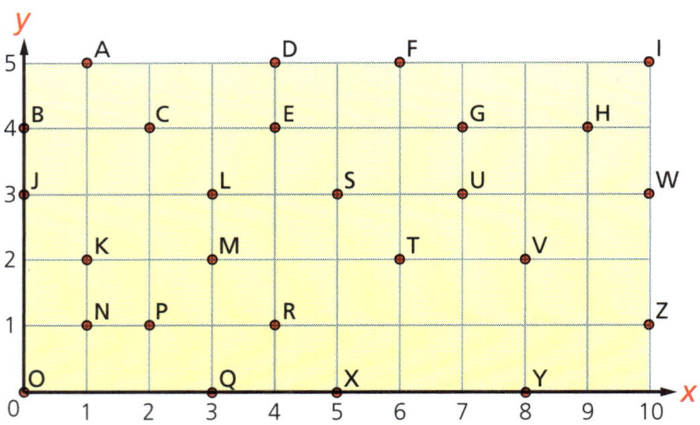

a (1, 1)		**b** (8, 2)	
c (9, 4)		**d** (2, 4)	
e (4, 5)		**f** (0, 4)	
g (8, 0)		**h** (6, 5)	
i (4, 1)		**j** (3, 0)	
k (0, 0)		**l** (1, 2)	

2 Write the coordinates of each point named.

a A	**b** E	**c** G	**d** J
e L	**f** M	**g** P	**h** S
i T	**j** U	**k** W	**l** Z

3 Write the coordinates of the point that is halfway between:

a T and V	**b** O and B	**c** Y and O
d L and U	**e** W and I	**f** N and S

4:21 The Number Plane – Four Quadrants

CONCEPT

- On a number line, negative numbers are placed to the left of zero.

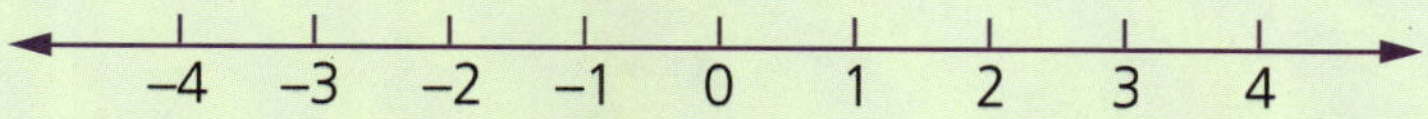

- On a vertical number line, negative numbers are placed below zero.
- We can use negative numbers on a number plane as well, since we are using both a horizontal and a vertical number line.

A is the point (**–1**, **2**). **B** is the point (**–1**, **0**). **C** is the point (–**2, –1**).
D is the point (**0**, **–2**). **E** is the point (**2**, **–2**). (**0**, **0**) is called the origin.

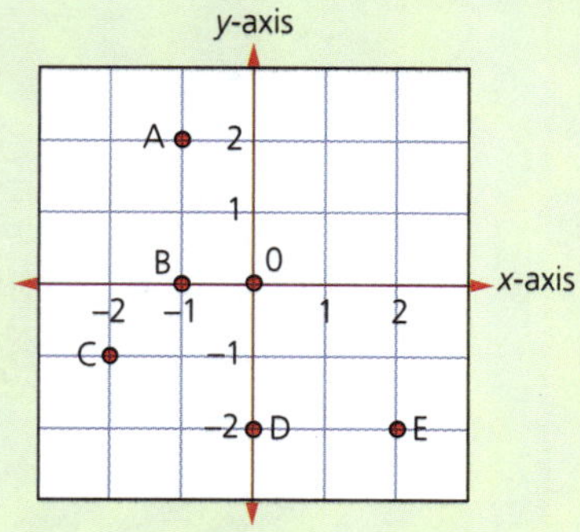

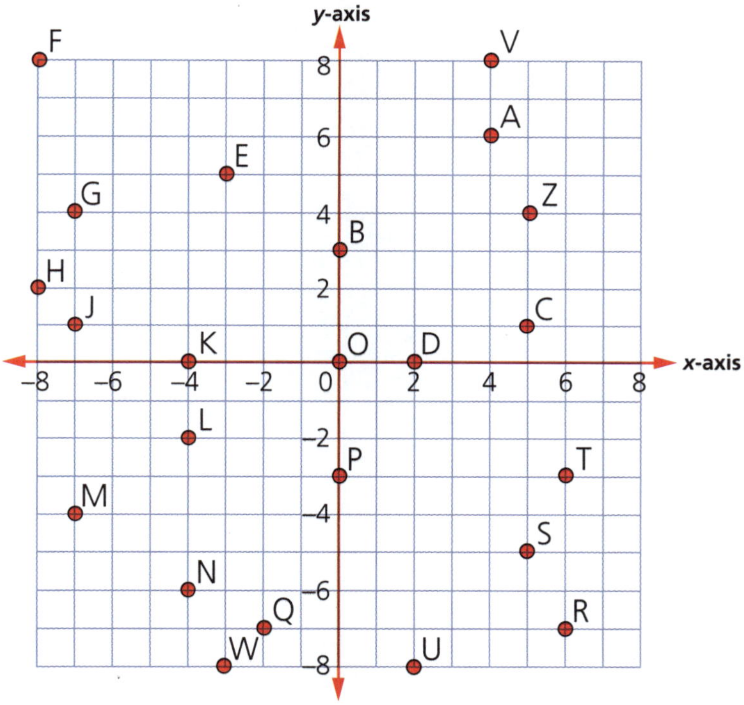

1 Write the letter used to name each point.

a (4, 6)		**b** (5, 1)	
c (–4, –2)		**d** (0, –3)	
e (0, 0)		**f** (6, –3)	
g (–8, 2)		**h** (–7, –4)	
i (–2, –7)		**j** (5, –5)	

2 Write the coordinates of each point.

a B		**b D**	
c E		**d N**	
e F		**f G**	
g J		**h K**	
i R		**j U**	

3 Write the coordinates of the point that is halfway between:

a (4, 6) and (4, 8)		**b** (–8, 2) and (–8, 8)	
c (–4, –2) and (–4, 0)		**d** (–4, –2) and (–4, –6)	
e (6, –3) and (6, –7)		**f** (2, 0) and (2, –8)	
g (–7, –4) and (–7, 4)		**h** (5, 4) and (–7, 4)	

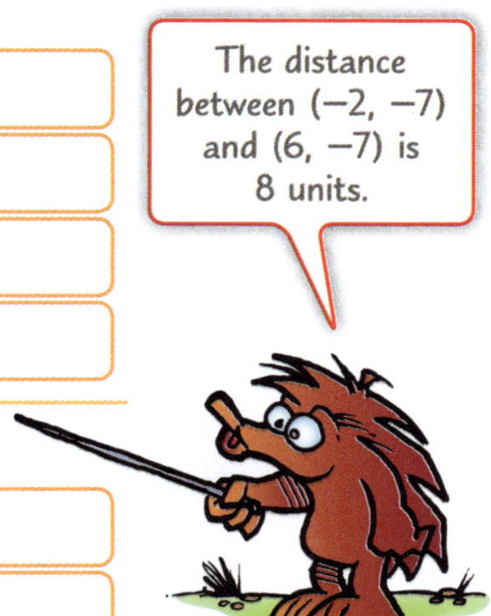

4 Write the distance in grid units, between:

a (0, 0) and (2, 0)		**b** (5, 1) and (5, 4)	
c (–8, 8) and (4, 8)		**d** (5, 1) and (–7, 1)	
e (–3, 5) and (–3, –8)		**f** (–6, –2) and (–1, –2)	

4:22 Using Coordinates

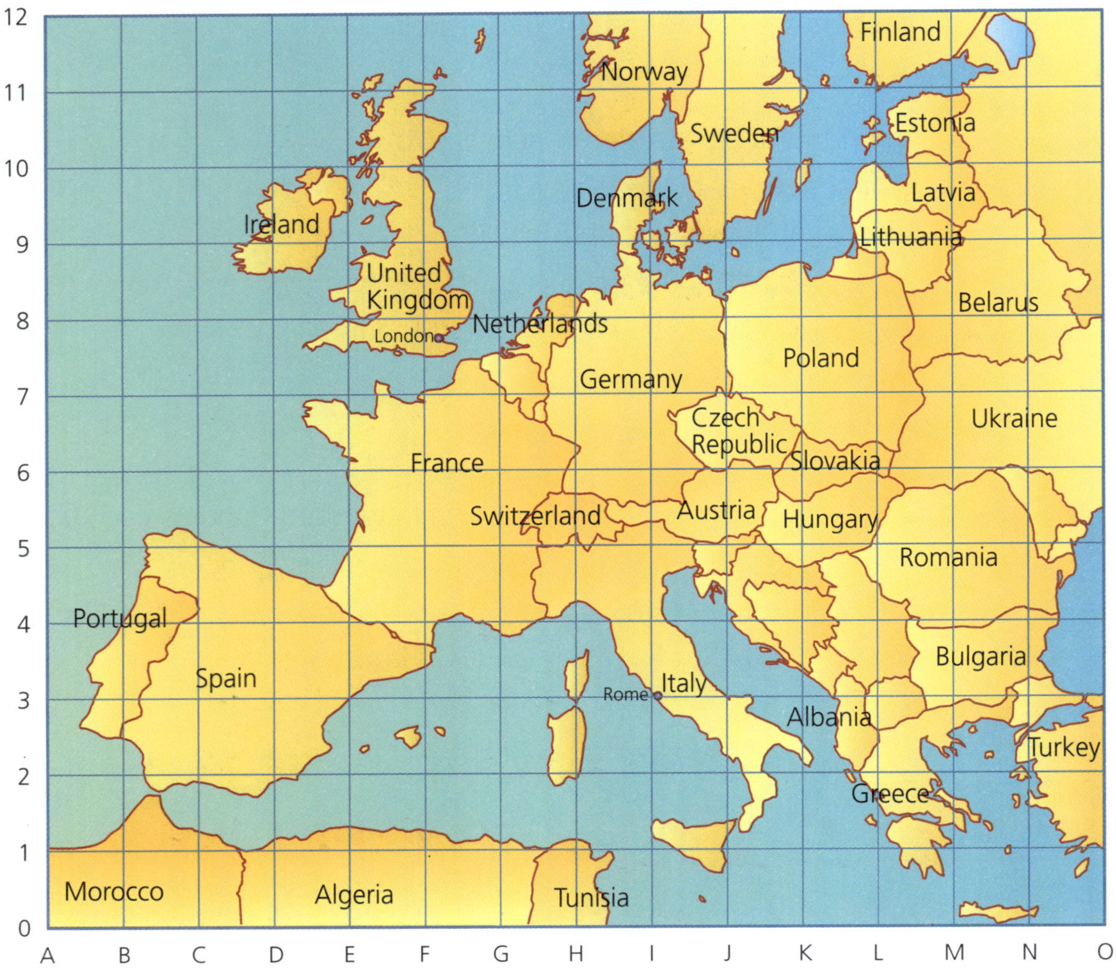

Scale
1:25 000 000

Grid reference coordinates refer to the intersection of lines.

1. Write down the country that has coordinates:
   a C3 ______ b J10 ______ c I4 ______ d B1 ______
2. Which coordinates should be used for:
   a Portugal? ______ b Greece? ______ c Bulgaria? ______ d Hungary? ______
3. Give four sets of coordinates for Poland. ______
4. Write all sets of coordinates that could be used for Germany. ______
5. Use the scale of the map to give the approximate distance across these countries, at the widest point (west to east). Give your answers correct to the nearest 100 km.
   a Bulgaria ______ b Romania ______ c Poland ______
6. Give the approximate distance from north to south of these countries at their longest point. Give your answers correct to the nearest 100 km.
   a Latvia ______ b Spain ______ c Romania ______
7. Estimate, to the nearest 100 km, the shortest distance by sea from London to:
   a Norway ______ b Spain ______ c Rome ______ d Finland ______

4:23 Using Maps

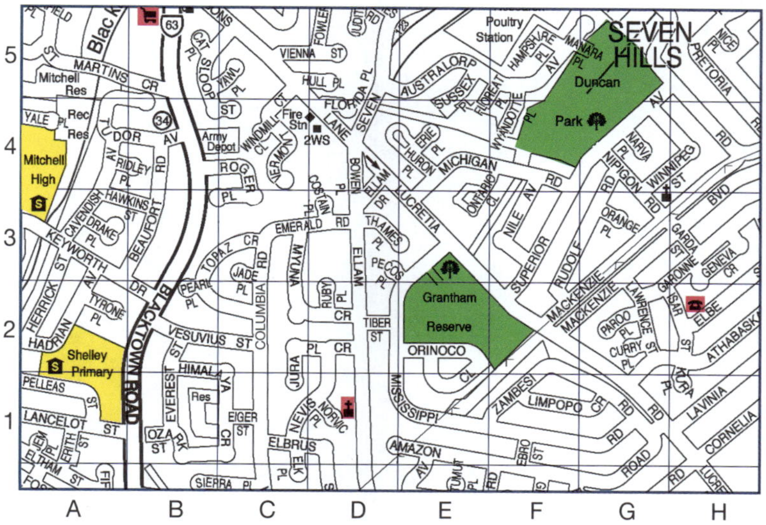

Legend
- park
- places of worship
- school
- shops
- telephone

Scale
5 cm = 1 km
1 : 20 000

1 **a** Estimate the distance by road from Shelley Primary School (A1) to the intersection of Superior Av and Nipigon Rd (G4).

b List the roads you would use to travel from Tyrone Pl (A2) to Thames Pl (D3).

c Estimate the time it would take to walk from Lancelot St (A1) to Pretoria Rd (H5).

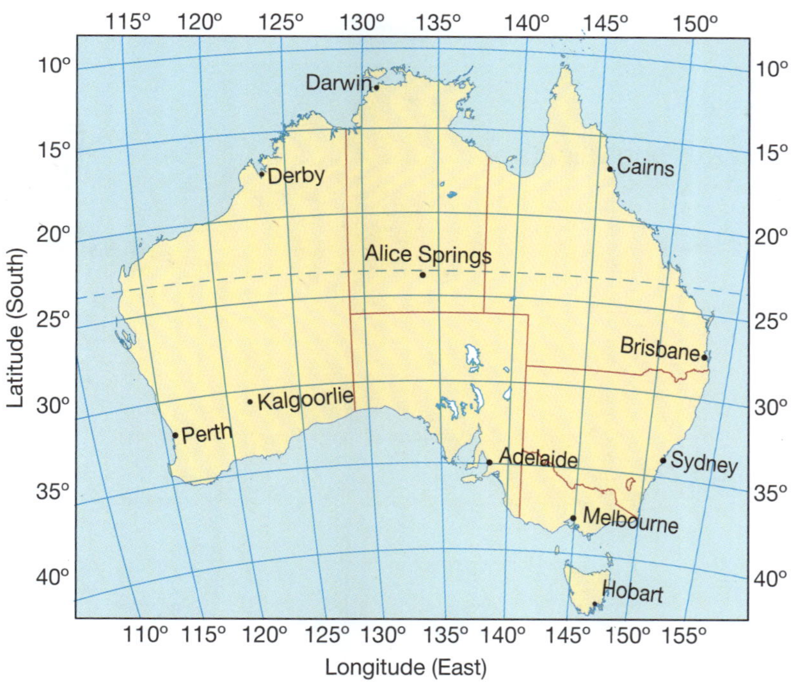

Scale
1 cm = 475 km

2 Estimate, to the nearest 50 km, the distance from:

a Alice Springs to Brisbane

b Perth to Brisbane

c Darwin to Melbourne

My plane flies at 600 km/h. Estimate, to the nearest hour, the time it would take me to fly from:

d Alice Springs to Brisbane

e Perth to Brisbane

f Darwin to Melbourne

3 Using the map in Question 1, write directions to travel from Cornelia Rd (H1) to Paroo Pl (G2).

ACTIVITY

- Take turns to give directions using the map in Question 1.

4:24 Transformations

1 Complete each pattern, then describe how it was made.

a

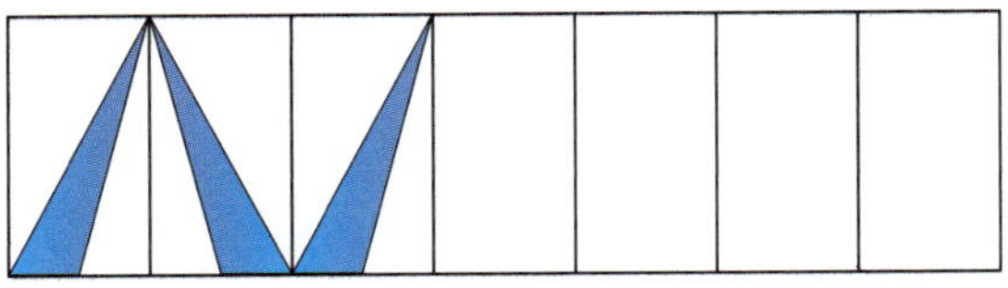

b

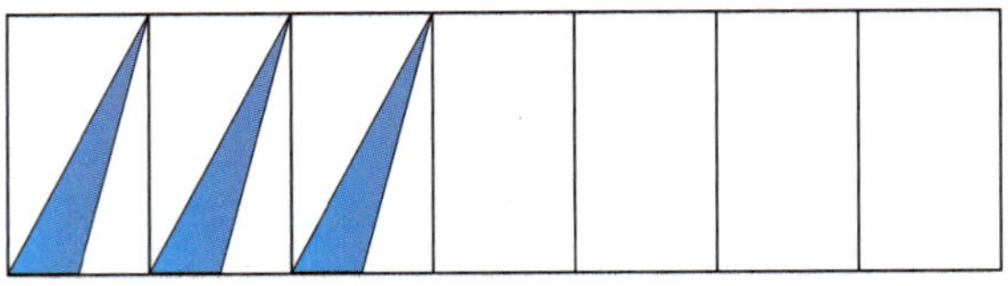

c

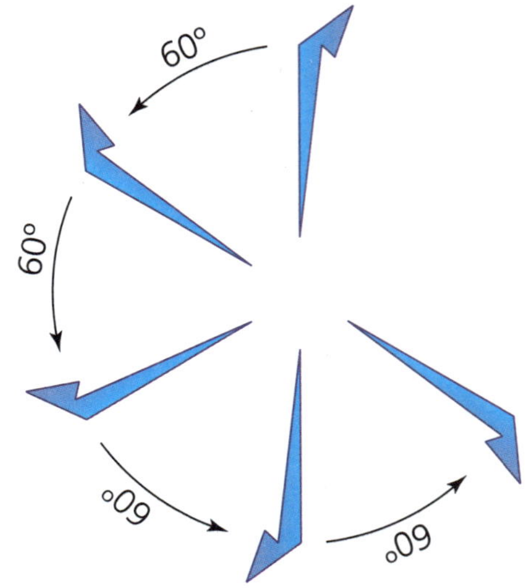

2 Question 1**c** may help you to answer these.

What single transformation is the same as (is equivalent to):

a 2 rotations of 60°?

b 3 rotations of 60°?

c 4 rotations of 60°?

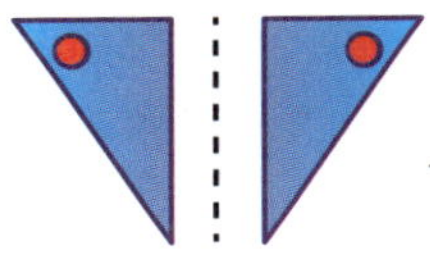

d a rotation of 40°, then a rotation of 20°?

e translations of 2 cm right then 3 cm right?

f two reflections about the same line?

g a reflection about **A** then a reflection about **B**?

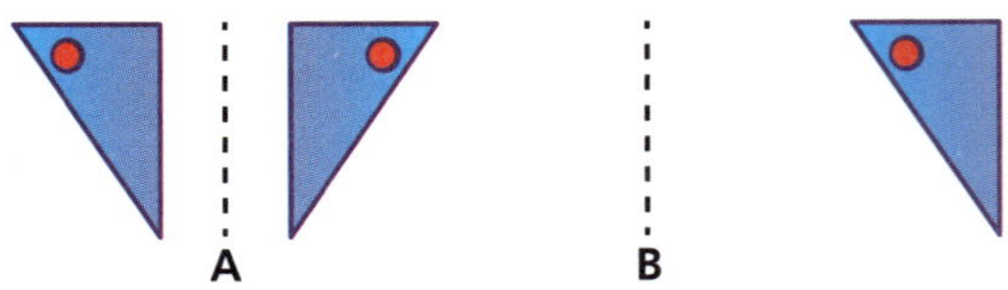

Computer Patterns

ICT

- Use 'Copy', 'Paste' and 'Turn' features to make these patterns.
- Design a school logo.

 • *AUSTRALIAN SIGNPOST MATHS 6* • ISBN 978 1 4886 2193 2

4:25 Transformations

1. Describe the change that takes place as **A** moves to **B**. What is the same? What is different?

	The Change	Description
a	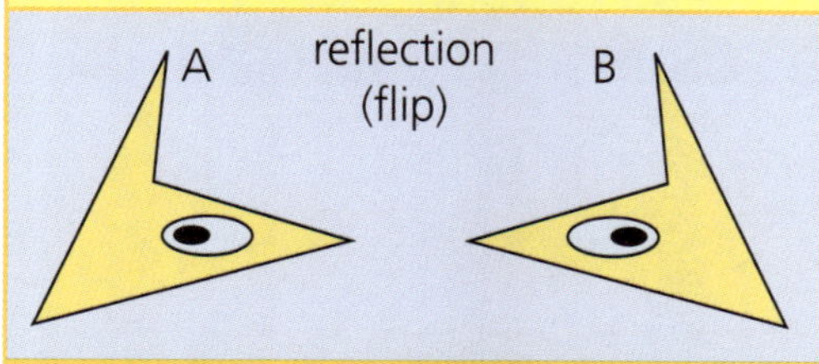 A reflection (flip) B	
b	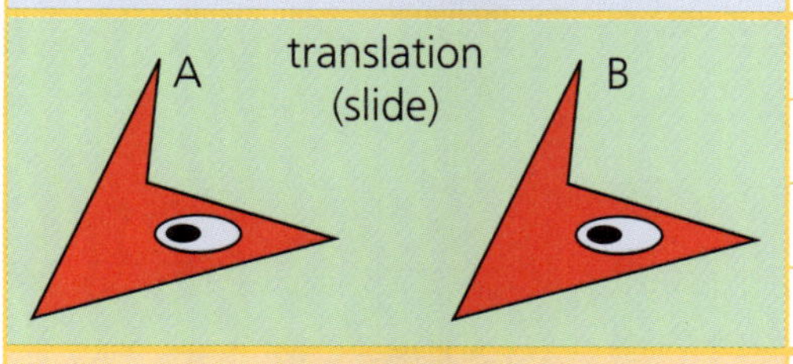 A translation (slide) B	
c	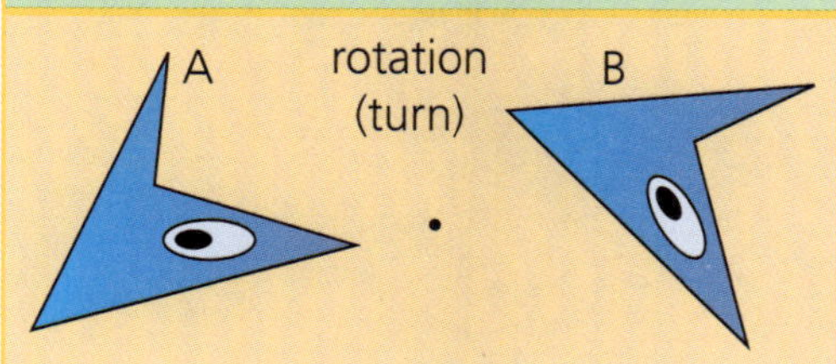 A rotation (turn) B	

2. Describe each transformation (for example, 'a rotation of 90° about a point').

	The Change	Description
a	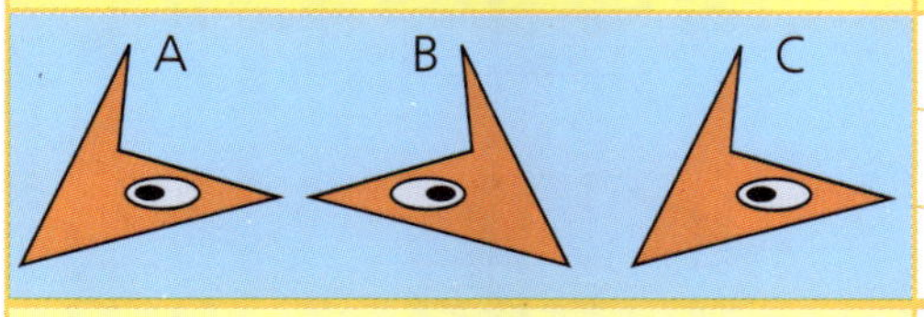 A B C	A to C:
		A to B:
		B to C:
b	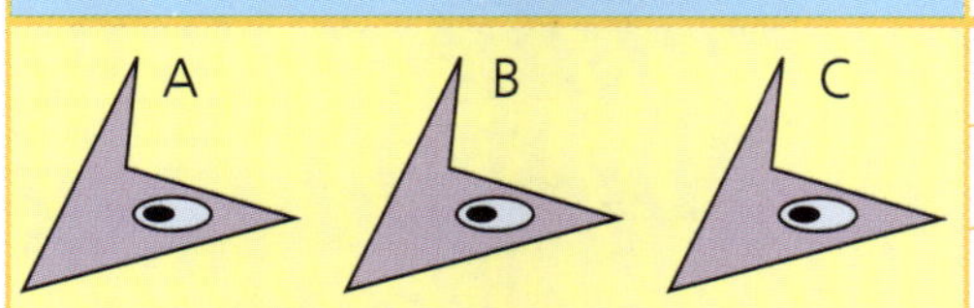 A B C	A to C:
		A to B:
		B to C:
c	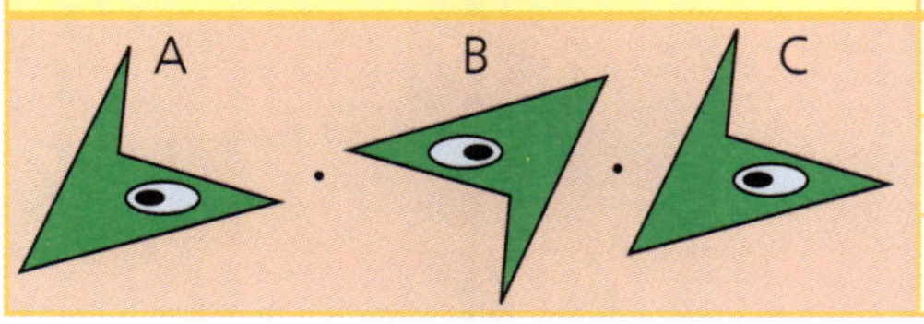 A B C	A to C:
		A to B:
		B to C:

d Is going from A to B, then from B to C, the same as going from A to C in one step?

Using Computers to Copy a Pattern

ICT

3. Use a drawing program to copy this pattern. List the transformations used.

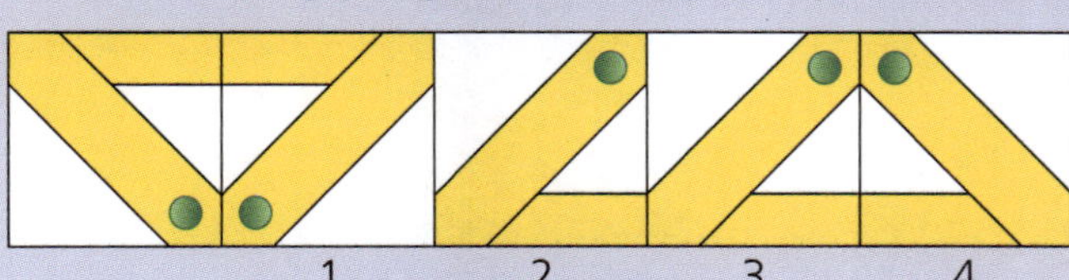

4:26 Making Patterns on a Computer

1 Heather made a design on a computer and then made three different patterns. She used **flip** (reflection) in one, **slide** in one and **turn** (rotation) in the other. Which is which?

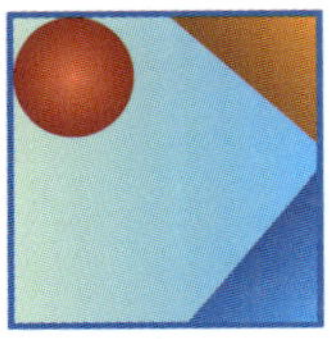

Heather's Design

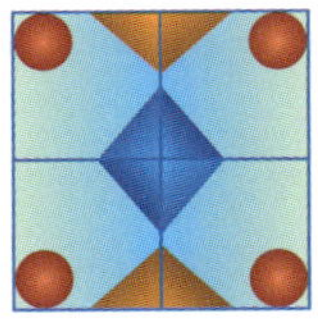

a

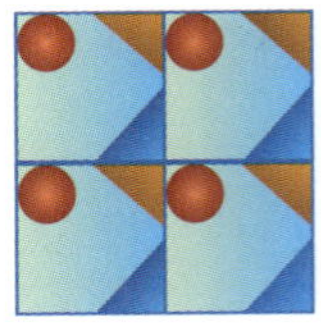

b

c

2 Design your own tile on a computer. Create three patterns of your own as Heather has done.

My Tile

a flip **b** slide **c** turn

3 On a computer make these squares: 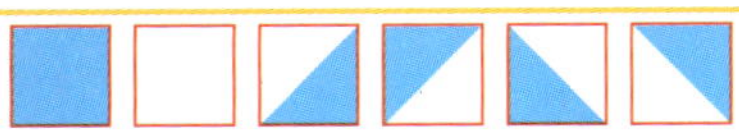

Join 4 of the squares to make this basic design: 

a **Flip** each strip about its lower edge to get the next line of the pattern. Put all strips together.

Basic design 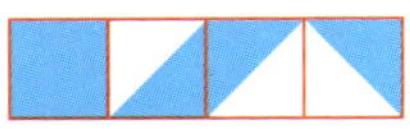

Flip 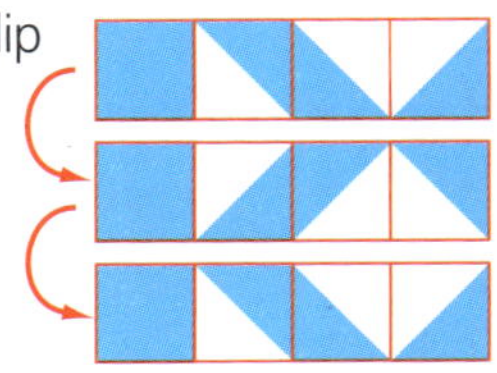

Your design 

b **Slide** the design one square to the left, moving the first square to the end. Put all strips together.

Basic design 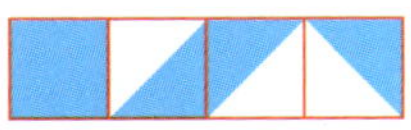

Slide 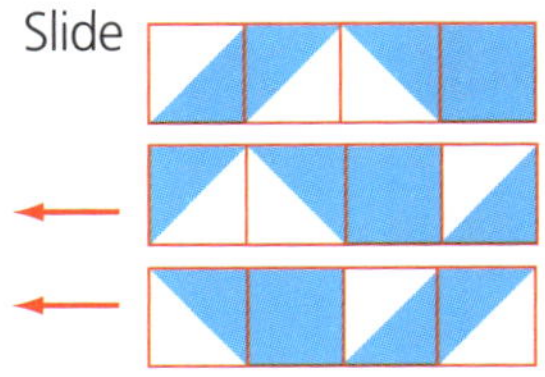

Your design 

c **Turn** each square of the strip clockwise through 90° (about each square's centre). Put all strips together.

Basic design 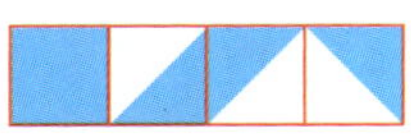

Turn 

Your design 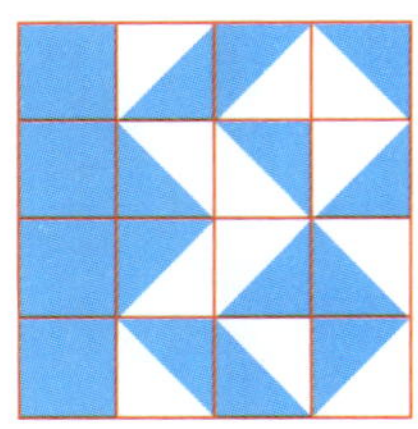

4 On a computer, make your design and use it to make patterns below.

Choose 4 of these squares. 

My Basic Design

a flip **b** slide **c** turn

 • *AUSTRALIAN SIGNPOST MATHS 6* • ISBN 978 1 4886 2193 2

4:27 Constructing Regular Shapes

CONCEPT

Drawing a Regular Decagon

1 Draw a circle.
Mark the centre.

2 Divide the 360° at the centre by 10.
360° ÷ 10 = 36°

3 Extend the lines to the circle.
Join the points on the circle to get a decagon.

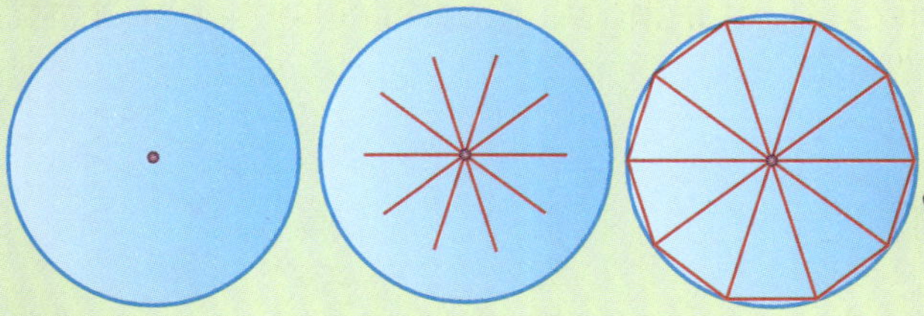

1 By dividing the angle at the centre into equal parts, draw inside the circle:

a a regular pentagon

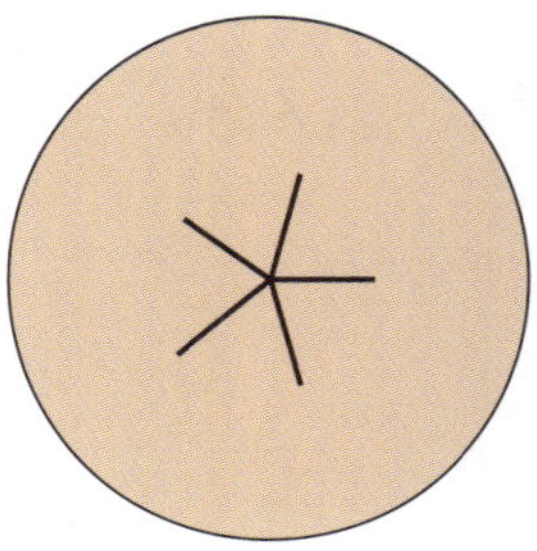

360° ÷ 5 = 72°, so angles at the centre are 72°.

b a regular hexagon

360° ÷ 6= ☐°, so angles at the centre are ☐°.

c a regular octagon

☐

2 Match each description with a diagram and a name.

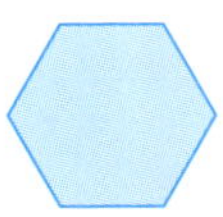

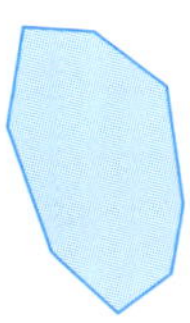

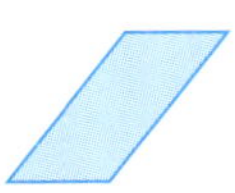

- opposite sides equal, angles measure 90°, 2 diagonals
- 6 equal sides, 6 axes of symmetry, 9 diagonals
- opposite sides parallel, opposite angles equal, 2 diagonals
- 8 sides, angles different sizes, 24 diagonals

- regular hexagon
- rectangle
- irregular octagon
- parallelogram
- regular pentagon
- right-angled triangle
- irregular pentagon
- trapezium

- 5 sides, 5 diagonals, sides not equal
- 5 equal sides, 5 axes of symmetry, 5 diagonals
- 3 sides, one angle is 90°, no diagonals
- 4 sides, one pair of parallel sides, 2 diagonals

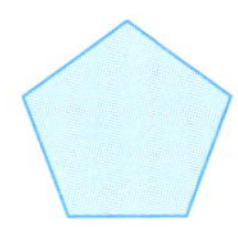

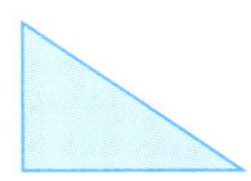

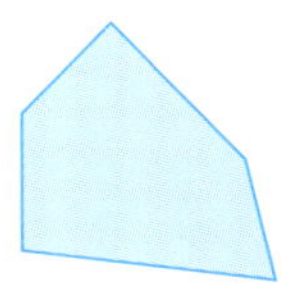

 • *AUSTRALIAN SIGNPOST MATHS 6* • ISBN 978 1 4886 2193 2

4:28 Drawing and Recognising 3D Objects

CONCEPT

We can draw many 3D objects by starting with parallel lines that cross.

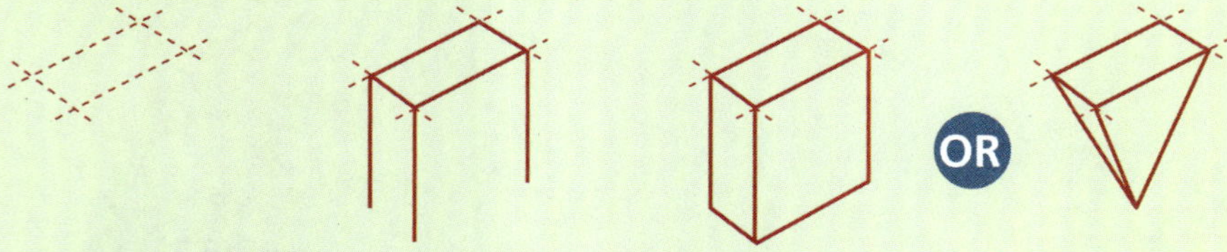

If we start with an oval, we can draw a cylinder or a cone.

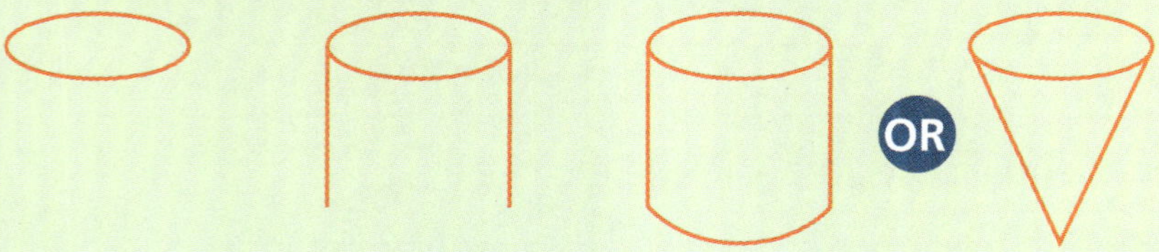

1 How are the rectangular prism and rectangular pyramid above alike? How are they different?

2 On your own paper, and using the methods above, draw:

a two prisms **b** two pyramids **c** two cylinders

3 This nail is made of part of a sphere, a cylinder and a cone. Name the 3D objects, or parts of 3D objects, that are part of the items below.

a

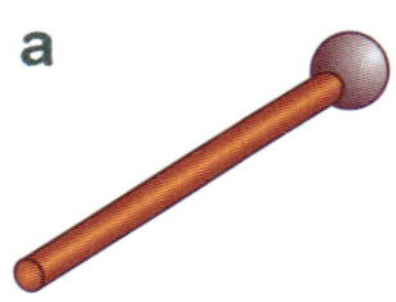

drumstick

b

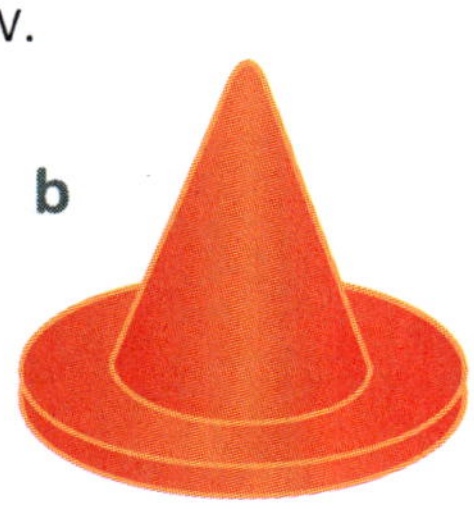

traffic cone

c

vase

d

water bottle

e

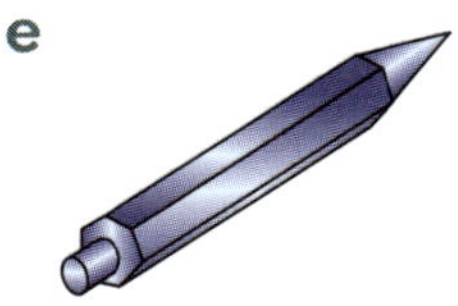

propelling pencil

ACTIVITY

- Write, as headings in a table, 3D objects found in everyday items.
- In the table, list items in which each 3D object is found.

Cone	Cylinder	Sphere	Prism
nail	nail	nail	

 • *AUSTRALIAN SIGNPOST MATHS 6* • ISBN 978 1 4886 2193 2

4:29 Cones, Cylinders and Spheres

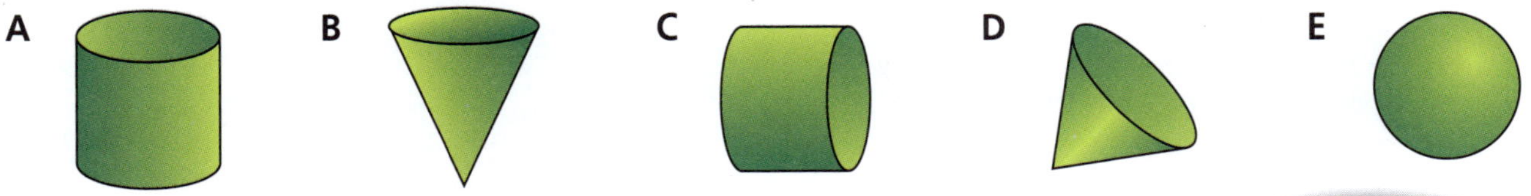

1 Give the name of:

a shape **C**

b shape **E**

2 a Which cone has the larger base?

b Which cone has the greater height?

c Which cylinder has the greater height?

d Which cylinder has the larger base?

e Which of these 3D objects rolls the best?

f How does a cone roll?

g How does a cylinder roll?

3 Complete the table below.

Shape	Number of Edges	Number of Surfaces	Number of Corners	Does It Roll?	Does It Have a Curved Surface?
cone					
cylinder					
sphere					
cube					

4 a Is it easier to stack cones or cylinders?

b Stack ten of each type of coin mentioned in the table.

- Order the stacks from the highest (1) to lowest (6) and fill in that column.
- Order the stacks from widest (1) to narrowest (6) and fill in that column.
- Write in the value of each stack.

Stack	Height Order	Width Order	Value
5c			
10c			
20c			
50c			
$1			
$2			

ACTIVITY

- Investigate the shapes you can form by stacking cones and cylinders.
- Make two cones from thin cardboard.
  - Join the bases of the cones.
  - Place this shape on a set of non-parallel rails.
  - See how far you can blow the shape uphill along rails using one breath.

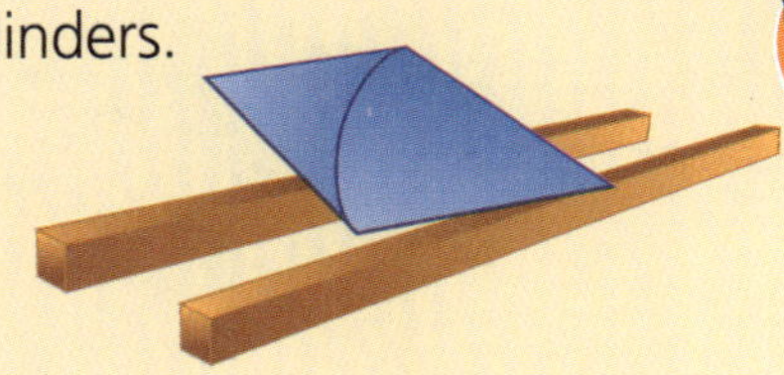

 • *AUSTRALIAN SIGNPOST MATHS 6* • ISBN 978 1 4886 2193 2

5:01 Reading Graphs

1 Our family uses a lot of petrol, especially during our annual holiday.

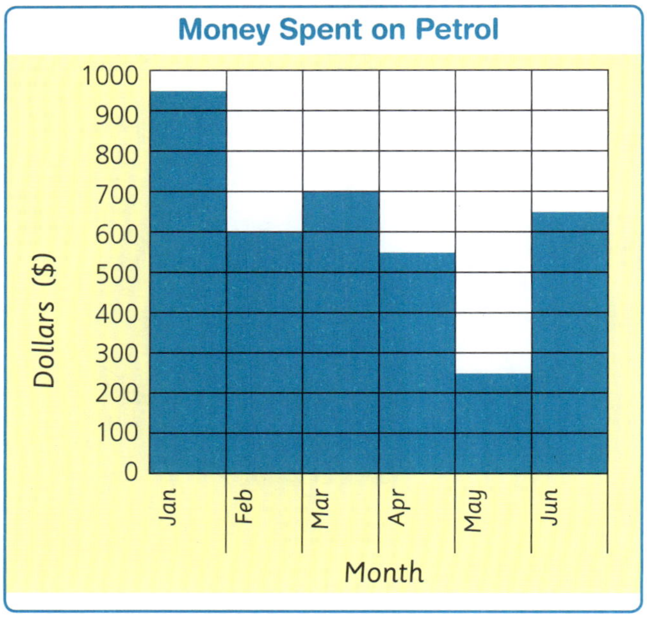

a In which month were we probably on our holiday?

b In which month did we spend $550 on petrol?

c What was the total money spent in April and June?

d What was the difference in money spent in January and February?

e In which two months did we spend a total of:

i $800?

ii $1 650?

2 Throughout the year, the coach of the under-12 hockey team kept this tally.

Goals Scored (卌 = 5)

Aziz	卌 |
Erin	卌 卌 |
David	卌 卌 卌 卌
Chris	卌 卌 卌 卌 卌 卌 ||
Alicia	卌 卌 卌 |||
Brandon	卌 卌 |||

a Who scored the most goals?

b How many goals did Alicia score?

c How many goals did Brandon score?

d How many more goals did David score than Aziz?

e Who scored 20 goals?

f How many goals altogether did the team score during the year?

3 Two dice were thrown 250 times and the total of each throw was recorded on this tally sheet.

a Why did the score of 1 not occur?

Scores of Thrown Dice

					卌					
				||||	卌	||||				
			|||	卌	卌	卌	||||			
		||	卌	卌	卌	卌	卌			
	|	卌	卌	卌	卌	卌	卌	卌		
	卌	卌	卌	卌	卌	卌	卌	卌	|||	
||	卌	卌	卌	卌	卌	卌	卌	卌	卌	||
卌	卌	卌	卌	卌	卌	卌	卌	卌	卌	卌
2	3	4	5	6	7	8	9	10	11	12

b Use the tally sheet to complete the bar graph below.

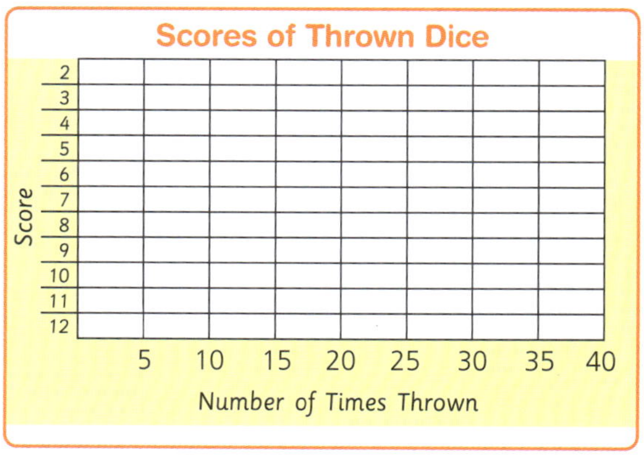

5:02 Line Graphs

1 The results of our weekly dictation test are graphed here. We were given the same dictation passage for 8 weeks in a row.

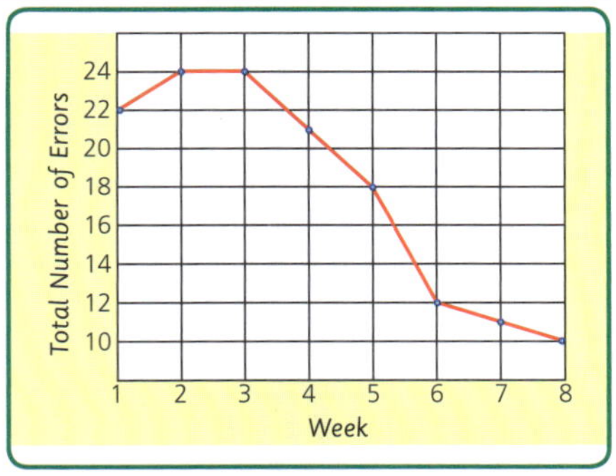

a The vertical axis does not start at zero. What is the least number of errors that can be shown on this graph?

b How many errors were made in week 4?

c When were the greatest number of errors made?

d How many more errors were there in week 1 than in week 8?

e How many errors were there altogether?

f On which week was the greatest improvement shown?

g Why do you think the results improved so much over time?

h How many errors do you think would be made in week 9?
Why?

i Write down some advantages of using a line graph.

2 This graph shows the monthly sales of copiers.

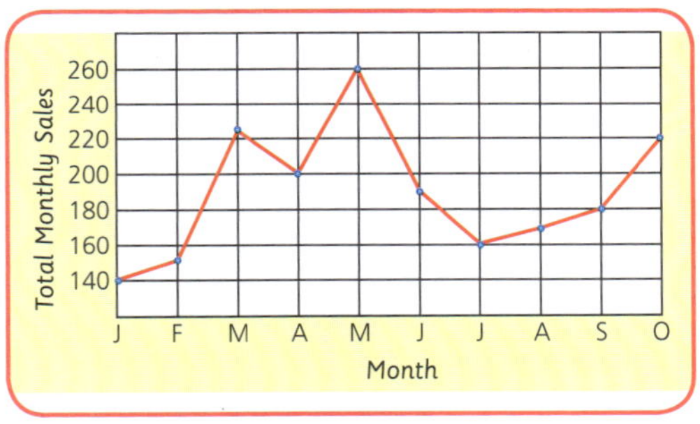

a In which month were 152 copiers sold?

b How many sales were made in the last three months altogether?

c How many more copiers were sold in May than in July?

d How many more sales were made in the eighth month than in the first month?

e Which three-month period had the most sales?

f How many sales would you predict for November?

3 Some of Di's bank balances for last year were:

April	$100	May	$200
June	$200	July	$300
August	$350	September	$375
October	$325	November	$250

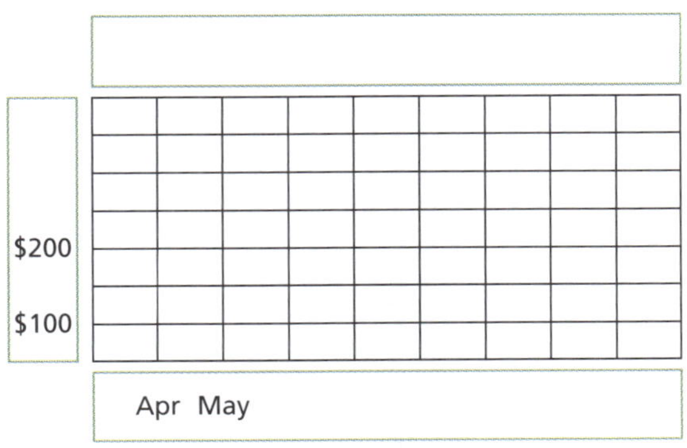

a Complete the labels and draw a line graph to show this information.

b What is most likely to have happened to the balance in December? Why?

5:03 Tables and Graphs

Politician's Rating	I. Speakalot			U. Cantrustme		
	% Men	% Women	% of Total	% Men	% Women	% Total
Very good	22	20	21	8	12	10
Good	46	34	40	22	30	26
Unsure	18	36	27	20	10	15
Poor	12	9	10·5	40	36	38
Very poor	2	1	1·5	10	12	11

Opinion polls are often reported in newspapers. The data collected is categorical data.

1. What percentage of men rated I. Speakalot as very good? ☐
   What fraction is this? ☐

2. What percentage of women rated U. Cantrustme as good? ☐ What fraction is this? ☐

3. What percentage of people rated I. Speakalot as poor? ☐

4. What percentage of people rated U. Cantrustme as:
   a very good or good? ☐ **b** poor or very poor? ☐

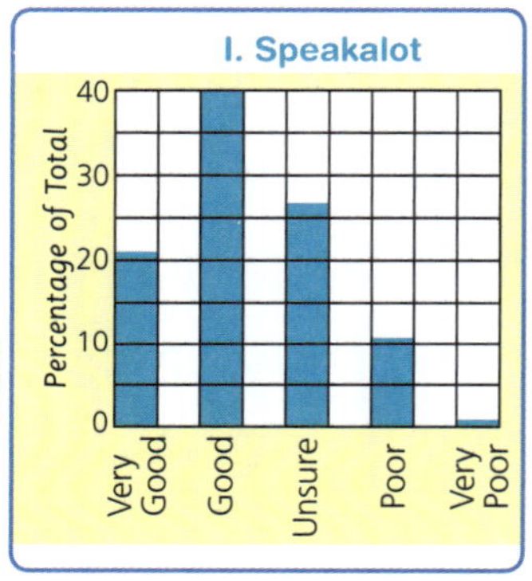

5. Which politician was the more popular? ☐

6. Was I. Speakalot more popular with men or women? ☐

7. If these figures came from asking 1 000 people their opinions about each politician, how many people said that U. Cantrustme was very good? ☐

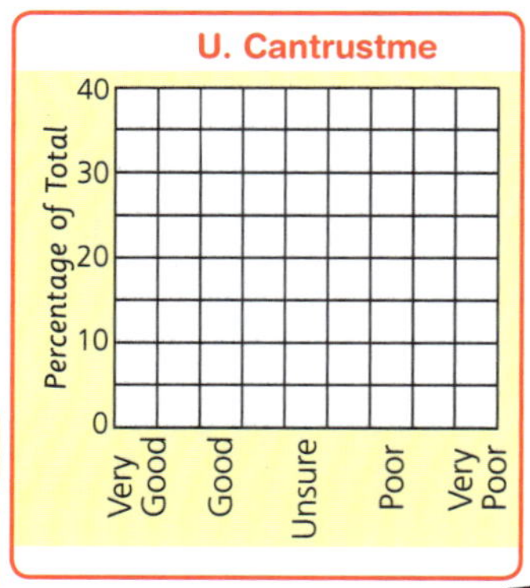

8. **a** Which part of the table above has been used to draw the I. Speakalot graph? ☐
   b Complete the second graph using the table.
   c Which was the most common response for I. Speakalot? ☐
   d Which was the least common response for U. Cantrustme? ☐

FUN SPOT

9. Pierre, Rachel and Kelly play a new version of handball. For a court they need two joined shapes from the space shown.
   a How many different courts could they choose? ☐
   b How many courts can be used at the same time? ☐
   c If **three** joined shapes were needed for a court, how many different courts could they choose? ☐

 • *AUSTRALIAN SIGNPOST MATHS 6* • ISBN 978 1 4886 2193 2

5:04 Reading Graphs and Tallies

1 An aid agency provided meals for the poor in May.

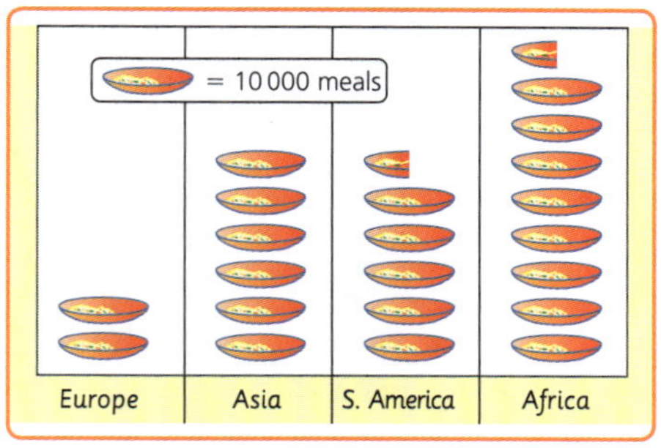

a How many meals were provided in:

i South America? ☐ **ii** Africa? ☐

b How many meals were provided altogether for these places in May? ☐

c Why do you think no name has been used for the horizontal axis? ☐

d Do you think that the bowl of food is a good choice to stand for meals provided? ☐

Can you think of another picture that could be used? ☐

e Give two reasons which may have led to Africa receiving more meals than any other continent.

☐

f Name the three continents not listed. ☐

2 Heather made a tally of animals seen during our trip.

Tally of Animals Seen	
Rabbit	𝍸 𝍸 𝍸 𝍸 |||
Dingo	𝍸 |
Possum	𝍸 𝍸
Wallaby	𝍸 𝍸 𝍸 ||||
Kangaroo	𝍸 𝍸 𝍸 𝍸 ||
Echidna	|||

a Which animal was seen 22 times? ☐

b How many sightings were made? ☐

c How many more rabbits were seen than dingoes? ☐

d In what ways is the tally like a picture graph?

☐

3

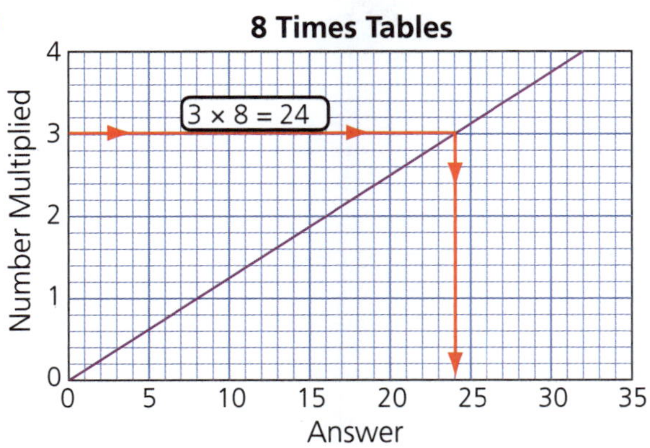

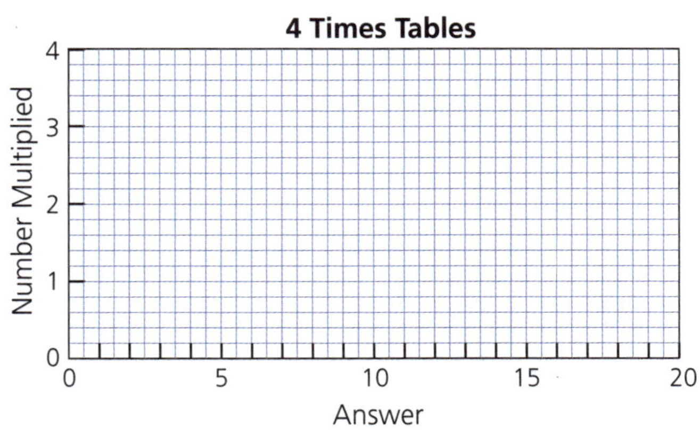

a Use the 8 Times Tables graph to find the answers to these as accurately as you can.

i 2×8 ☐ **ii** $1\frac{1}{2} \times 8$ ☐ **iii** $\frac{3}{4} \times 8$ ☐ **iv** $3\frac{1}{4} \times 8$ ☐

b On the grid to the right, draw a graph of the 4 Times Tables.

c Use the 4 Times Tables graph to find the answers to these as accurately as you can.

i 3×4 ☐ **ii** $3{\cdot}5 \times 4$ ☐ **iii** $1{\cdot}5 \times 4$ ☐ **iv** $2\frac{1}{4} \times 4$ ☐

5:05 The Probability of Events

CONCEPT

0% means 'no chance'.

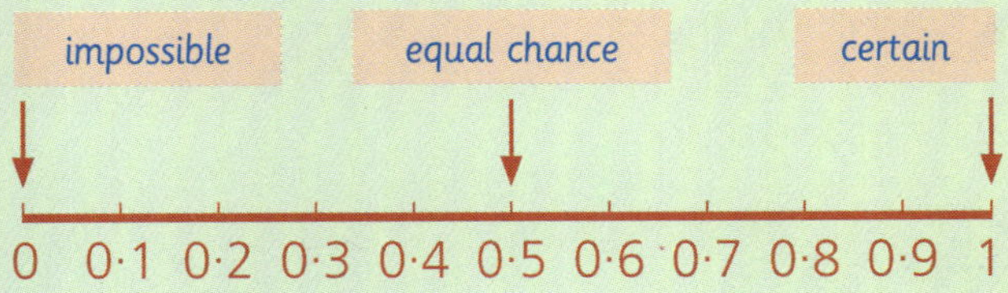

100% means 'certain'.

Unlikely events fall between 0% and 50%.
Likely events fall between 50% and 100%.

1 Use the scale 0% to 100% to rate the chance of the following events happening.

	Event	Chance
a	My first toss of a coin will be a tail.	
b	My first throw of a dice will be a three.	
c	I will be one year old next birthday.	
d	I will not go to school on Australia Day.	
e	I will have homework next week.	

≑ means 'is approximately equal to'.

The chance of rolling a five

$= \frac{1}{6}$

$= 1 \div 6$

$≑ 0.166....$

$≑ 17\%$

2 Use the scale 0 to 1 to write the probability, as a decimal, of each spinner landing on green.

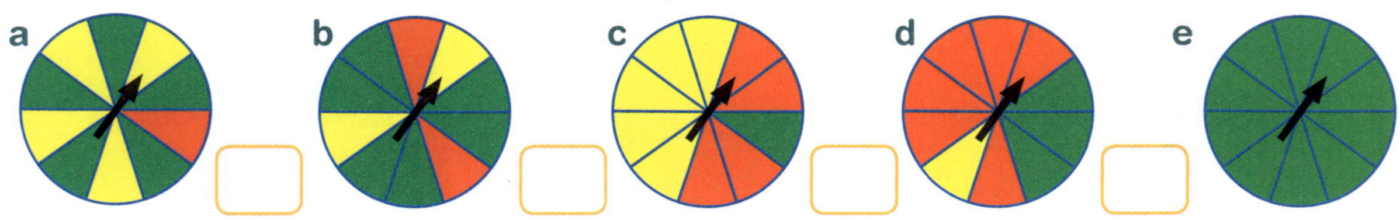

3 Use the scale 0 to 1 to write the probability, as a decimal, of each spinner landing on yellow.

a ____ b ____ c ____ d ____ e ____

4 Use the scale 0 to 1 to write the probability, as a decimal, of each spinner landing on red.

a ____ b ____ c ____ d ____ e ____

ACTIVITY

- Make a list of three events that may happen next week.
  Use the scale of 0 to 1 to rate the probability of each happening.

1 ____ ____
2 ____ ____
3 ____ ____

5:06 Ordering the Probability of Events

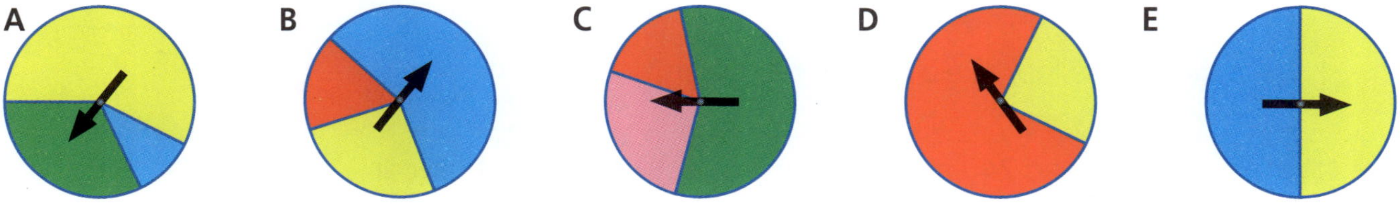

1 Which spinner has the most chance of landing on:

a green? ☐ **b** yellow? ☐ **c** red? ☐ **d** blue? ☐

2 Which spinner has more than a 50% chance of landing on:

a red? ☐ **b** blue? ☐ **c** yellow? ☐ **d** green? ☐

3 Which spinner has between a 0% and 50% chance of landing on:

a green? ☐ **b** blue? ☐ **c** pink? ☐ **d** red? ☐

4 Use the scale 0 to 1 to rate the probability of spinning each colour on this spinner.

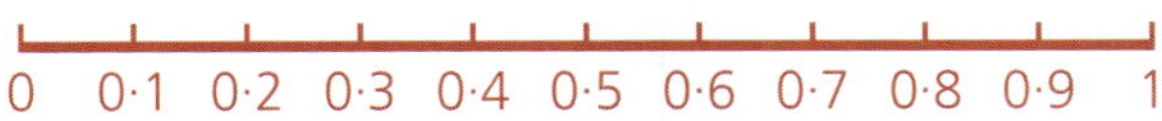

Each part is one tenth of the circle.

a (green) ☐ **b** (red) ☐ **c** (yellow) ☐ **d** (blue) ☐ **e** (pink) ☐

5 Jake listed all the possible totals for a pair of dice on this table.

Dice One (columns), **Dice Two** (rows)

	1	2	3	4	5	6
1	2	3	4	5	6	7
2	3	4	5	6	7	8
3	4	5	6	7	8	9
4	5	6	7	8	9	10
5	6	7	8	9	10	11
6	7	8	9	10	11	12

Which total is:

a most likely? ☐ **b** least likely? ☐

Which totals could be expected once in every:

c 36 throws? ☐ **d** 18 throws? ☐

e 12 throws? ☐ **f** 9 throws? ☐

g 6 throws? ☐

 • *AUSTRALIAN SIGNPOST MATHS 6* • ISBN 978 1 4886 2193 2

5:07 Divided Bar and Sector Graphs

1 Mercury, water and oil were used to fill this 1 000 mL jug.

Liquid in the Jug

a What fraction of the jug contains:

i mercury? ii water? iii oil?

b In the jug, what is the volume of:

i mercury? ii water? iii oil?

c What is the title of the graph?

d Does the picture look like a divided bar graph?

2 Naomi's class surveyed 240 people. Each chose their preferred flavour.

How could your class carry out this survey?

Flavour Preferred

The results have been graphed.

a What kind of graph is this?

b What is the title of the graph?

c What fraction of people chose:

i vanilla? ii strawberry?

d What percentage chose:

i vanilla? ii strawberry?

e How many people chose:

i vanilla? ii strawberry? iii chocolate or mango?

3 We have 100 books on the library shelves of our class. This is a graph of the number of each type of book.

Books in the Class Library

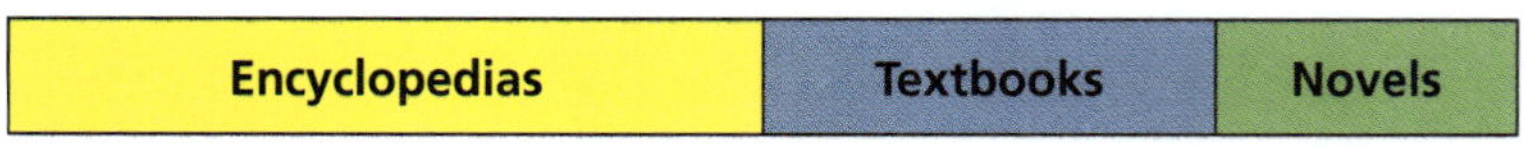

a Which category has the greatest number of books?

b What is the length of the graph in millimetres?

c In our library, what is the number of: i novels? ii encyclopedias?

d Would we get the same graph if the mass of the books in each category were measured?

Explain your answer.

e On the rectangle below, draw your estimate of the graph of the total mass of the books in each category.

5:08 Drawing Sector Graphs (Pie Charts)

CONCEPT

At the Party

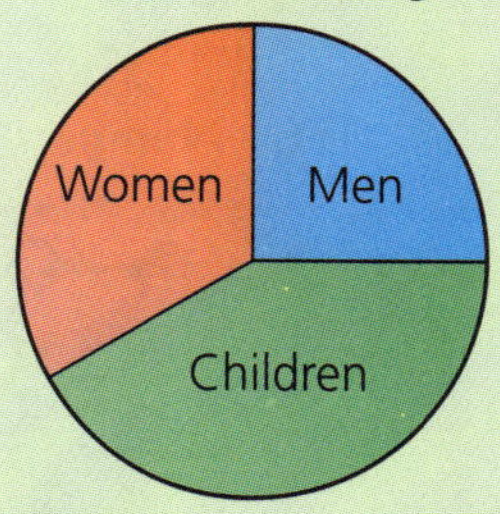

30 men, 40 women, and 50 children attended the party.

- The whole circle represents all of the data.
- The angles at the centre allow us to compare the categories.
- An angle of 360° is all the way around the centre of the graph.

If there are 120 people, then one person would be represented by $\frac{1}{120}$ of 360°. 360 divided by 120 = 3.

Each person is represented by 3°.

Category	Men (30)	Women (40)	Children (50)
Angle at the centre	Angle = 30 x 3° = 90°	Angle = 40 x 3° = 120°	Angle = 50 x 3° = 150°

1 In the half-hour after waking, Naomi spent 8 minutes making her bed, 12 minutes dressing and 10 minutes eating breakfast. Draw the graph.

The 30 minutes after waking is represented by 360°.

360° divided by 30 = 12°.

Each minute is represented by 12° at the centre.

Calculating the angles at the centre of the graph:

Making the Bed	Dressing	Eating Breakfast
Angle = 8 x 12° =	Angle = 12 x 12° =	Angle = 10 x 12° =

Naomi's Actions

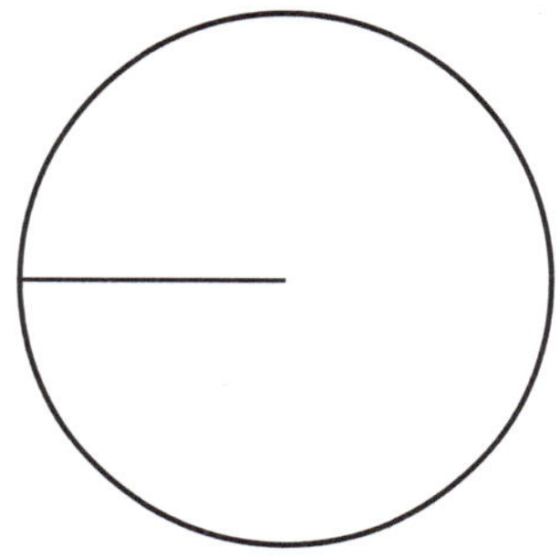

2 In one hour of lunchtime, Luke spent 15 minutes eating his lunch, 20 minutes playing handball and 25 minutes talking to friends.

The 60 minutes of lunchtime is represented by 360°.

360° divided by ____ = ____.

Each minute is represented by ____ at the centre.

Calculating the angles at the centre of the graph:

Eating Lunch	Playing Handball	Talking to Friends
Angle = ____ × ____ = ____	Angle = ____ × ____ = ____	Angle = ____ × ____ = ____

Luke's Lunchtime

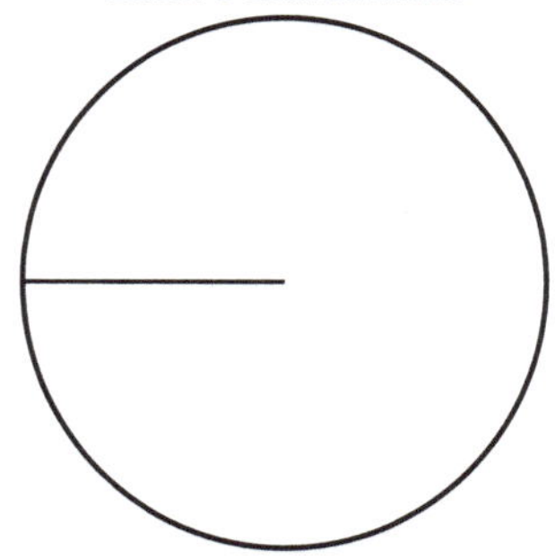

Using Technology to Construct a Sector Graph

ICT

3 At the sale, 15% of animals sold were horses, 57% were cows and 28% were goats. On a computer or your own paper, draw a sector graph that shows the animals sold.

 • *AUSTRALIAN SIGNPOST MATHS 6* • ISBN 978 1 4886 2193 2

5:09 Using Samples

1 This two-way table shows the results of a survey conducted with 50 boys and 50 girls. If one of those surveyed is chosen at random, find the probability that the person will be:

	Boys	Girls
like cheese	30	40
don't like cheese	20	10
like Vegemite	25	20
don't like Vegemite	25	30

$$\text{Probability} = \frac{\text{Number}}{\text{Total Number}}$$

a a girl ☐ **b** a man ☐

c a girl who does not like cheese ☐

d a boy who likes cheese ☐

e a child who does not like cheese ☐

f If this is a sample of a school that has 500 boys and 500 girls, estimate the number of students who don't like cheese. ☐

g Use the table data to complete the graph.

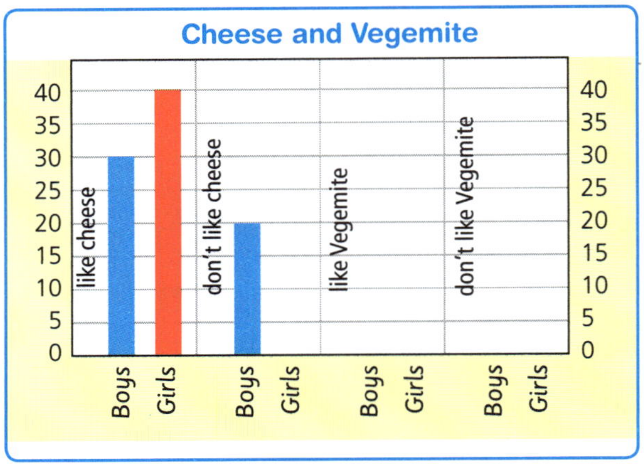

2 Petra recorded the number and flavour of ice creams sold in 15 minutes at lunchtime.

Vanilla	Chocolate	Strawberry	Lemon
24	33	19	12

Use the results in the table to predict how many:

a vanilla ice creams sold in 1 hour ☐ **b** strawberry ice creams sold in 1 hour ☐

c Explain why the results would be different if we conducted the survey at another time.

☐

3 One hundred marbles were placed in a bag and the following sample was pulled out at random.

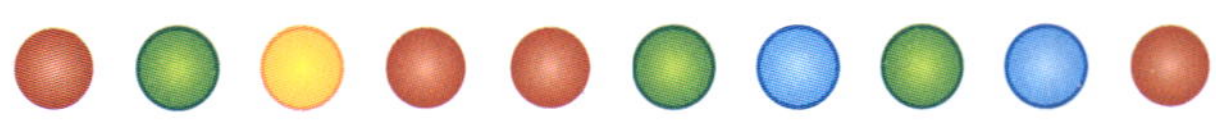

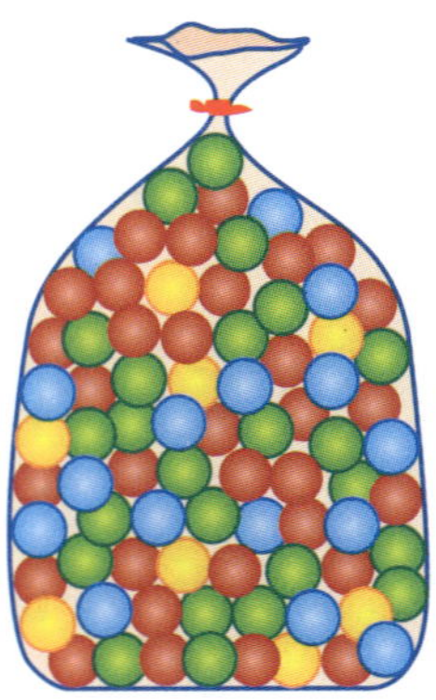

Use this sample to predict how many marbles of each colour were in the bag.

a green ☐ **b** red ☐ **c** yellow ☐ **d** blue ☐

e If the marbles were returned and a different 10 marbles chosen, how would they resemble the first lot of marbles?

☐

☐

 • *AUSTRALIAN SIGNPOST MATHS 6* • ISBN 978 1 4886 2193 2

5:10 Dot Plots

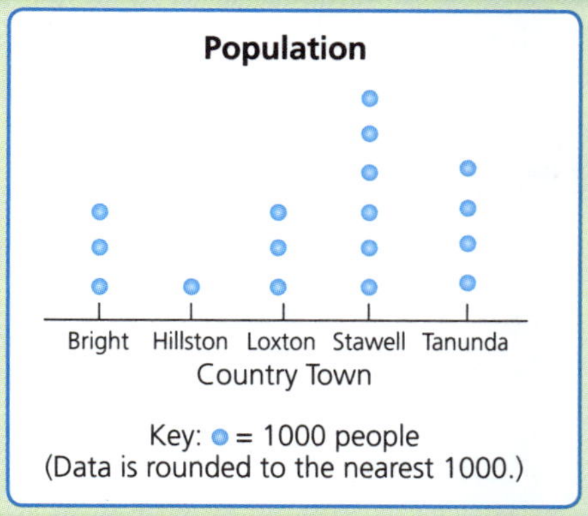

The information in this table was used to make this dot plot.

Population of Country Towns (to the nearest thousand)					
Country Town	Bright	Hillston	Loxton	Stawell	Tanunda
Population	3000	1000	3000	6000	4000

One dot represents 1 000 people.

Figures have been rounded to the nearest 1 000.

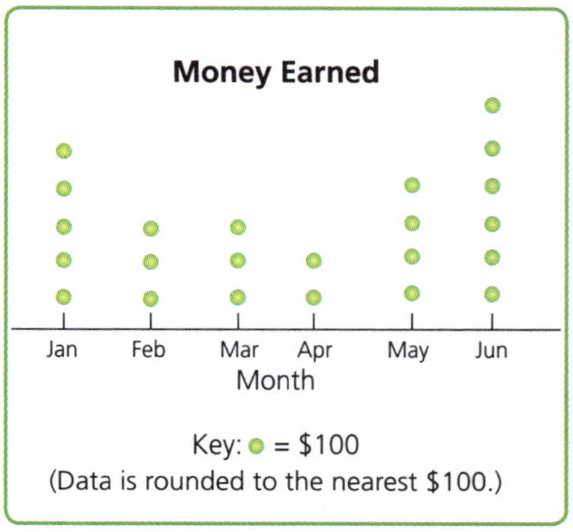

1 **a** How much money was earned in February?

b How much money was earned in April and May?

c When was the most money earned?

d When was the least money earned?

e Use this dot plot to complete the table.

Money Earned						
Month						
Money Earned						

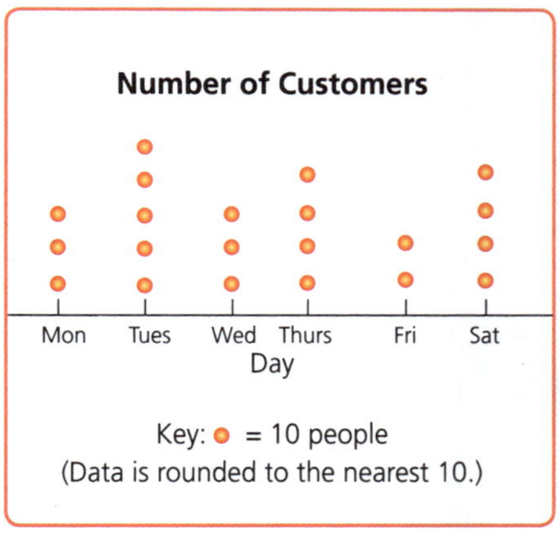

2 **a** How many customers were recorded for Thursday?

b How many customers were recorded for Monday?

c Which day has the most customers recorded?

d Which day has the least customers recorded?

e Use this dot plot to complete the table.

Customers						
Day						
Number of Customers						

3 **a** Use this information to complete the dot plot.

Money Earned				
Month	March	April	May	June
	$2906	$4392	$1896	$879
Rounded to the Nearest $1000				

b Ask a partner questions about this dot plot.

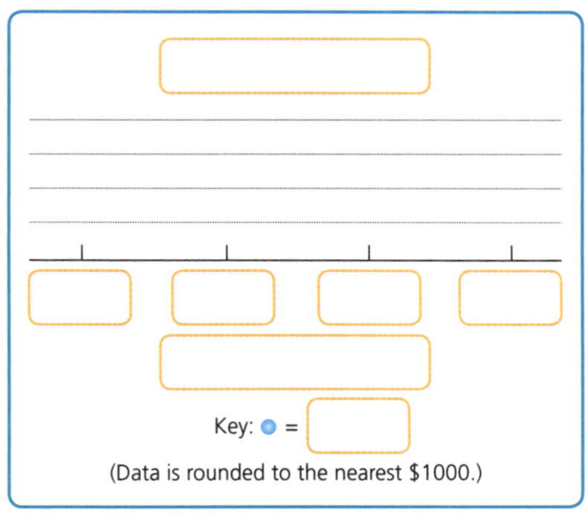

 • *AUSTRALIAN SIGNPOST MATHS 6* • ISBN 978 1 4886 2193 2

5:11 Misleading Displays

CONCEPT

We wanted to find the percentage of people in the world who were hungry.

Hungry People in the World	
Country	**% Hungry**
Australia	6%
NZ	5%
UK	7%
US	7·5%

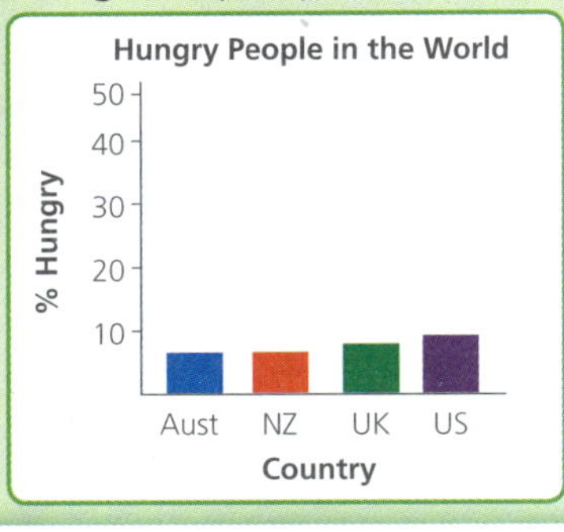

1 Explain why the graph above is a misleading graph.

2 Explain and discuss how these displays below could be misleading.

a

I am comparing money spent.

This is an example of a broken axis.

b

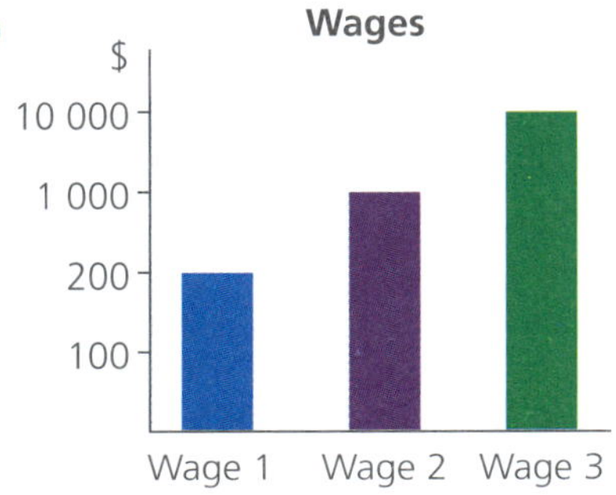

I am comparing wages.

This is an example of a non-linear scale.

c

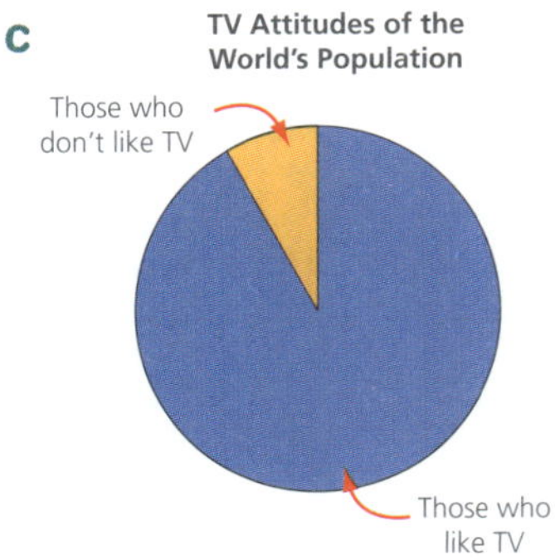

I asked people I know.

This sector graph does not represent the entire population.

INVESTIGATION

3 Consider and discuss these statements. The best attitude is ____.

A

B

C

 • *AUSTRALIAN SIGNPOST MATHS 6* • ISBN 978 1 4886 2193 2

5:12 Misleading Displays

CONCEPT

We asked these two students if they liked mathematics.

Liked Maths	2
Did Not Like Maths	0

Conclusion = 100% of people like mathematics.

Do you like maths?

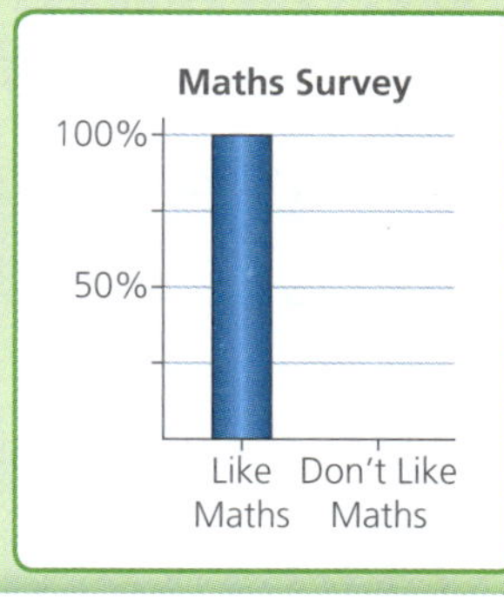

1. Explain why the graph above is a misleading graph.

2. Explain how each graph below could be misleading.

a

b

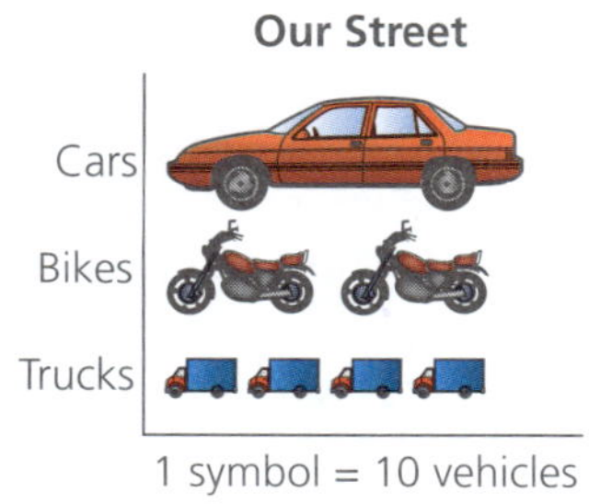

c

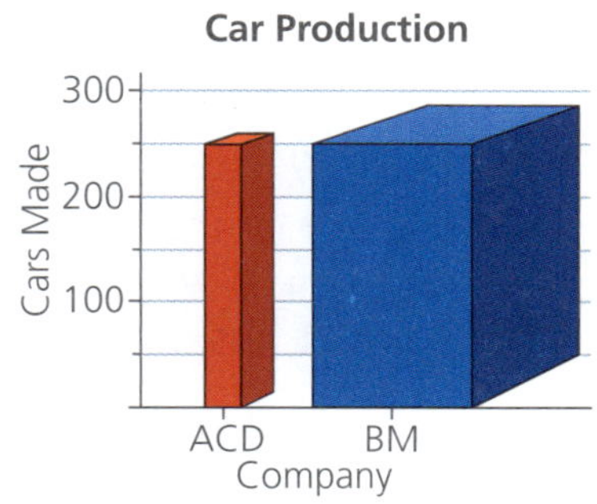

Collecting and Displaying Data

ACTIVITY

- Tick the things that are important if you are taking and reporting on a survey.

where information is collected		the pen you use	
how many people are asked		the scale used on graphs	
the age of people asked		the way you draw graphs	
the type of graph used		the questions on the survey	
the colour of your car		the conclusions you make	

A sample is part of a specific population.

- Discuss your answers.

5:13 Possible Outcomes to Experiments

1 **a** List the possible outcomes when a coin is tossed.

b What is the chance of tossing a head on a coin?

c Toss a coin 30 times. What fraction of the time did you toss a head?

2 **a** List the possible outcomes when a dice is rolled.

b What is the chance of rolling a 6 on a dice?

c Roll a dice 30 times. What fraction of the time did you roll a 6?

3 List all of the possible names that could be made using Marcus or Alan as a first name, Peter or John as a second name and Pieris as a last name.

4 100 tickets were sold in a raffle. Ruth bought 3 tickets. Write her chance of winning as a percentage and as a fraction.

5 A spinner has 5 red, 3 blue and 2 yellow sections. Diane wins when yellow is spun, Alan when blue is spun and Tom when red is spun.
If the spinner is spun, what is the chance that the winner is:

a Tom? **b** Alan? **c** Diane?

d Who is most likely to win when the spinner is spun?

6 4 coins are tossed at the same time. Do this 40 times.

ACTIVITY

Keep a tally of the results.

Alana wins if 4 heads are tossed.
Rachel wins if 3 heads are tossed.
Naomi wins if 2 heads are tossed.
Heather wins if 1 head is tossed.
Rhonda wins if no heads are tossed.

Winner	Tally	Total
Alana		
Rachel		
Naomi		
Heather		
Rhonda		

From your results, who is most likely to win?

 • *AUSTRALIAN SIGNPOST MATHS 6* • ISBN 978 1 4886 2193 2

5:14 Chance Using Two Dice

Two dice are to be rolled.

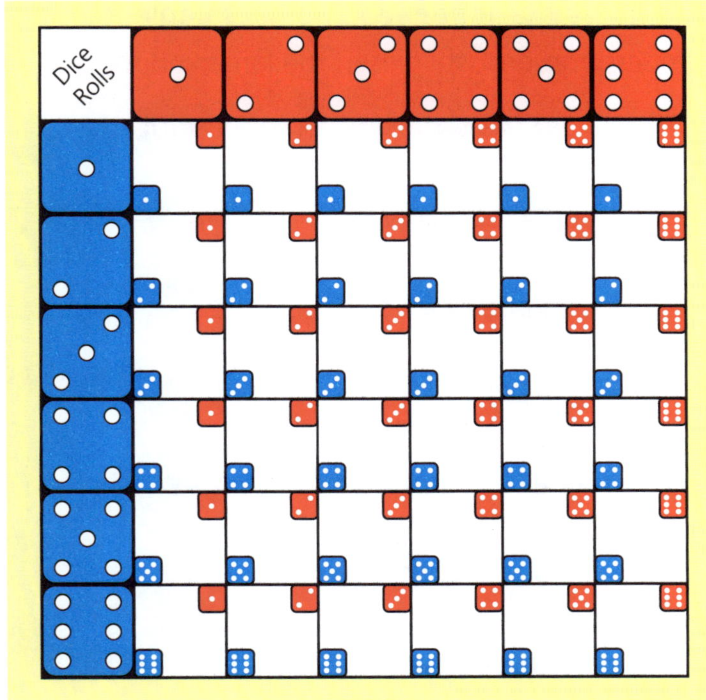

1 List all possible totals of the dice.

2					

2 Write the number of ways we could roll a total of:

a 2 ☐ **b** 3 ☐
c 4 ☐ **d** 5 ☐
e 6 ☐ **f** 7 ☐
g 8 ☐ **h** 9 ☐
i 10 ☐ **j** 11 ☐
k 12 ☐

3 How many different outcomes are shown on the diagram above? ☐

4 Write, as a fraction, the chance that the dice will have a total of:

a 2 ☐ **b** 11 ☐ **c** 10 ☐ **d** 5 ☐
e 9 ☐ **f** 7 ☐ **g** 8 ☐ **h** 6 ☐

Which Numbers are Rolled Most Often?

ACTIVITY

- Using two dice, follow the instructions in the flowchart.
- Use your results to draw a column graph.

Rolling Two Dice

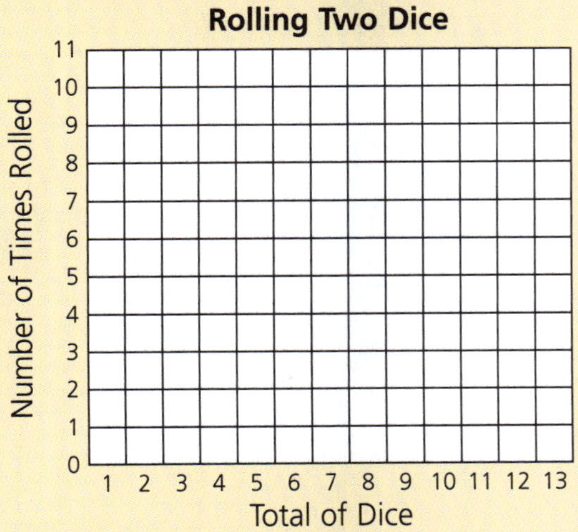

- Compare your results with those of others.

Number of Rolls
Tally:

Total of Dice	
2	
3	
4	
5	
6	
7	
8	
9	
10	
11	
12	

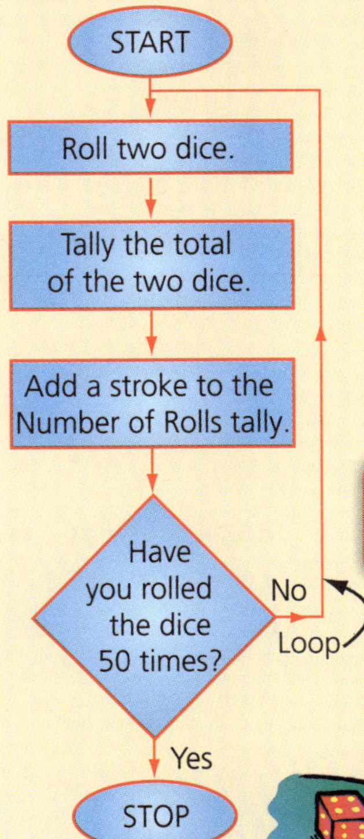

5:15 Using Samples

We can use the information from a **sample** to estimate numbers and probabilities in the larger population.

- The larger the sample, the more useful it is.

CONCEPT

1 Mia conducted a survey of 50 children to find their favourite fruit.

Apples	Bananas	Kiwi Fruit	Peaches
14	21	6	9

Use the table to predict how many children in a group of 100 would prefer:

a peaches ☐ **b** apples ☐ **c** kiwi fruit ☐ **d** bananas ☐

Use the table to predict how many children in a group of 1 000 would prefer:

e kiwi fruit ☐ **f** bananas ☐ **g** peaches ☐ **h** apples ☐

2 Fifty children from a school were surveyed to find which coloured pencil they believed they sharpened most frequently.

	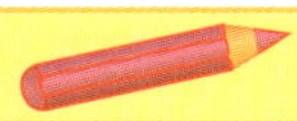	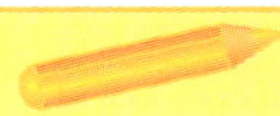	
10	5	25	10

Use their results in the table to predict how many pencils of each colour would be the results if 100 children were surveyed.

a blue ☐ **b** green ☐ **c** red ☐ **d** yellow ☐

Use the scale 0% to 100% to estimate the probability of the following responses from this group:

e red ☐ **f** yellow ☐ **g** blue ☐ **h** green ☐

i yellow or green ☐ **j** red or blue ☐ **k** not red ☐

l Do you think that asking 50 children would give helpful results? ☐

m Should we ask all children from one school year or some from each year? Why or why not?

____

____

Carry Out a Survey

ACTIVITY

- Ask 50 students, "Which coloured pencil do you think you use most often: green, red, yellow or blue?"
- Keep a tally.
- Compare your results with those in Question 2.
- Conclusion: ______________________________

Green	Red	Yellow	Blue

 • *AUSTRALIAN SIGNPOST MATHS 6* • ISBN 978 1 4886 2193 2

5:16 Collecting Information

When carrying out a **census**, everybody in the chosen population is asked the questions.
When using a **sample**, only a part of the chosen **population** is asked the questions.

CONCEPT

1 Ask 10 people from one class at school to choose the fruit they like best from apples, oranges, bananas or pears.

Complete the table below, then draw graphs to display the results.

Choice of Fruit (Survey 1)			
Fruit Chosen	Tally	Total	Percentage
Apples (A)			
Oranges (O)			
Bananas (B)			
Pears (P)			

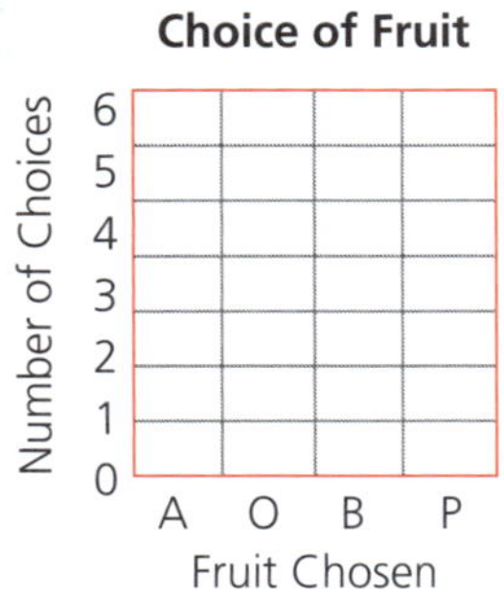

Which fruit was most popular? [ ]

Which fruit was least popular? [ ]

2 Repeat the survey above, asking 10 different people.

Choice of Fruit (Survey 2)			
Fruit Chosen	Tally	Total	Percentage
Apples (A)			
Oranges (O)			
Bananas (B)			
Pears (P)			

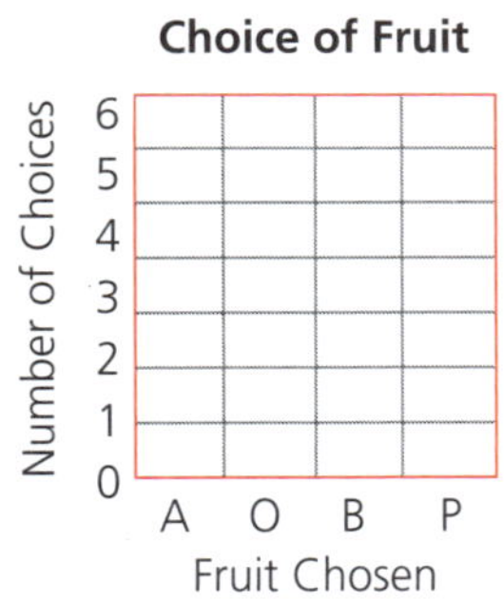

Were the results of Survey 2 the same as for Survey 1? [ ]

3 Would it have been better to carry out a census of the class? [ ]

4 Write a report of your surveys above, mentioning the method you used to collect the responses and the results you obtained.

 • *AUSTRALIAN SIGNPOST MATHS 6* • ISBN 978 1 4886 2193 2

5:17 Repeating an Experiment

INVESTIGATION

Experiment Toss a plastic bottle top 50 times to see whether it is more likely to land 'open end up' or 'open end down'.

Trial 1 (Carry out the experiment.)

How the Top Fell	Tally	Number	Fraction	Percentage
Open end up			$\frac{\quad}{50}$	$\frac{\quad}{100}$ or ____ %
Open end down			$\frac{\quad}{50}$	$\frac{\quad}{100}$ or ____ %
	Total			

Trial 2 (Repeat the experiment.)

How the Top Fell	Tally	Number	Fraction	Percentage
Open end up			$\frac{\quad}{50}$	$\frac{\quad}{100}$ or ____ %
Open end down			$\frac{\quad}{50}$	$\frac{\quad}{100}$ or ____ %
	Total			

Combine the data above. Include all data from Trial 1 and Trial 2.

How the Top Fell	Total in Trial 1 and Trial 2	Fraction	Percentage
Open end up		$\frac{\quad}{100}$	$\frac{\quad}{100}$ or ____
Open end down		$\frac{\quad}{100}$	$\frac{\quad}{100}$ or ____
Total			

To change $\frac{24}{50}$ to hundredths, multiply the numerator and denominator by 2.

1. Which result is more likely: 'open end up' or 'open end down'? ____________

2. Which would give the best estimate of the true probability of tossing the lid 'open end up'?
   Would it be Trial 1, Trial 2, or Trial 1 plus Trial 2? ____________
   Explain your answer.
   ____
   ____
   ____
   ____

5:18 Chance: Expected Results

INVESTIGATION

Calculate the expected results for each experiment first, then complete the experiment

1 **Experiment 1:** Randomly select a counter from a container with three red counters and one blue counter. Record the colour taken, then return the counter to the container. Do this 40 times.

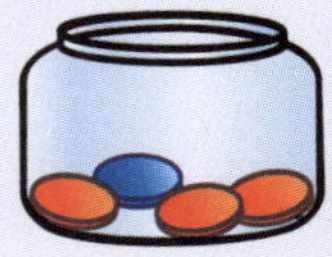

Calculated Probability	
Chance of red =	
Chance of blue =	

Experimental Results		
	Tally	Number out of 40
Red		
Blue		

a Out of 40, what fraction of red counters would you expect to take out?

b Out of 40, what fraction of blue counters would you expect to take out?

c What fraction of the counters taken (out of 40) were red?

d What fraction of the counters taken (out of 40) were blue?

e Was the expected (or calculated) result close to the result of your experiment?

f Is it possible to select a red counter 40 out of 40 times? Is it likely?

g When we calculate the probability, what are we really calculating?

h Compare the results of your experiment with the results of other students in your class. Did anyone take red 30 out of 40 times?

2 **Experiment 2:** Toss a coin 20 times.

Calculated Probability	
Chance of heads =	
Chance of tails =	

Experimental Results		
	Tally	Number out of 20
Heads		
Tails		

a What is the percentage probability of tossing a head?

b How many heads did you expect to toss out of 20?

c How many heads did you toss? Discuss your results.

5:19 Tree Diagrams

Tree diagrams are used to show all possible outcomes from a simple experiment.

For example, this tree diagram shows how many different outfits can be made from 3 different tops and 4 different pairs of shorts.

Tree Diagram

12 different outfits can be made.

1 For the outfits above, how many outfits had:

a green shorts? ☐ **b** a yellow top? ☐ **c** blue shorts and top? ☐

If an outfit is chosen at random, what is the chance that it has:

d blue shorts? ☐ **e** a blue top? ☐ **f** any part blue? ☐

2 Draw a tree diagram to show how many different:

a sandwiches could be made using either brown or white bread and one filling of either meat (M), cheese (C) or jam (J).

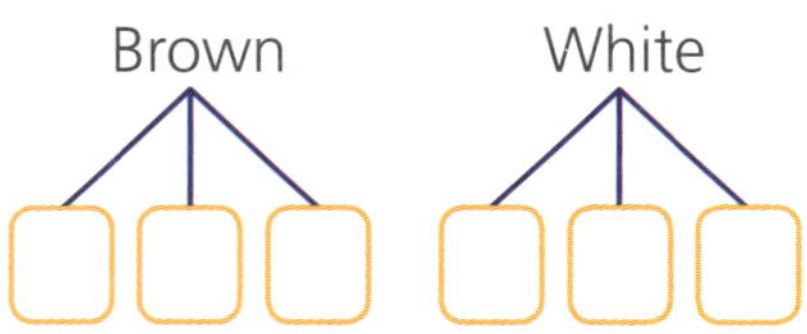

b outfits could be made from three hats and two dresses.

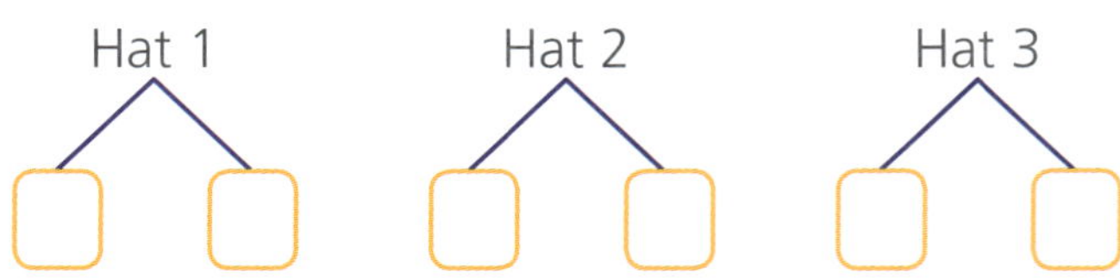

For the sandwiches above, how many sandwiches had:

c brown bread? ☐ **d** meat? ☐ **e** cheese or jam? ☐

If one of the sandwiches is chosen at random, what is the chance that it has:

f white bread? ☐ **g** jam? ☐ **h** white bread or jam? ☐

3 A coin is tossed, and a spinner showing **A**, **B**, **C** and **D** is spun.

a How many different outcomes are there? ☐

What is the chance that the outcome will have:

b a head? ☐

c a head and **C**? ☐

d an **A**? ☐

e an **A**, **B** or **C**? ☐

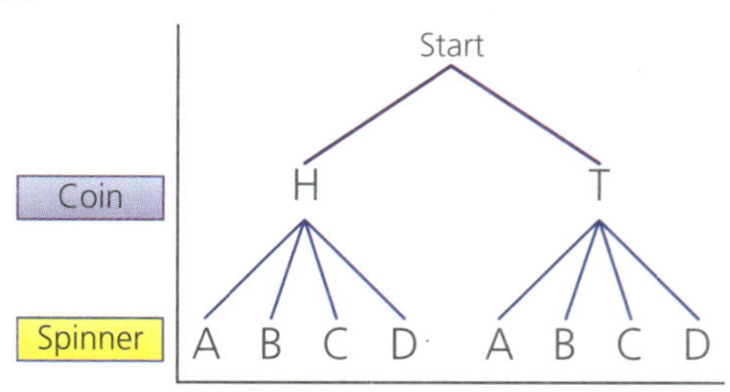

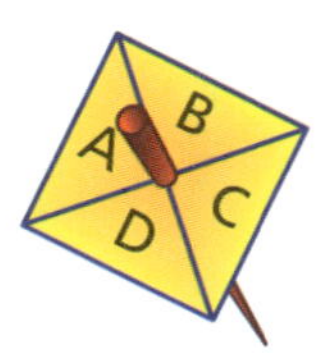

f Does the diagram look like a tree? (Turn it upside down.) ☐

5:20 Likelihood of an Outcome

CONCEPT

Question: If 3 children are born tomorrow, what is the chance that they will all be boys?

We assume that a boy and a girl are equally likely. Discuss this assumption.

Method 1: Toss 3 coins 100 times and keep a tally.

The result '3 heads' stands for '3 boys'. If the table shows the final result, we could say the chance of 3 boys is about 11%.

3 heads = 3 boys

3 heads	11	1 head	36
2 heads	40	0 heads	13

Method 2: Use a tree diagram.

This tree diagram lists all possibilities.

The 3 linked boxes along the top stand for 3 boys (bbb).

- One out of the eight possibilities stands for '3 boys'.
- The chance that 3 boys are born is $\frac{1}{8}$ or $12\frac{1}{2}\%$ or 0·125.

1st child	2nd child	3rd child	Result
boy	boy	boy	bbb
		girl	bbg
	girl	boy	bgb
		girl	bgg
girl	boy	boy	gbb
		girl	gbg
	girl	boy	ggb
		girl	ggg

1 Using the results in the table for Method **1** above, what is the chance of:

a 2 boys and a girl? ______ **b** 1 boy and 2 girls? ______

2 Using Method **2** above, what is the chance of:

a 2 boys and a girl? $\frac{\square}{\square}$ or ______ **b** 3 girls? $\frac{\square}{\square}$ or ______

3 **True** or **false**?

a To find the chance that 2 boys are born out of 2 children, we could toss two coins many times to see what fraction of the time two heads occur. ______ Why or why not?

______

______

b On a dice, a 6 is harder to roll than a 3. ______ Why or why not?

______

______

  ISBN 978 1 4886 2193 2

5:21 Planning a Good Survey

Mr Adams believed that students in the school liked to watch tennis more than all other sports combined.

He asked his students to carry out a survey to see if this was true.

- Bill decided to ask the 5 students in the boys' tennis team. Since 4 of them chose tennis, he reported that 80% of students preferred to watch tennis.
- Sam asked the members of the girls' hockey team. None of them chose tennis, so she reported that 0% of students preferred to watch tennis.
- Mary asked all of the students in Kindergarten. 2 out of 50 preferred to watch tennis. She reported that 4% preferred to watch tennis.
- Callum asked 4 students (2 boys and 2 girls) from every class in the school. 17 out of 56 said they preferred to watch tennis. He reported that about 30% preferred to watch tennis.

1 Comment on the method used by each student above. Would the results be reliable?

a Bill

b Sam

c Mary

d Callum

2 Which student would have collected the most reliable results?

3 Imagine you are in Mr Adams's class. Explain how you would carry out the survey.

 • *AUSTRALIAN SIGNPOST MATHS 6* • ISBN 978 1 4886 2193 2

5:22 Unusual Graphs

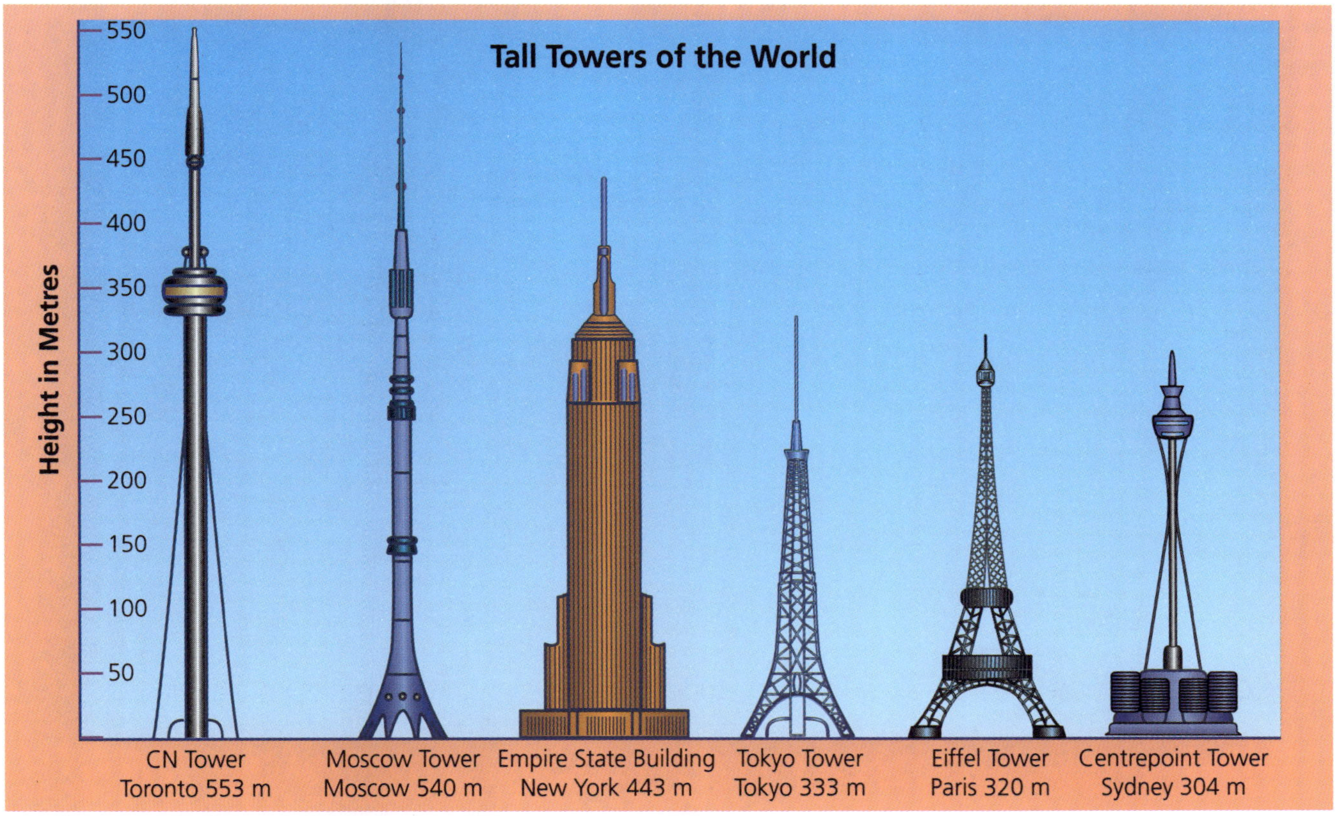

The towers are drawn to scale.

1. What type of graph does this one most resemble?
2. How much shorter is the Centrepoint Tower than the:
   a CN Tower? **b** Moscow Tower?
3. Which of these towers has the widest base?
   How wide is it?

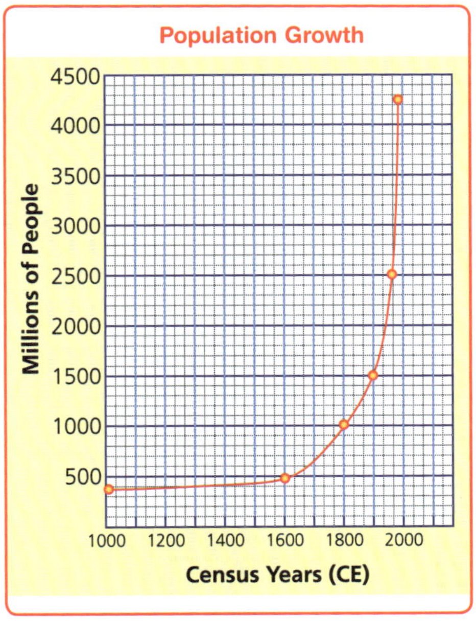

4. What population was recorded in 1983?
5. Was the population less than 500 million in the year 1000?
6. When did the population reach 1 500 million?
7. What is the difference in recorded populations for 1800 and 1983?
8. Between which two year readings did the population increase the most?
9. Between which two year readings was the increase of population least?
10. Over what period of time does this graph record population growth?
11. Use the internet to find the current population, correct to the nearest 100 million.

 • *AUSTRALIAN SIGNPOST MATHS 6* • ISBN 978 1 4886 2193 2

6 Answers

1:01

1. a 49 760 621 b 83 132 549
2. a 4 hundreds of thousands b 3 millions
   c 6 tens of thousands d 9 tens of millions
   e 7 millions f 5 millions
3. a 26 349 721; 43 296 714; 62 419 637
   b 56 811 769; 63 497 624; 65 375 670
   c 17 634 658; 32 693 475; 41 623 912
4. a 30 000 000 b 30 000 000 c 40 000 000
5. a

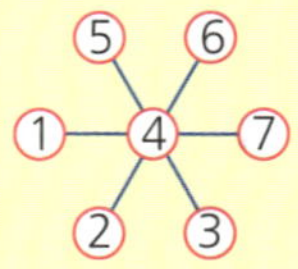

b

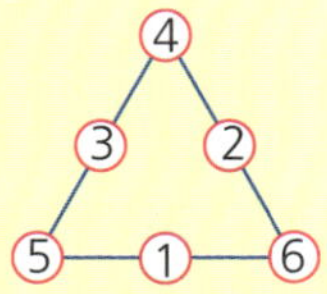

c

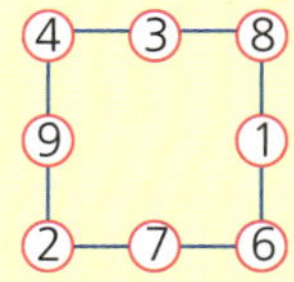

1:02

1. a 25 b 49 c 36 d 16 e 9 f 64
2. a 9 b a square
   c

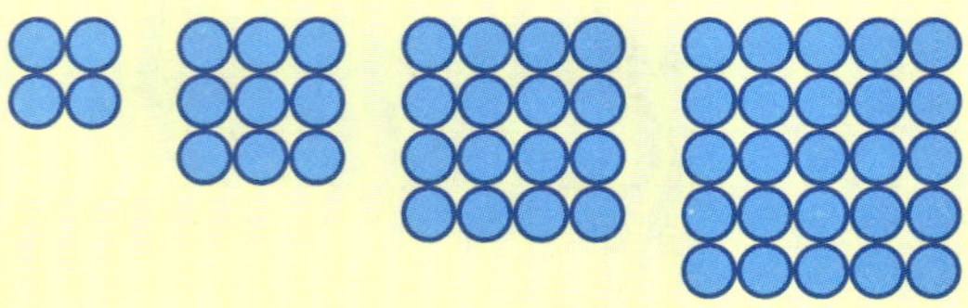

3. 1, 4, 9, 16, 25, 36, 49, 64, 81, 100
4. Answers can vary. (Some are 121, 144, 169, 196, 225, 256 and 289.)
5. Counters representing square numbers can be arranged in the shape of a square.

1:03

25 = 5 × 5, 36 = 6 × 6

1. a 25 b 36 c 49 d 64 e 81
   f 100 g 4 h 16 i 49 j 9
   k 36 l 100 m 1 n 64
2. a + 11, + 13, + 15, + 17, + 19
   b 121, 144 c 4, 16, 36, 64, 100
3. a 3 squared b 5 squared c 4 squared d 6 squared
4. a 1, 25, 5 b 1, 9, 3 c 1, 49, 7 d 1, 4, 2
   e 1, 16, 2, 8, 4

1:04

1. a 30% b 80% c 90% d 50% e 75%
   f 10% g 60% h 30%
2. a 70% b 20% c 10% d 50% e 25%
   f 90% g 40% h 70%
3. a

0·25	$\frac{25}{100}$	25%

b

0·35	$\frac{35}{100}$	35%

c

0·65	$\frac{65}{100}$	65%

d

0·75	$\frac{75}{100}$	75%

e

0·15	$\frac{15}{100}$	15%

f

0·55	$\frac{55}{100}$	55%

g

0·90	$\frac{90}{100}$	90%

h

0·40	$\frac{40}{100}$	40%

i

0·80	$\frac{80}{100}$	80%

1:05

1. a 80% b 60% c 70% d 40% e 30%
   f 50% g 100% h 90%
2. a 20% b 40% c 30% d 60% e 70%
   f 50% g 0% h 10%
3. a

$\frac{25}{100}$	0·25	25%

b

$\frac{55}{100}$	0·55	55%

c

$\frac{75}{100}$	0·75	75%

d

$\frac{95}{100}$	0·95	95%

e

$\frac{65}{100}$	0·65	65%

f

$\frac{45}{100}$	0·45	45%

g

$\frac{9}{10}$	0·9	90%

h

$\frac{3}{10}$	0·3	30%

i

$\frac{7}{10}$	0·7	70%

j

$\frac{4}{10}$	0·4	40%

k

$\frac{5}{10}$	0·5	50%

l

1	1·0	100%

4. a

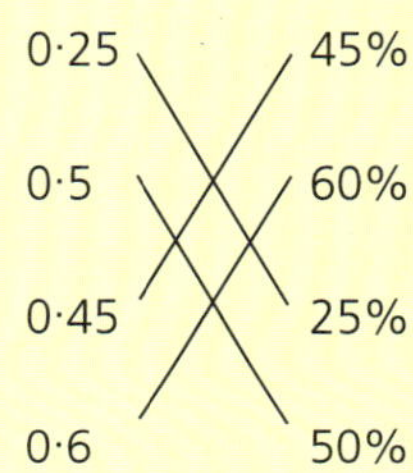

b

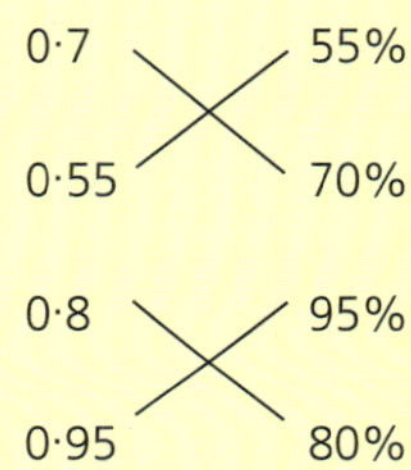

c

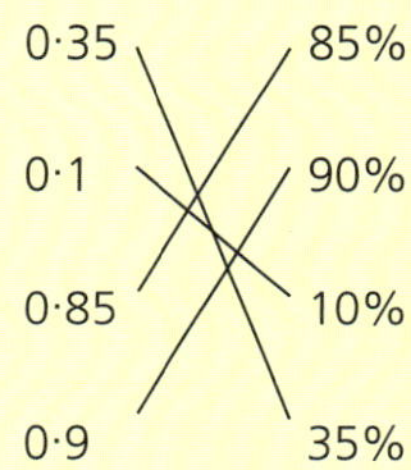

d

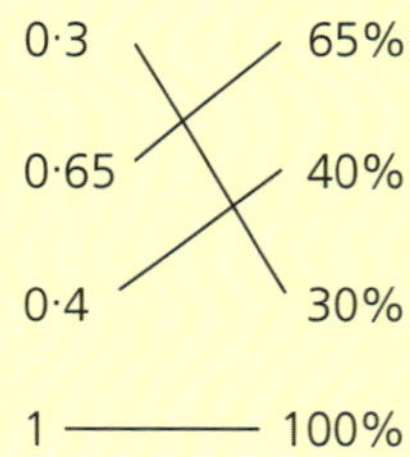

1:06

1. a 30% b 90% c 75% d 60% e 25%
   f 40% g 65% h 35% i 50% j 10%
2. a 15% b 45% c 85% d 55% e 95%
   f 80% g 20% h 70% i 10% j 5%
3. a 0·25 b 0·4 c 0·9 d 0·75 e 0·1
   f 0·35 g 0·25 h 0·65 i 0·05 j 0·15

4 a
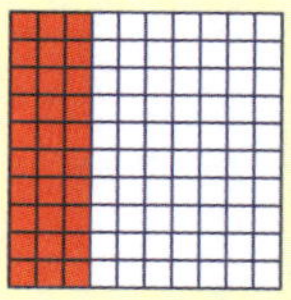
0·3 = 30%

b
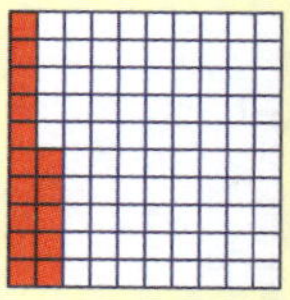
0·15 = 15%

c
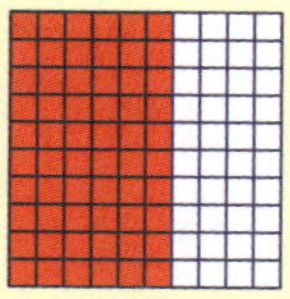
0·6 = 60%

d
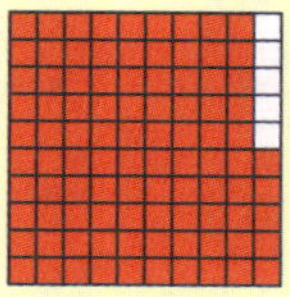
0·95 = 95%

e
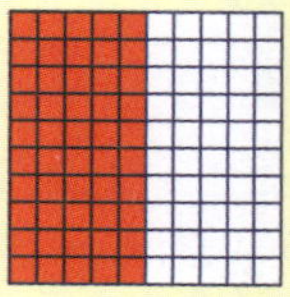
0·5 = 50%

f
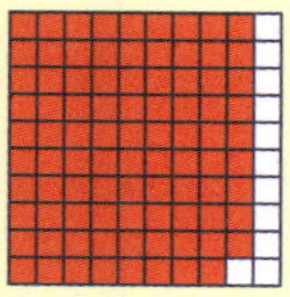
0·89 = 89%

g
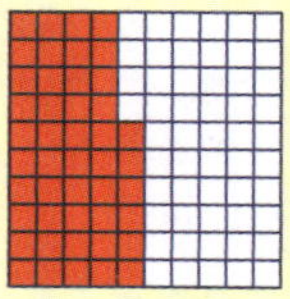
0·46 = 46%

h
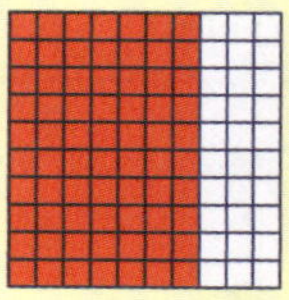
0·7 = 70%

i
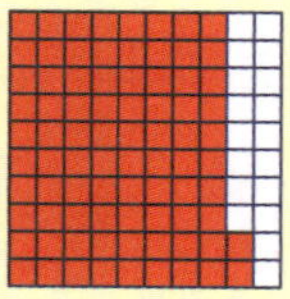
0·82 = 82%

j
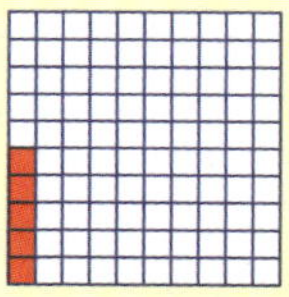
0·05 = 5%

1:07

1 **a** 37 952 **b** 96 831 **c** 62 475
d 89 354

2 **a** $(6 \times 10^3) + (4 \times 10^2) + (9 \times 10^1) + 1$
b $(2 \times 10^4) + (7 \times 10^3) + (2 \times 10^2) + (4 \times 10^1) + 5$
c $(7 \times 10^4) + (8 \times 10^3) + (3 \times 10^2) + (1 \times 10^1) + 9$
d $(4 \times 10^4) + (5 \times 10^3) + (6 \times 10^2) + (2 \times 10^1) + 8$

3

	Ten thousands	Thousands	Hundreds	Tens	Ones
a	7	9	2	3	4
b	4	6	7	9	3
c	3	5	6	8	6
d	8	3	5	6	2

4 **a** 64 958 **b** 96 743 **c** 372 598
d 785 461 **e** 159 356

1:08

1 **a** −10°C **b** 0° C **c** −20° C
d −14° C **e** −6° C **f** −2° C

2 **a** 2 **b** −1 **c** −3 **d** −11 **e** −7
f −10 **g** −8 **h** −5

3 **a** −4 **b** −2 **c** 8 **d** 3 **e** 1
f −5 **g** 5 **h** −1 **i** 7 **j** −6
k −3 **l** 0

1:09

1 a
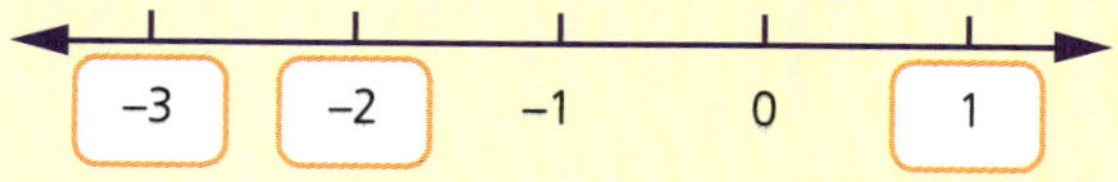

b
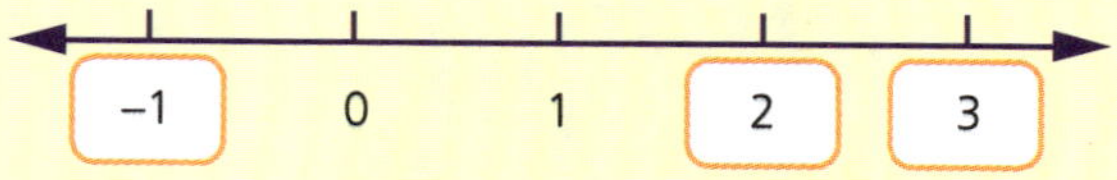

c
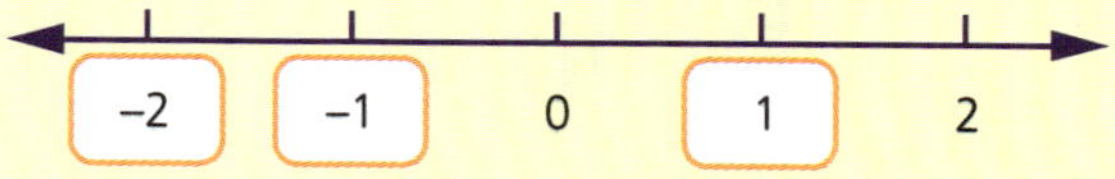

d
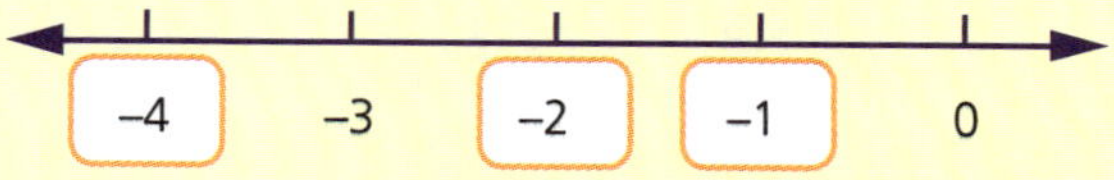

2 a
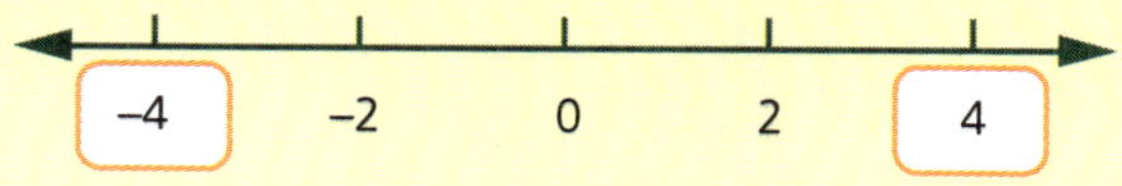

b
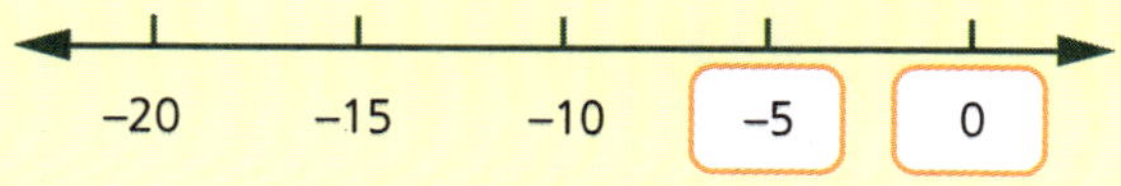

c
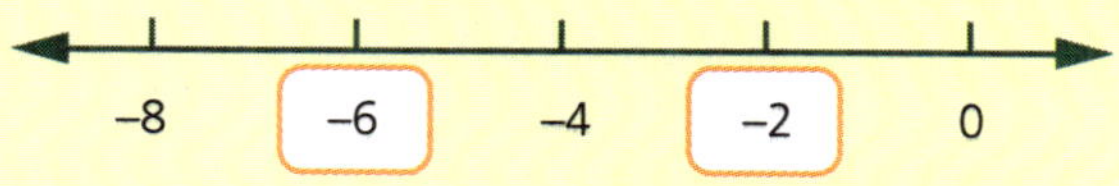

d
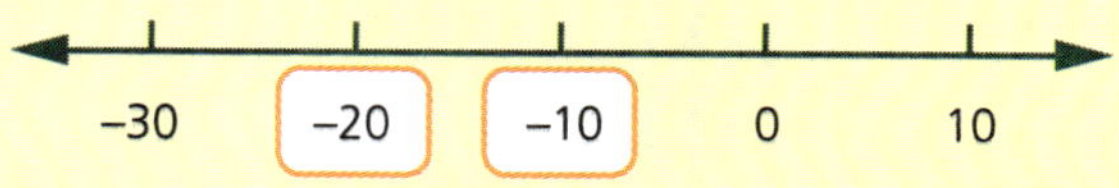

3 a
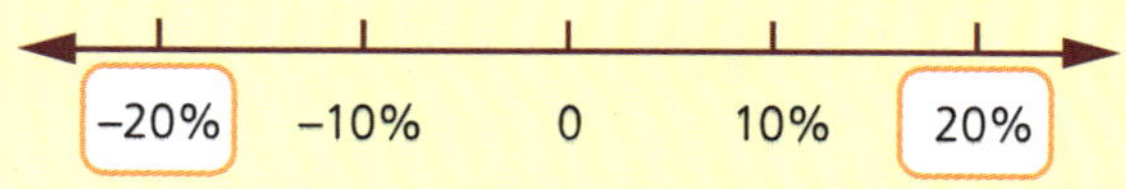

b
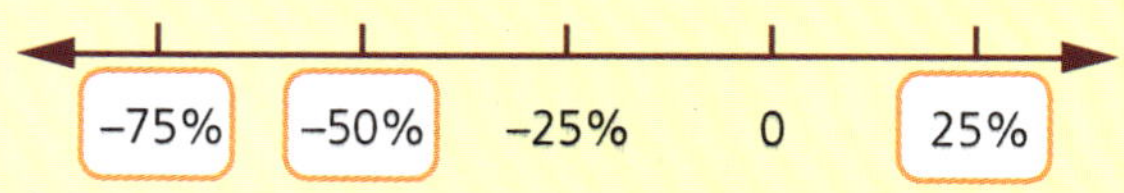

c
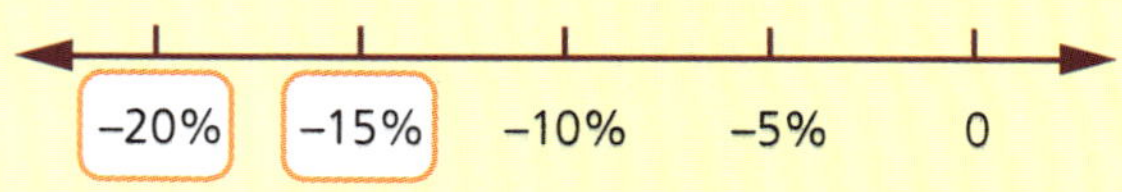

d
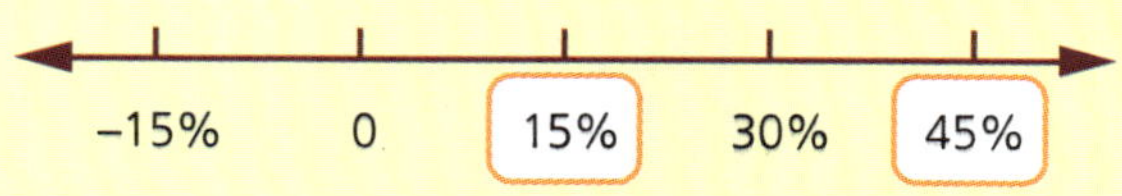

1:10

1 a $\frac{7}{4}, 1\frac{3}{4}$ b $\frac{11}{6}, 1\frac{5}{6}$ c $\frac{8}{5}, 1\frac{3}{5}$ d $\frac{11}{8}, 1\frac{3}{8}$ e $\frac{14}{5}, 2\frac{4}{5}$ f $\frac{13}{6}, 2\frac{1}{6}$

2 a $1\frac{1}{5}$ b $2\frac{1}{5}$ c $1\frac{4}{5}$ d $2\frac{3}{5}$ e $3\frac{1}{5}$ f $2\frac{2}{5}$
g $3\frac{2}{5}$ h $1\frac{2}{5}$ i $1\frac{4}{5}$ j $1\frac{3}{5}$

3 a $\frac{9}{5}$ b $\frac{13}{5}$ c $\frac{17}{5}$ d $\frac{6}{5}$ e $\frac{12}{5}$ f $\frac{16}{5}$
g $\frac{7}{5}$ h $\frac{11}{5}$ i $\frac{19}{5}$ j $\frac{8}{5}$

4 a $1\frac{1}{4}$ b $1\frac{3}{10}$ c $1\frac{1}{8}$ d $1\frac{1}{6}$ e $2\frac{1}{4}$ f $1\frac{7}{10}$
g $1\frac{3}{8}$ h $1\frac{1}{12}$ i $1\frac{5}{8}$ j $2\frac{3}{4}$ k $2\frac{5}{6}$ l $1\frac{5}{12}$

1:11

1 a $1\frac{3}{5}$ b $1\frac{5}{6}$ c $1\frac{1}{3}$ d $1\frac{62}{100}$ e $2\frac{1}{3}$
f $2\frac{3}{5}$ g $2\frac{1}{6}$ h $1\frac{24}{100}$

2 a $\frac{3}{2}$ b $\frac{9}{4}$ c $\frac{7}{4}$ d $\frac{11}{4}$ e $\frac{18}{5}$
f $\frac{45}{8}$ g $\frac{47}{5}$ h $\frac{87}{8}$ i $\frac{151}{10}$ j $\frac{152}{10}$
k $\frac{153}{10}$ l $\frac{154}{10}$

3 a 1·2, 1·4, 1·6
b 0·7, 0·9, 1·1
c 2, 2·5, 3
d 2·4, 2·5, 2·6
e 7·9, 8·2, 8·5
f 10·5, 11·0, 11·5

4 a 2, $2\frac{1}{2}$, 3
b 1 or $\frac{8}{8}$, $1\frac{2}{8}$ or $\frac{10}{8}$, $1\frac{4}{8}$ or $\frac{12}{8}$
c $\frac{7}{6}$ or $1\frac{1}{6}$, $\frac{9}{6}$ or $1\frac{3}{6}$, $\frac{11}{6}$ or $1\frac{5}{6}$
d $\frac{7}{5}$ or $1\frac{2}{5}$, $\frac{9}{5}$ or $1\frac{4}{5}$, $\frac{11}{5}$ or $2\frac{1}{5}$
e $\frac{8}{10}$, $\frac{10}{10}$ or 1, $\frac{12}{10}$ or $1\frac{2}{10}$
f $\frac{4}{3}$ or $1\frac{1}{3}$, $\frac{5}{3}$ or $1\frac{2}{3}$, $\frac{6}{3}$ or 2

1:12

1 a −1, 2 b −3, 1 c −10, 30 d −10, 10
e −8, 2 f −5, 5 g −15, 5 h −4, 2

2 a
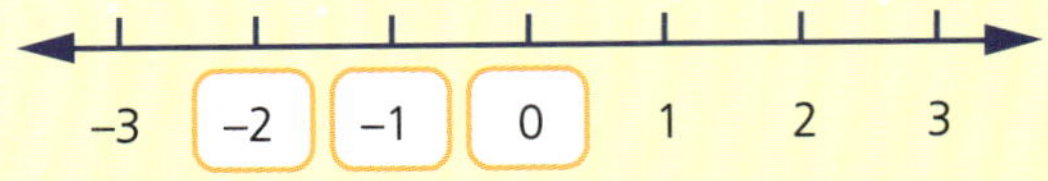

b
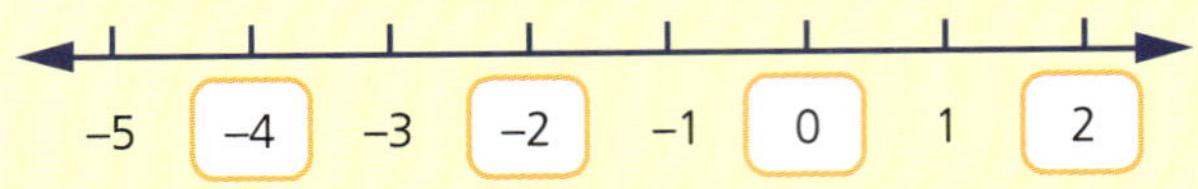

c
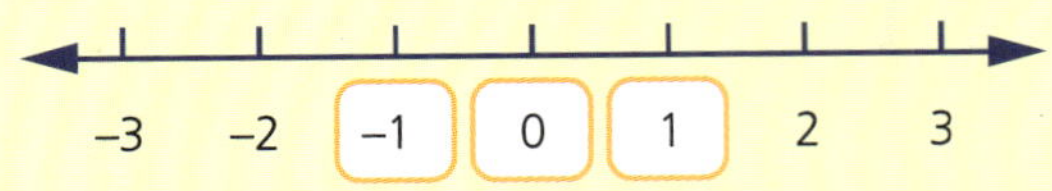

d
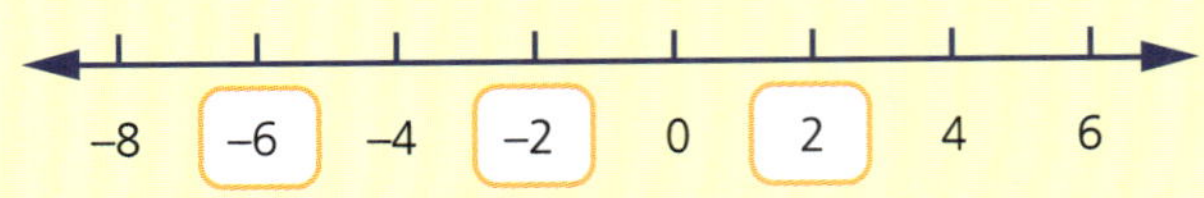

3 a −2°C b −5°C c −5°C
d −5°C e −10°C f −7°C

4 a −2, 1, 4, 5 b −4, −1, 0, 1
c −4, −3, −2, −1 d −3, −2, 2, 3

5 a 0 b −2 c −1 d +3 (or 3)

1:13

1 a +50 b −10
c −3 d +30
e −14 f −100

2 a +8 b −4
c −14 d −15
e −11 f −4
g −2 h −150

3 a 30°C b 20°C c 0°C
d −10°C e −20°C

1:14

1 a −2 b −20 c −4 d −16
e −13 f −7 g −11 h −1

2 a 2°C b −2°C c −5°C d 0°C e −3°C f −11°C

3 a \$5 b −\$4 c \$0 d −\$17
e −\$20 f −\$32 g −\$59 h −\$5
i −\$39 j −\$89

4 a −2 and 6 b −7 and 1
c −18 and −10

5 a −7 b −9 c −17 d −14 e −34
f −62 g −54 h −56 i −73 j −86

1:15

1 a

Number of Hexagons	1	2	3	4	5	6
Number of Sides	6	12	18	24	30	36

b Multiply by 6 c 54

2 a

Number of Circles	1	2	3	4	5	6
Number of Stars	4	8	12	16	20	24

b Multiply by 4 c 80

3

1	1 + 2	1 + 2 + 3	1 + 2 + 3 + 4	1 + 2 + 3 + 4 + 5	1 + 2 + 3 + 4 + 5 + 6
1	3	6	10	15	21

4 a 9, 18, 27, 36
b 99, 198, 297, 396
c 1, 121, 12321, 1234321

1:16

1 a 2 b 2 c 3 d 4 e 3 f 3
g 5 h 6
2 a 12 b 8 c 6 d 4 e 3 f 2
3 a 4 b 3 c 4 d 5 e 6 f 4
g 7 h 7
4 Answers will vary.
a 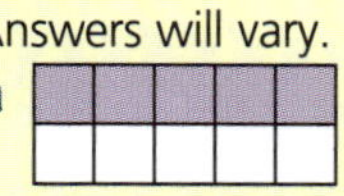 $\frac{1}{2}$ of 10 will be covered.
b 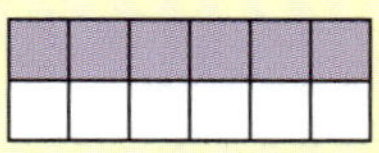 $\frac{1}{2}$ of 12 will be covered.
c 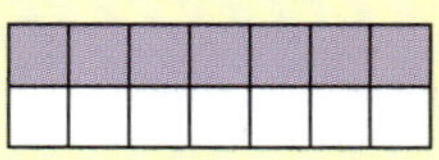 $\frac{1}{2}$ of 14 will be covered.
d $\frac{1}{2}$ of 14 is greater.

1:17

1 a 6 b 10 c 9 d 12
2 a 18 b 6 c 16 d 20 e 8 f 15
g 10 h 9 i 21
3 a 20 b 6 c 16 d 54 e 15 f 15
g 21 h 25 i 84 j 35 k 28 l 24
4 a 30 cards b 60 sandwiches c 117 emails d No

1:18

1 a $\frac{5}{12}$ b $\frac{7}{12}$ c $\frac{5}{12}$ d $\frac{7}{12}$
e $\frac{9}{12}$ or $\frac{3}{4}$ f $\frac{7}{12}$ g $\frac{10}{12}$ or $\frac{5}{6}$ h $\frac{9}{12}$ or $\frac{3}{4}$
i $\frac{12}{12}$ or 1 j $\frac{5}{12}$ k $\frac{4}{12}$ or $\frac{1}{3}$ l $\frac{6}{12}$ or $\frac{1}{2}$
2 a $\frac{7}{8}$ b $\frac{9}{10}$ c $\frac{3}{4}$ d $\frac{5}{6}$
e $\frac{8}{10}$ or $\frac{4}{5}$ f $\frac{3}{6}$ or $\frac{1}{2}$ g $\frac{5}{8}$ h $\frac{8}{10}$ or $\frac{4}{5}$
i $\frac{8}{8}$ or 1 j $\frac{6}{4}$
3 a $\frac{1}{4}$ b $\frac{4}{8}$ or $\frac{1}{2}$ c $\frac{3}{5}$
d $\frac{1}{6}$ e $\frac{7}{10}$ f $\frac{2}{6}$ or $\frac{1}{3}$
g $\frac{2}{8}$ or $\frac{1}{4}$ h $\frac{4}{6}$ or $\frac{2}{3}$ i $\frac{7}{12}$
j $\frac{1}{5}$
4 a $\frac{5}{8}$ b $\frac{2}{6}$ or $\frac{1}{3}$ c $\frac{8}{10}$ or $\frac{4}{5}$
d $\frac{3}{8}$ e $\frac{2}{5}$ f $\frac{8}{10}$ or $\frac{4}{5}$
g $\frac{4}{6}$ or $\frac{2}{3}$ h $\frac{2}{12}$ or $\frac{1}{6}$ i $\frac{5}{8}$

1:19

1 a $\frac{5}{6}$ b $\frac{9}{10}$ c $\frac{7}{8}$ d $\frac{11}{12}$ e $\frac{4}{5}$ f $\frac{1}{4}$
g $\frac{3}{10}$ h $\frac{3}{5}$ i $\frac{1}{3}$ j $\frac{1}{6}$ k $\frac{5}{8}$ l $\frac{7}{12}$
2 a $\frac{4}{8}$ b $\frac{3}{6}$ c $\frac{1}{3}$ d $\frac{2}{4}$ e $\frac{2}{6}$ f $\frac{3}{8}$
g $\frac{6}{10}$ h $\frac{2}{5}$ i $\frac{5}{12}$ j $\frac{3}{10}$ k $\frac{2}{5}$ l $\frac{4}{12}$
3 a $2\frac{1}{2}$ b $1\frac{2}{3}$ c $1\frac{5}{6}$ d $2\frac{4}{5}$ e $3\frac{9}{10}$
f $3\frac{11}{12}$ g $2\frac{7}{8}$ h $3\frac{3}{4}$ i $1\frac{1}{4}$
4 a $\frac{1}{6}$ b $\frac{5}{6}$ c $\frac{2}{3}$ or $\frac{8}{12}$ d $\frac{4}{6}$ or $\frac{8}{12}$
e $\frac{4}{6}$ or $\frac{2}{3}$ f $\frac{6}{12}$
5 a true b true c false d true

1:20

1 a $\frac{1}{5}$ b $\frac{2}{5}$ c $\frac{3}{5}$ d $\frac{8}{10}$ e 1 or $\frac{1}{1}$ f $\frac{2}{10}$
2 a $\frac{1}{3}$ b $\frac{2}{3}$ c 1 or $\frac{1}{1}$ d $\frac{2}{6}$ e $\frac{4}{6}$ f 1 or $\frac{1}{1}$
3 a true b true c false d false
e true f true
4 a true b true c true
d false e false f true
5 a $\frac{1}{5} > \frac{1}{10}$ b $\frac{3}{5} < \frac{8}{10}$
c $\frac{6}{10} > \frac{2}{5}$ d $\frac{5}{10} < \frac{5}{5}$
6 a $\frac{1}{3} > \frac{1}{6}$ b $\frac{3}{3} > \frac{4}{6}$
c $\frac{2}{6} < \frac{2}{3}$ d $\frac{5}{6} > \frac{1}{3}$
7 a 4 b 3 c 9 d 6 e 8
8 a $\frac{1}{3}$ b $\frac{3}{4}$ c $\frac{1}{2}$ d $\frac{3}{4}$ e $\frac{9}{12}$ f $\frac{1}{2}$
9 a 0·25 b 0·75 c 0·5 d 0·333…

1:21

1 a $\frac{2}{12}$ b $\frac{10}{12}$ c $\frac{6}{12}$ d $\frac{4}{6}$ or $\frac{8}{12}$ e $\frac{2}{6}$ or $\frac{1}{3}$ f $\frac{2}{3}$ or $\frac{8}{12}$ g $\frac{5}{6}$ h $\frac{3}{6}$

2 a true b true c false d true e false f true g true h false

3 a
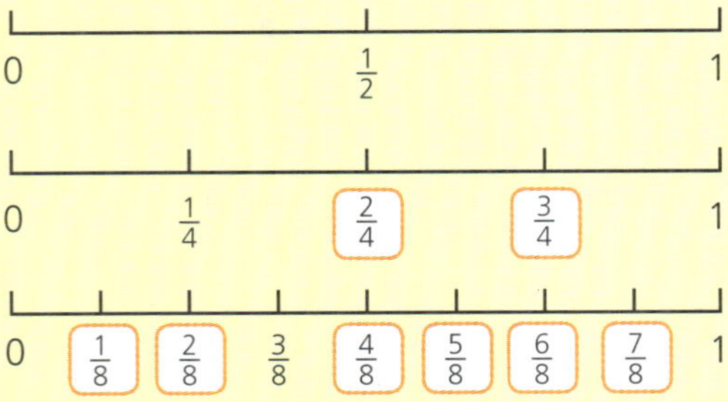

b
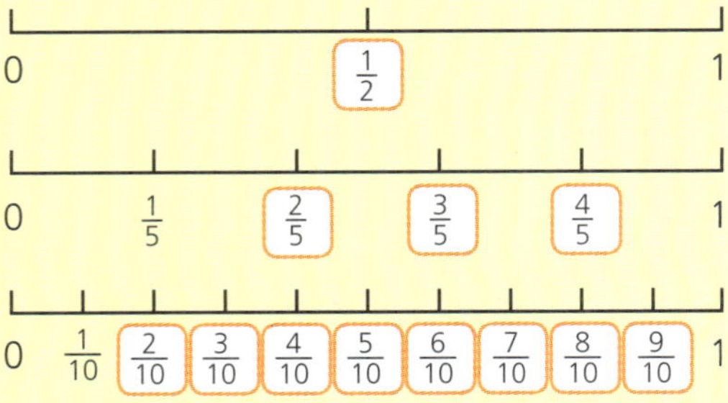

4 a true b true c false d true e false f true g true h false i true j true k true l false

5 a $\frac{4}{5}$ b $\frac{1}{4}$ c $\frac{4}{10}$ d $\frac{3}{4}$ e $\frac{2}{10}$ f $\frac{3}{5}$ g $\frac{1}{2}$ h $\frac{6}{10}$

1:22

1 a $\frac{6}{10}$ b $\frac{3}{12}$ c $\frac{4}{12}$ d $\frac{2}{12}$ e $\frac{8}{12}$ f $\frac{16}{20}$ g $\frac{9}{12}$ h $\frac{10}{12}$

2 a $\frac{1}{2}$ b $\frac{1}{2}$ c $\frac{1}{4}$ d $\frac{1}{2}$ e $\frac{3}{4}$ f $\frac{2}{5}$ g $\frac{3}{4}$ h $\frac{3}{5}$

3 a $\frac{12}{20}$ b $\frac{4}{8}$ c $\frac{4}{40}$ d $\frac{8}{16}$ e $\frac{4}{20}$

4 a $\frac{5}{10}$ b $\frac{5}{15}$ c $\frac{20}{25}$ d $\frac{10}{15}$ e $\frac{15}{20}$

5 a $\frac{2}{5}$ b $\frac{1}{3}$ c $\frac{2}{4}$ d $\frac{3}{6}$ e $\frac{1}{5}$

6 a $\frac{2}{10}$ or $\frac{1}{5}$ b $\frac{5}{6}$ c $\frac{2}{3}$ d $\frac{1}{5}$ e $\frac{15}{50}$ or $\frac{3}{10}$

1:23

1 a $\frac{5}{8} + \frac{2}{8} = \frac{7}{8}$
b $\frac{7}{8} - \frac{4}{8} = \frac{3}{8}$
c $\frac{6}{8} - \frac{2}{8} = \frac{4}{8}$ or $\frac{1}{2}$
d $\frac{3}{8} + \frac{2}{8} = \frac{5}{8}$

2 a $\frac{4}{10} + \frac{4}{10} = \frac{8}{10}$ or $\frac{4}{5}$
b $\frac{9}{10} - \frac{5}{10} = \frac{4}{10}$ or $\frac{2}{5}$
c $\frac{10}{10} - \frac{6}{10} = \frac{4}{10}$ or $\frac{2}{5}$
d $\frac{3}{10} + \frac{5}{10} = \frac{8}{10}$ or $\frac{4}{5}$

3 a $\frac{3}{12} + \frac{8}{12} = \frac{11}{12}$
b $\frac{8}{12} - \frac{6}{12} = \frac{2}{12}$ or $\frac{1}{6}$
c $\frac{3}{12} + \frac{3}{12} = \frac{6}{12}$ or $\frac{1}{2}$
d $\frac{9}{12} - \frac{3}{12} = \frac{6}{12}$ or $\frac{1}{2}$

4 a $\frac{5}{10}$ or $\frac{1}{2}$ b $\frac{3}{12}$ or $\frac{1}{4}$ c $\frac{9}{12}$ or $\frac{3}{4}$ d $\frac{2}{10}$ or $\frac{1}{5}$

1:24

1 a $\frac{5}{6}$ b $\frac{5}{8}$ c $\frac{5}{10}$ or $\frac{1}{2}$ d $\frac{7}{8}$ e $\frac{8}{10}$ or $\frac{4}{5}$ f $\frac{8}{10}$ or $\frac{4}{5}$ g $\frac{9}{10}$

2 a $\frac{3}{8}$ b $\frac{5}{10}$ or $\frac{1}{2}$ c $\frac{1}{12}$ d $\frac{1}{10}$ e $\frac{3}{12}$ or $\frac{1}{4}$ f $\frac{3}{6}$ or $\frac{1}{2}$ g $\frac{3}{10}$ h $\frac{2}{12}$ or $\frac{1}{6}$

3 a $\frac{9}{10}$ b $\frac{2}{9}$ c $\frac{11}{12}$ d $\frac{3}{8}$

1:25

1 9 (oranges)
2 $\frac{1}{10}$
3 $\frac{7}{10}$
4 $\frac{4}{7}$ (includes father/includes parents)
5 936 (students)
6 12 (away)
7 $10
8 $2.00
9 25 (litres have been used)
10 15 (hours)
11 12 (cartons left)
12 $12·60
13 230 (cards)
14 50 (students liked maths best)
15 120 (litres)
16 10 (used green the least)

1:26

1 a $18 b $40 c $20 d $200

2 a 4 b 25 c 45 d 10 e 32 f 15 g 40 h 27 i 75 j $6 k $8 l $6 m $9 n $1 o $28 p $8 q $15 r $12

3 a 54 b 128 c 75

1:27

1 a $7 b $8 c $15 d $8 e $18 f $5 g $6 h $11 i $7

2 a 9 boys b 6 girls c 27 women d 12 sheep e 14 pens f 12 dogs

		a	b	c	d	e	f	g	h
3	Price	$30	$450	$180	$300	$32	$120	$650	$900
	Discount	$6	$45	$45	$15	$8	$12	$130	$45
	Discount Price	$24	$405	$135	$285	$24	$108	$520	$855

4 **a** $31.50 **b** $58.50

5 **a** $42 \div 24 = \frac{\cancel{2}\times\cancel{3}\times 7}{\cancel{2}\times 2\times 2\times\cancel{3}} = 7 \div 4 = 1\frac{3}{4}$

b $78 \div 30 = \frac{\cancel{2}\times\cancel{3}\times 13}{\cancel{2}\times\cancel{3}\times 5} = 13 \div 5 = 2\frac{3}{5}$

1:28

1

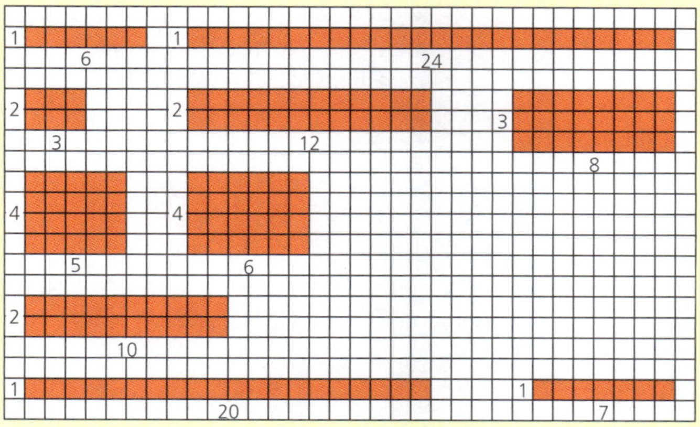

2 **a** 1, 6, 2, 3 **b** 1, 24, 2, 12, 3, 8, 4, 6
c 1, 7 **d** 1, 20, 2, 10, 4, 5

3 **a** 7 **b** 6, 24 and 20
c 2, 3, 5, 7, 11, 13, 17, 19

1:29

1 **a** 1, 11 **b** 1, 8, 2, 4 **c** 1 **d** 1, 9, 3
e 1, 15, 3, 5 **f** 1, 13
g 1, 24, 2, 12, 3, 8, 4, 6
h 11 and 13 **i** 8, 9, 15, 24

2

1	1			
(2)	1	2		
(3)	1	3		
4	1	4	2	
(5)	1	5		
6	1	6	2	3

(7)	1	7				
8	1	8	2	4		
9	1	9	3			
10	1	10	2	5		
(11)	1	11				
12	1	12	2	6	3	4

3 **a** 2, 3, 5, 7, 11, 13, 17, 19, 23, 29
b 4, 6, 8, 9, 10, 12, 14, 15, 16, 18, 20, 21, 22, 24, 25, 26, 27, 28

4

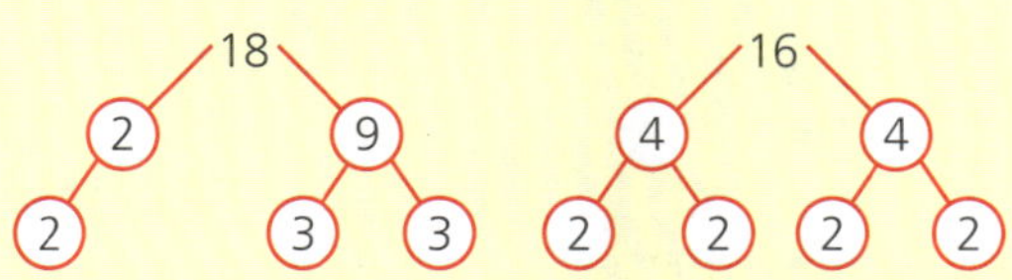

a $18 = 2 \times 3 \times 3$
$= 18$

b $16 = 2 \times 2 \times 2 \times 2$
$= 16$

2:01

1 **a**

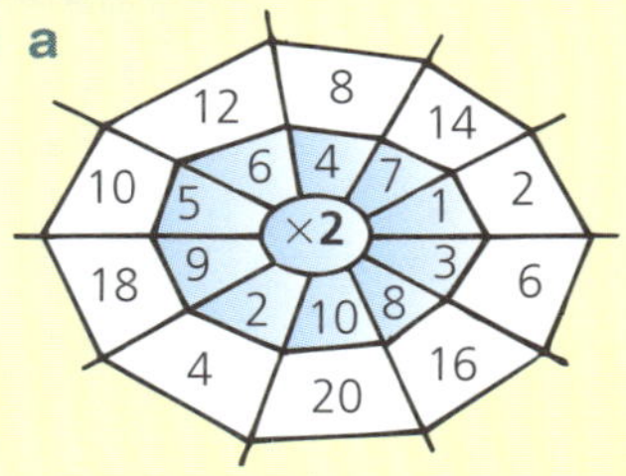

b

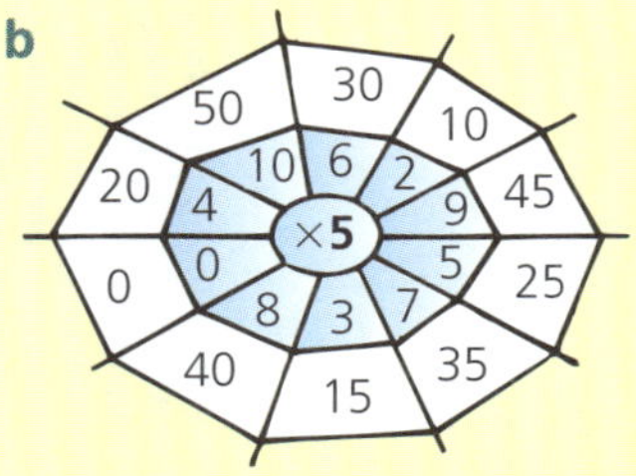

c

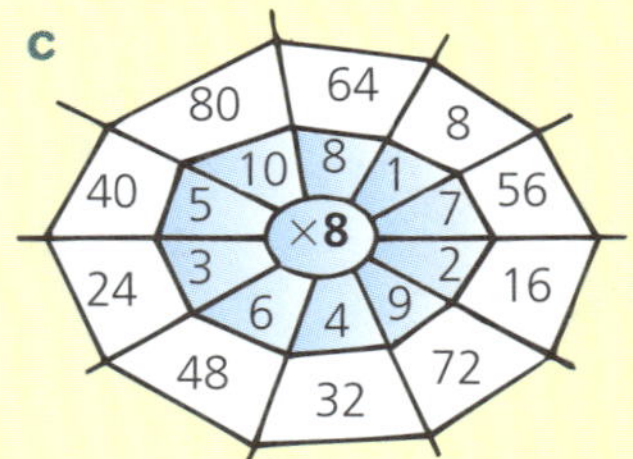

d

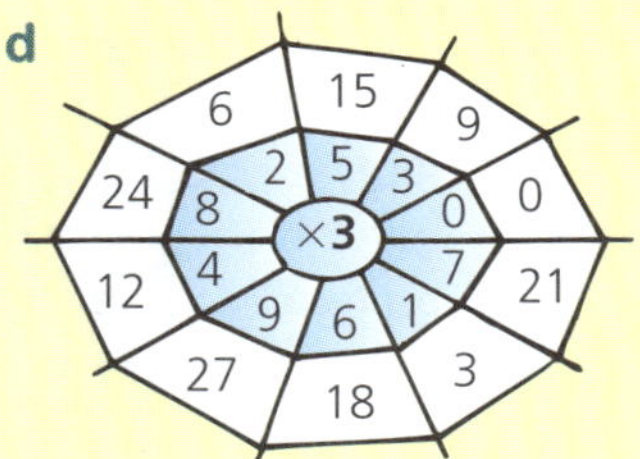

e

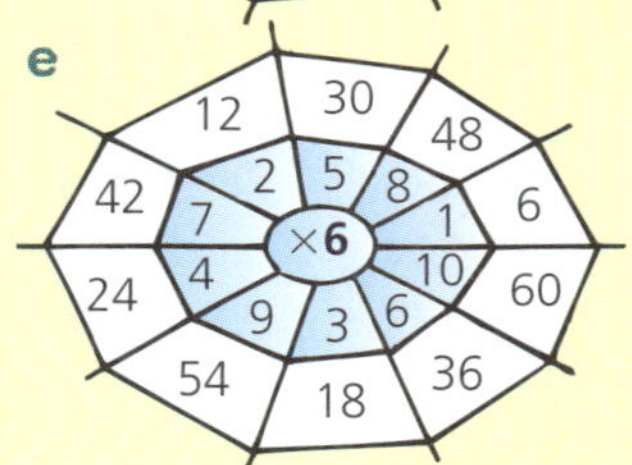

f

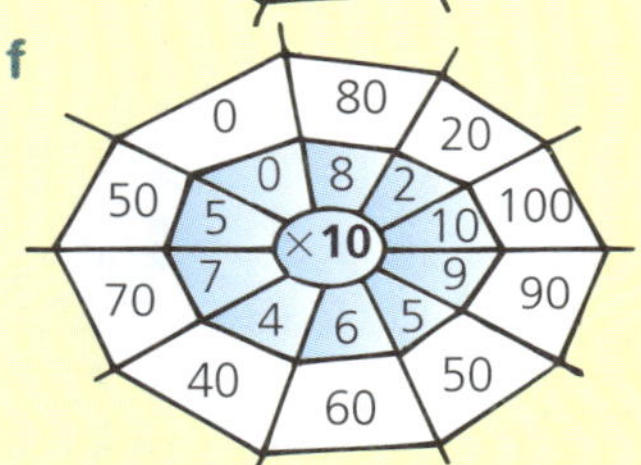

g

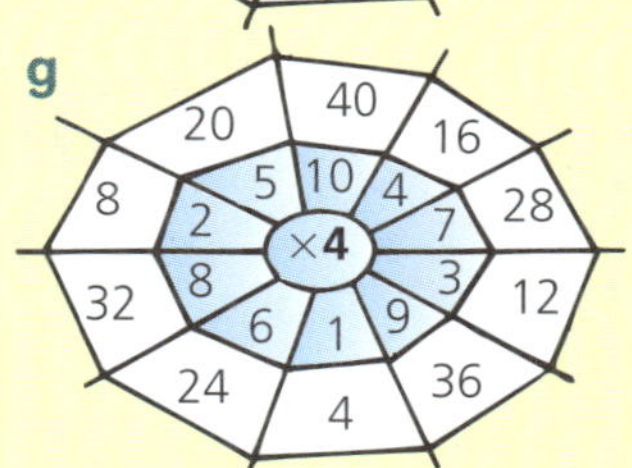

h

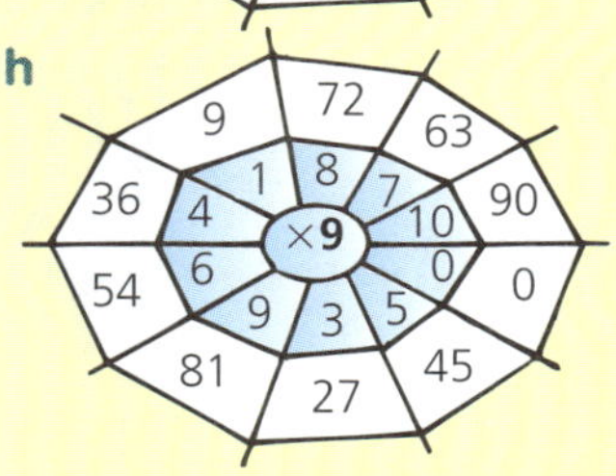

i

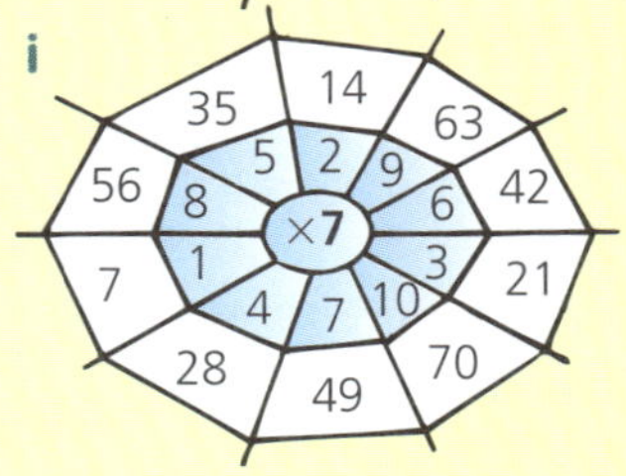

2 **a** 1, 18, 2, 9, 3, 6 **b** 1, 49, 7
c 1, 24, 2, 12, 3, 8, 4, 6 **d** 1, 13

3 **a** 4, 8, 12, 16, 20, 24, 28, 32, 36, 40, 44
b 9, 18, 27, 36, 45, 54, 63, 72, 81, 90, 99

4 **a** 9 **b** 64 **c** 100 **d** 7 **e** 1 **f** 5

5 **a** false **b** true

2:02

1 **a** 6 turtles **b** 4 turtles **c** 8 turtles **d** 3 turtles

2 **a** 4 r 4 **b** 3 r 3

3 **a** 8 groups **b** 6 groups

4 a 2 groups, 2 left over b 3 groups, 0 left over
c 5 groups, 10 left over
5 a 11 and 1 left b 12 and 2 left
c 14 and 1 left
6 a 56 ÷ 8 = 7, 56 ÷ 7 = 8 b 54 ÷ 6 = 9, 54 ÷ 9 = 6
c 48 ÷ 6 = 8, 48 ÷ 8 = 6 d 63 ÷ 9 = 7, 63 ÷ 7 = 9
7 a 9 b 7 c 8 d 2 e 4 f 7
g 8 h 8
8 a 7 b 3 c 3 d 4 e 5 f 10
g 4 h 5 i 6 j 5 k 5 l 6
m 6 n 1 o 9 p 3 q 8 r 9
s 7 t 7 u 8

2:03

1 a 905 b 671 c 650 d 841
2 a 7099 b 5800 c 6605 d 6564
3 a 8187 b $99.10 c 8274 d $53.92
4 a 4851 b 9050 c 7315 d 7496
5 a 9455 b 9010 c 9000
6 a 4252 b 7217

2:04

1 a 837 b 641 c 433 d 265
2 a 407 b 455 c 207 d 713
3 a 4271 b 4351 c 279 d 2639
4 a 916 b 2085 c 3591
5 a 730 b 5550 c 2876
6 a 4876 b 2319 c 64
7 a 1560 − 1494 = 66 b 4200 − 3285 = 915 m

2:05

1 a 34 b 3 c 34 d 2 e 10 f 16
g 30 h 3 i 60
2 a 25 b 17 c 19 d 70 e 17 f 26
g 20 h 52 i 66
3 a 24 b 15 c 3 d 18 e 18 f 12
g 2 h 4 i 2
4 a 59 b 14 c 3 d 35 e 6 f 27
g 7 h 26 i 40

2:06

1 a 25 b 33 c 0 d 16 e 4 f 8
g 38 h 38 i 24 j 10 k 50 l 420
2 a 16 b 72 c 16 d 18 e 18 f 18
g 83 h 68 i 26 j 4
3 a 50 b 250 c 170 d 14 e 8 f 3
g 44 h 45 i 21 j 1
4 a 50 b 50 c 12 d 12 e 1 f 1
g 2 h 2
5 a 1, 3, 6, 10, 15 b 21, 28, 36, 45, 55

2:07

1 a 8 tens b 28 tens c 45 tens
d 2 hundreds 8 tens e 4 hundreds 5 tens
f 35 hundreds g 32 hundreds
h 3 thousands 2 hundreds i 42 thousands
2 a 240 b 360 c 300 d 490 e 540
f 600 g 2000 h 4800 i 4200 j 2400
k 56 000 l 30 000 m 18 000 n 50 000 o 72 000
3 a 200 toothpicks b 4800 stamps c 21 000 words
d $42 000 e $28 000 f 42 000 L
g 60 000 L h 40 000 seeds; 32 000 grew; 8000 did not grow

2:08

1 a 91 b 90 c 105 d 129 e 135
f 212 g 440 h 603 i 288
2 a 640 b 822 c 832 d 833 e 912
f 2092 g 2439 h 4298 i 3024 j 3570
k $18.96 l $14.60 m $49.92 n $14.91 o $75.60

524, 542

2:09

1 a 4 b 2 c 4 d 4 e 3 f 8
g 5 h 6 i 4 j 3 k 5 l 7
2 a 56 b 50 c 42 d 64 e 313 f 334
g 146 h 150 i 269 j 98 k 389 l 609
3 a 56 b 13 c 23 d 26 e 33 f 23
g 19 h 38
4 a 128 b 416 c 718 d 147 e 627 f 928
g 415 h 502
5 a 270 b 106 c 36

2:10

1 148 L 2 $152 3 1108 m
4 $382 5 512 tins 6 916 km

2:11

1 a 8 r 4 b 6 r 3 c 5 r 5 d 6 r 1
2 a 12 b 11 c 42 d 21
3 a 19 b 14 c 17 d 12
4 a 214 b 88 c 165 d 55
5 a 121 r 2 b 63 r 7 c 179 r 1 d 93 r 1
6 a 37 b 58 r 3 c 80 d 61 r 4
7 a 204 b 107 c 207 d 107
8 a 150 r 1 b 120 r 2 c 225 d 140 r 2
9 a 38 mm b $190

2:12

1 a 5 r 5 b 7 c 8 r 5 d 8 r 2 e 9 r 3
f 15 g 14 h 18 i 15 j 29

2 a 242 b 120 c 111 d 101 e 211
f 126 g 133 h 134 i 234 j 122
3 a 156 r 3 b 175 c 150 r 2 d 128 r 4 e 33
f 60 g 35 h 74 i 30 j 72
k 30 r 3 l 107 m 300 r 2 n 107 o 30 r 3
4 a 59 filled b 60 trips c 25 cars
d 134 mL e $96

2:13

1 a $1\frac{1}{4}$ slices b $1\frac{1}{2}$ slices c $1\frac{3}{4}$ slices
2 a $2\frac{1}{3}$ homes b $3\frac{2}{3}$ homes c $7\frac{2}{3}$ homes
3 a $4\frac{3}{5}$ b $3\frac{4}{5}$ c $17\frac{1}{2}$ d $14\frac{2}{3}$ e $5\frac{7}{10}$ f $7\frac{3}{4}$
g $5\frac{5}{7}$ h $7\frac{8}{9}$ i $11\frac{2}{8}$ j $14\frac{4}{6}$ k $18\frac{3}{5}$ l $23\frac{3}{4}$
4 a $\frac{3}{5}$ b $\frac{2}{4}$ or $\frac{1}{2}$ c $\frac{7}{10}$ d $\frac{1}{3}$ e $\frac{5}{8}$
5 a $93\frac{2}{3}$ b $84\frac{3}{5}$ c $20\frac{7}{9}$ d $29\frac{9}{10}$ e $124\frac{1}{6}$
f $112\frac{7}{8}$ g $130\frac{2}{7}$ h 175

2:14

1 a 25 b 8 c 6 d 46 e 521
f 1968 g $22\frac{1}{2}$ (or 22·5) h $12\frac{1}{4}$ (or 12·25)
i 1406 j 555 k $18.20 l $7.90
2 a 90 km/h b 71 km/h c 274 km/h
d 2·1 cm/min e 900 km/day f 335 m/s

2:15

1 a 61 500 b 47 121 c 65 750 d 80 803
e 270 050 f 412 230 g 734 981 h 622 100
2 a 48 655 b 117 980 c 100 650 d 96 935
e 887 450 f 441 610 g 734 412 h 912 000
3 a 515 634 b 198 217 c 801 702
4 Answers will vary.
a about 17 500 b about 35 000
c about 32 500

2:16

1 a 47 180 b 53 630 c 42 676 d 24 359
e 15 000 f 24 338 g 55 904 h 21 863
2 a 165 945 b 327 169 c 539 783
d 773 616 e 484 473 f 365 000 g 451 431
3 a $75 100 b $298 075 c 75 000 m^2
d 27 000 apples e 215 500

2:17

1 a 5864 b 8577 c 1904 d 7599
e 22 954 f 11 984 g 57 045 h 46 359
2 a 71 927 b 43 276 c 30 666 d 576
e 69 153 f 39 057 g 52

3 a 1155 km b 615 km
c 2579 ice blocks d 1825 m
e $1610, $34 390 f 2528 m

2:18

1 a $48.37 b $124.47 c $249.25 d $395.14
2 a $11.75 b $18.45 c $35.72 d $204.75
e $126.42 f $187.74 g $4.32 h $340.60
3

	a	b	c	d	e
Total	$24.12	$27.39	$27.99	$25.85	$29.75
Rounded off	$24.10	$27.40	$28.00	$25.85	$29.75
Change	$5.90	$2.60	$2.00	$4.15	$0.25

2:19

1 Steven travelled 416 km. His sister travelled 378 km. Difference is 38 km.
2 a 223 km b 312 km c 478 km
d 270 km e 355 km f 356 km
3 a 1380 km b 138 L c $220.80
4 336 km, average speed 56 km/h
5 Cowra and Orange

2:20

1 a 30 b 0 c 11 d 6 e 80 f 3
g 2 h 23 i 50 j 0 k 88 l 136
2 a 26 b 49 c 16 d 150 e 49 f 37
g 10 h 48 i 100 j 58 k 16 l 18
3 a (16 − 3) × 2 = 26 b (6 − 3) × 2 = 6
c 10 × (6 + 4) = 100 d 6 × (3 + 5) = 48
e 50 ÷ (5 + 5) = 5 f 70 ÷ (10 − 3) = 10
g 27 − (14 − 8) = 21 h (4 + 3) × 12 − 2 − 10 = 72
i 100 × (40 − 20 − 10) = 1000
4 a

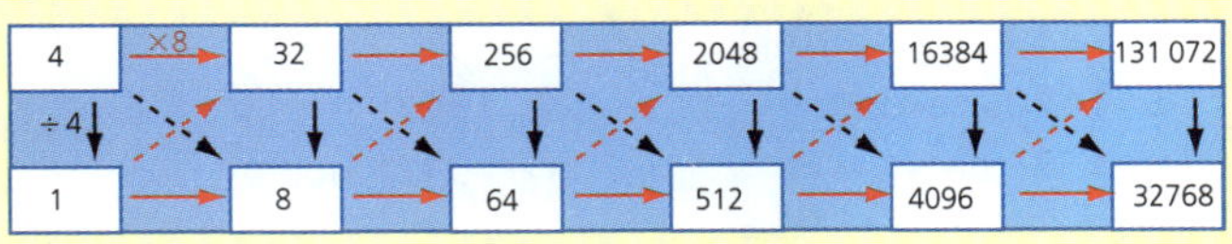

b black broken arrows are × 2
c red broken arrows are × 32

2:21

1 a 1 b 1 c 0·1 d 1
e 1 f 0·1 g 1 h 0·1
i 1
2 a 2 b 2 c 2 d 2
e 2 f 2
3 a 3 b 4 c 3 d 4·1
e 3 f 2·1
4 10, 1, 0·1, 0·01, 0·001
5 Calculator check

6 a $17.50 b $20 c $11.70
d $6.90 e $13.80

2:22

1 a 3·791 b 3·727 c 8·147
d 11·898 e 6·57 f 9·422
g 13·189 h 9·449 i 14·525
j 5·377 k 11·591 L l 11·429 m
m 9·72 km n 10·421 L o 8·803 m
2 Answers will be checked.
3 a 0·136 b 0·203 c 0·284
d 0·42 e 0·339 f 0·487

2:23

1 a 15·86 b 19·6 c 24·42 d 24·57 e 115·83
f 29·91 g 20·22 h 68·46 i 28·05 j 89·31
2 a 17·785 b 25·641 c 20·596 d 40·095
e 60·719 f 27·625 g 26·857 h 55·232
i 111·005 j 84·833
3 a 3·625 L b 10·15 km c 60·41 cm
d 4·875 kg e 11·35 m

2:24

1 a 8·2 b 5·47 c 15·79 d 12·383
e 2·7 f 57·78 g 2·62 h 17·726
i 276·256 j 78·613 k 15·786 l 27·566
2 a 1·507 km b 23·199 L c 2·189 kg d 0·933 m

2:25

1 a 8076 b 7300 c 7548 d 9228 e 9400
2 a 24 750 b 33 668 c 20 468 d 24 836 e 12 711
f 65 450 g 39 720 h 22 155 i 13 540 j 35 880
3 a 56 000 b 85 050 c 60 250 d 81 702 e 82 940
f 73 596 g 90 000 h 69 235 i 75 000 j 94 890
4 a 14 400 seconds b 10 080 minutes
c 4383 days d 70 128 hours

2:26

1 a 0·6 b 0·8 c 0·9 d 0·8
e 3 f 1·5 g 5·4 h 2·8
i 5·6 j 8·1 k 6·4
2 a 1·5 b 0·7 c 0·8 d 1·4
e 2·7 f 5·6 g 3·6 h 3
i 4
3 a 3·2 b 7·2 c 30·5 d 10·8
e 14 f 37 g 46 h 18
i 58·4 j 45·5 k 28·2 l 74·7
4 a 42·5 km b 46 m c 5·6 L
d 14 million dollars

5 a 1157·1 L b 250·2 m c 671·4 kg
d 382·5 L e 355·2 kg f 627·2 t

2:27

1 a 6·39 b 26·88 c 77·76 d 184·56 e 322·56
2 a $52.30 b $74.22 c $151.68
d $928.13 e $3 207.33
3 a $13.50 b $26.00 c $30.00 d $17.70
e $22.05 f $30.00 g $29.40
4 a 6 b 6·148 c 7·914 d 10·2 e 18·4 f 18·2
5 a 110·4 cm b 26·915 kg c 1·95 L

2:28

1 a 12·4 b 43·4 c 77·6 d 138·20
e 329·31 f 256·50 g 455·28 h 321·6
i 765·27 j 303·44 k 372·8
2 a 12·984 b 5·859 c 14·345 d 28·882
e 30·438 f 9·220 g 28·672 h 8·690
i 18·063 j 18·630
3 a $95.55 b $135.25 c $274.80 d $482.52
e $585.21 f $1 812.24
4 a 1552 L b 1 337·6 m c 151·375 kg
d 1 313·2 cm e 3075·6 g f 400·74 km

2:29

1 a 528 b 723 c 739 d 459
e 1 187 f 1 439
2 a 579 r 2 b 677 r 1 c 877 r 6 d 165 r 1
e 18 277 f 28 224 g 13 579 h 17 315
3 a $1 437 b 347 runs c 491 books
d 532 people e $367 f 375 mL
g 273 books
4 a 7204 b 6312 c 3090 d 4275 e 8660
f 6414 g 3700 h 4368 i 9774

2:30

1 a 606 b 802 c 708 d 207
e 1 005 f 2 040 g 2 090 h 1 021
i 1 350 j 2 071 k 1 209 l 1 007
2 a 540 r 1 b 400 r 1 c 700 r 4 d 600 r 2
e 1 200 r 6 f 1 035 r 1 g 1 500 r 2 h 1 402 r 4
i 10 020 j 12 022 k 10 305 l 11 302
3 a 2 183 sheep b $1 052 c 625 mL
d 2 704 weeks, 52 years

2:31

1 a 835 r 2 b 186 r 5 c 364 r 8 d 901 r 4
e 784 f 238 g 508 h 411
i 814 r 5 j 900 k 380 r 6 l 720

2 a 1638 b 4560 c 3724 d 8600
e 3635 r 3 f 4786 r 4 g 1005 r 6 h 7134 r 6
i 21632 j 13005 r 6 k 18650 l 37103 r 9

3 a 800 cm b 5685 cm c 1868 cm d 784 cm

4 a 693 b 510 c 6127 d 5600

5 3600 boxes

6 $816.50

2:32

1 a 647·2 b 2523·1 c 86509·2
d 442·933 e 173·9 f 5094·61
g 37527·7 h 2980·5 i 3700·1

2 a 77291·4 b 157·3 c 779·31
d 84291·3 e 910·033 f 8793·5
g 2106·8 h 643·375 i 40357·1
j 351·13 k 9612·1 l 9427·4

3 a 2·714 b 5·6336 c 7·2456
d 9·067 e 87·794 f 6·304
g 6·3119 h 4·62731 i 8·56094

4 a 1·53483 b 9·746 c 7·2563
d 8·84366 e 73·508 f 8·3251
g 8·05603 h 8·449 i 5·6239
j 7·2634 k 3·6217 l 83·673
m 8·0446 n 7·238 o 8·469

2:33

1 a 3500 b 960 c 736 d 2400

2 a 350 b 96 c 70 d 240

3 a 400 b 4 c 90 d 500 e 22
f 21 g 20 h 119

4 a 20·6 b 355·2 c 33·7 d 134 e 18·2
f 147·7 g 26·1 h 18·7 i 182·9 j 53·7
k $70.10 l $33.70 m $95.40 n $78.90 o $52.70

5 a 6 b 95 c 163 d 57 e 140

2:34

1 a ■ = $1\frac{1}{2}$ b ■ = 1·2 c ▲ = $8\frac{1}{4}$
d ▲ = 3·3 e ■ = 1·1 f ■ = 1

2 a F b F c T d T e F f T

3 a ■ + 3 b ■ − 7 c 5 × ■
d ■ + 2 e ■^{2} f $\frac{1}{5}$ × ■

4 a ■ + 4 = 11, ■ = 7 b ■ − 11 = 9, ■ = 20
c ■ − 7 = 6, ■ = 13 d ■ × 9 = 27, ■ = 3
e ■ ÷ 4 = 7, ■ = 28 f ■^{2} = 25, ■ = 5
g ■ + 8 = 14, ■ = 6 h ■ − 34 = 82, ■ = 116
i ■ ÷ 7 = 9, ■ = 63 j (15 + ■) ÷ 2 = 16, ■ = 17

2:35

1 a 0·4 b 0·3 c 0·1 d 0·4
e 0·3 f 0·7 g 0·7 h 0·9
i 0·6

2 a 4·2 b 3·1 c 1·1 d 1·6
e 0·42 f 0·31 g 0·11 h 0·16

3 a 0·402 b 0·268 c 0·201
d 0·012 e 0·008 f 0·006

4 a 3·1 b 24·6 c 0·35
d 0·103 e 0·714 f 0·238
g 2·48 h 0·003

5 a 26·314 m b 981·9 cm c 0·568 km
d 7·442 km e 2·568 L f 9·137 L

2:36

1 a T b T c F d T e T f T
g T h T i T

2 a 1·12 b 1·12 c 0·36 d 2·25
e 1·75 f 0·45

3 a 1·425 b 3·27 c 16·3 d 4·2 kg
e 20·75 km f 18·5 L g 1·275 m

4 a 0·375 b 6·775 c 5·12
d 0·125 e 99·75 f 68·55
g 1·35 h 21·15

5 a 12·56 m b 8·601 L c 0·026 km
d 3·095 km

2:37

1 a

3 × 8	24
3 × 80	240
3 × 800	2 400

b

2 × 3	6
2 × 30	60
2 × 300	600

c

6 × 4	24
6 × 40	240
6 × 400	2 400

d

4 × 7	28
4 × 70	280
4 × 700	2 800

e

8 × 6	48
8 × 60	480
8 × 600	4 800

f

7 × 9	63
7 × 90	630
7 × 900	6 300

2 a

30 × 8	240
30 × 80	2 400
30 × 800	24 000

b

20 × 4	80
20 × 40	800
20 × 400	8 000

c

50 × 7	350
50 × 70	3500
50 × 700	35 000

d

80 × 7	560
80 × 70	5 600
80 × 700	56 000

e

60 × 3	180
60 × 30	1 800
60 × 300	18 000

f

90 × 9	810
90 × 90	8 100
90 × 900	81 000

3 a

8 × 20	160
80 × 20	1 600
800 × 20	16 000
8000 × 20	160 000

b

70 × 3	210
70 × 30	2 100
70 × 300	21 000
70 × 3 000	210 000

c

4 × 80	320
40 × 80	3 200
400 × 80	32 000
4 000 × 80	320 000

d

3 × 90	270
30 × 90	2 700
300 × 90	27 000
3 000 × 90	270 000

e

50 × 6	300
50 × 60	3 000
50 × 600	30 000
50 × 6 000	300 000

Rule: Write down the zeros first, then multiply the numbers remaining.

4 **a** 800, 8000, 80 000, 800 000
b 900, 9000, 90 000, 900 000

2:38

1 **a** 350 **b** 150 **c** 780 **d** 510 **e** 700
f 300 **g** 1560 **h** 1020 **i** 1050 **j** 450
k 2340 **l** 1530

2 **a** 300 **b** 600 **c** 900 **d** 400 **e** 800
f 1200 **g** 800 **h** 1600 **i** 2400

3 **a** 800 **b** 900 **c** 600 **d** 400 **e** 1500
f 1200 **g** 1400 **h** 3000 **i** 2400 **j** 3600
k 5600 **l** 1800

4 **a** 30 000 **b** 32 000 **c** 35 000 **d** 64 000 **e** 28 000
f 35 000 **g** 24 000 **h** 72 000 **i** 12 000

5

×	60	90	300	800	400	600	500
50	3000	4500	15 000	40 000	20 000	30 000	25 000
200	12 000	18 000	60 000	160 000	80 000	120 000	100 000
700	42 000	63 000	210 000	560 000	280 000	420 000	350 000
900	54 000	81 000	270 000	720 000	360 000	540 000	450 000

2:39

1 **a** 1260 **b** 1120 **c** 3240 **d** 1290 **e** 2600
f 6440 **g** 2430

2 **a** 17 × 61 = (10 × 61) + (7 × 61)
b 16 × 95 = (10 × 95) + (6 × 95)
c 31 × 53 = (30 × 53) + (1 × 53)
d 46 × 66 = (40 × 66) + (6 × 66)
e 72 × 37 = (70 × 37) + (2 × 37)
f 29 × 48 = (20 × 48) + (9 × 48)
g 49 × 59 = (40 × 59) + (9 × 59)
h 72 × 64 = (70 × 64) + (2 × 64)

3 **a** 630 + 315 = 945 **b** 1480 + 592 = 2072
c 3680 + 552 = 4232 **d** 4050 + 162 = 4212
e 5250 + 375 = 5625 **f** 5280 + 792 = 6072
g 1560 + 260 = 1820 **h** 3840 + 192 = 4032
i 1400 + 196 = 1596 **j** 5040 + 252 = 5292

4 490 pages

2:40

1 **a** 448 **b** 1696 **c** 4743 **d** 2108
e 3124 **f** 1333 **g** 2068

2 **a** 1068 **b** 1431 **c** 3528 **d** 882
e 3515 **f** 2376 **g** 1825 **h** 2438

3 **a** 1008 mL **b** 704 g

2:41

1 **a** 552 **b** 1023 **c** 1944 **d** 2400 **e** 804
f 2173 **g** 1674 **h** 1386 **i** 450 **j** 3008
k 832 **l** 2277 **m** 868 **n** 3990

2 **a** $456 **b** cost $1944, sold for $3132, profit $1188

c

Item	Number of Items	Cost of One	Total Cost	Selling Price of One	Total From Sales	Total Profit
sugar bowl	65	$14	$910	$28	$1820	$910
oval tray	43	$22	$946	$35	$1505	$559
cup/ saucer	76	$28	$2128	$47	$3572	$1444

2:42

1 **a** 5577 **b** 4131 **c** 16 290 **d** 20 664
e 60 225 **f** 41 440 **g** 30 736 **h** 54 290
i 21 148 **j** 16 184 **k** 28 500 **l** 150 800

2 **a** 11 550 m, 46 200 m **b** 120 400 years

2:43

1 **a** ■ = 4, T **b** ■ = 100, T **c** ▲ = 8, T
d ▲ = 65, T **e** ■ = 82, T **f** ■ = 8, T
g ▲ = 77, T **h** ▲ = 21, T

2 **a** ■ = 3 **b** ■ = 21 **c** ▲ = 25
d ▲ = 21 **e** ■ = 199 **f** ■ = 10

3 **a** F **b** T **c** F **d** T **e** T **f** F
g T **h** T **i** T **j** T **k** T **l** T

2:44

1 **a** 154; 37 902; 89 366 **b** 732; 1092; 111 111
c 812; 4308; 17 224 **d** 415; 8400; 111 110
e 11 008; 23 000; 65 832; 614 016; 821 104; 7184
f 14 301; 91 422; 128 700 **g** 38 400; 75 830; 111 010

2 **a** 1, 24, 2, 12, 3, 8, 4, 6 **b** 1, 40, 2, 20, 4, 10, 5, 8
c 1, 105, 3, 35, 5, 21, 7, 15 **d** 1, 81, 3, 27, 9
e 1, 120, 2, 60, 3, 40, 4, 30, 5, 24, 6, 20, 8, 15, 10, 12

3 **a** 1, 2, 15, 3, 10, 5 and 6
b 1, 2, 12, 3, 8, 4 and 6
c 1, 2, 50, 4, 25, 5, 20 and 10

2:45

1 **a** 84, D **b** 14, U **c** 28, D
d 48, U **e** 6, U **f** 26, D
g 20, U **h** 22,U **i** 33, D

2 Parts **a**, **c** and **h** have wrong answers.

3 **a** 29 + 9 = 38 **b** 31 + 6 = 37
c 101 − 23 = 78 **d** 17 − 7 = 10
e 2 × 7 = 14 **f** 2 × 3 = 6

g 9 + 5 = 14 **h** 9 × 5 = 45
i 19 − 12 = 7 **j** 23 + 9 = 32

2:46

1. **a** 85·435 (Estimate = 85)
   b 71·055 (Estimate = 71)
   c 5·371 (Estimate = 5)
   d 38·266 (Estimate = 38)
2. Answers will vary.
   a half of 12 = 6 (or 0·5 × 10 = 5)
   b 40 × 2 = 80 **c** 20 × 1 = 20
   d 5 × 8 = 40 **e** half of 28 = 14 (or 30 × 0·5 = 15)
   f 4 × 7 = 28 **g** 10 × 7 = 70
   h 100 × 3 = 300 **i** 7 × 6 = 42
3. **a** 60 ÷ 3 = 20 **b** 30 ÷ 10 = 3 (or 32 ÷ 8 = 4)
   c 21 ÷ 7 = 3 **d** 8 ÷ 2 = 4
   e 15 ÷ 5 = 3 **f** 18 ÷ 2 = 9
   g 20 ÷ 4 = 5

2:47

1.

Question	Number of Decimal Places in Question	Calculator Answer	Number of Decimal Places in Answer
0·3 × 5	1	1.5	1
0·4 × 0·7	2	0.28	2
0·41 × 2	2	0.82	2
0·103 × 3	3	0.309	3
0·121 × 0·2	4	0.0242	4

2. **a** 2 **b** 1 **c** 3 **d** 1 **e** 2
   f 3 **g** 2 **h** 3 **i** 3 **j** 1
   k 4 **l** 4
3. **a** 1·6 **b** 2·4 **c** 0·27
   d 0·42 **e** 3·6 **f** 1·5
   g 0·36 **h** 0·06 **i** 0·4
   j 0·15 **k** 0·48 **l** 0·014
   m 0·007 **n** 0·008 **o** 0·21
4. **a** 0·86 **b** 1·1
   c 1·06 **d** 0·0214
   e 1·24 **f** 0·36
   g 0·0824 **h** 0·45
   i 21
5. **a** 0·24 kg **b** 0·36 cm^2 **c** 0·12 cm^2 **d** 1·2 kg

2:48

1. **a** 40 ÷ 5 = 8 **b** 45 ÷ 9 = 5
   c 24 ÷ 6 = 4 **d** 21 ÷ 7 = 3
   e 36 ÷ 9 = 4 **f** 40 ÷ 5 = 8
   g 20 ÷ 5 = 4 **h** 81 ÷ 9 = 9
   i 56 ÷ 8 = 7
2. Working will vary.
   a 18 ÷ 2 = 9 **b** 12 ÷ 6 = 2
   c 72 ÷ 9 = 8 **d** 54 ÷ 6 = 9
   e 54 ÷ 6 = 9 **f** 56 ÷ 7 = 8
   g 63 ÷ 7 = 9 **h** 30 ÷ 5 = 6
   i 14 ÷ 2 = 7
3. **a** 24 ÷ 6 = 4 **b** 45 ÷ 5 = 9
   c 88 ÷ 11 = 8 **d** 28 ÷ 4 = 7
   e 12 ÷ 6 = 2 **f** 81 ÷ 9 = 9
   g 40 ÷ 8 = 5 **h** 55 ÷ 5 = 11
   i 24 ÷ 8 = 3

2:49

1. **a** 1 717 217 **b** 9 486 251 **c** 12 611 000
   d 7 460 646 **e** 2 868 323 **f** 48 572 381
   g 8 134 000 000
2. **a** $347 337.50 **b** $926 188.91 **c** $647 900.00
   d $902 763.00
3. **a** 22 407 200 (including Territories)
4. Estimates will vary. (February 2010 is most likely.)

2:50

1. **a** 4 177 300 **b** 6 679 335 **c** 541 000
   d 14 443 660 **e** 47 902 000 **f** 50 814 000
   g 31 800 000 **h** 14 156 288
2. **a i** 2544 m **ii** 23 797 km
   b 163 638 400 km^2 **c** 356 524 200 (trees)
   d 619 000 000 (people) **e** 5 519 000 000 (people)
   f 148 140 000 km

2:51

1. **a** 15, 18 **b** 14, 15 **c** 21, 27 **d** 30, 34 **e** 11, 21
   f 20, 25 **g** 17, 24 **h** 11, 20 **i** 28, 30 **j** 10, 30
   k 11, 22 **l** 35, 40 **m** 5, 25 **n** 32, 38 **o** 50, 54
   p 48, 50 **q** 25, 30 **r** 41, 50
2. **a** 1·625 km **b** $1.63 (rounding up).
3.

1 8	L	2 E	3 S	4 S
I	■	5 6	0	0
6 L	E	6	S	■
L	■	S	■	■
7 S	8 0	■	9 h	10 E
■	11 I	12 S	■	E
■	13 L	0	■	L

2:52

1. a variegated one b $22.05
   c white, orange and variegated
2. 11, 10 posts
3. 8 hands
4. 8 posts
5. 17 days
6. B
7. B
8. C
9. A

3:01

1. A 24 mm, 2·4 cm B 49 mm, 4·9 cm
   C 66 mm, 6·6 cm D 92 mm, 9·2 cm
   E 118 mm, 11·8 cm
2. a 2·9 cm b 3·6 cm c 9·9 cm d 5·1 cm e 6·4 cm
   f 8·2 cm g 4·8 cm h 7·3 cm
3. a 39 mm b 72 mm c 91 mm d 45 mm e 68 mm
   f 24 mm g 53 mm h 106 mm
4. a 14 cm b 17 cm c 22 cm d 14 cm
5. a 12 cm b 19 cm

3:02

1. a 3000 m b 5000 m c 9000 m d 7000 m
   e 10000 m f 8000 m g 20000 m h 15000 m
   i 12000 m
2. a 2 km b 6 km c 1 km d 4 km e 10 km
   f 11 km
3. a km b m c cm d mm e km
   f km g m h m
4. a 386 km b 376 km c 491 km
   d 681 km e 877 km
5. a 500 m b 250 m c 750 m
   d 2100 m e 1500 m f 1700 m
   g 1750 m h 400 m i 6250 m
   j 7300 m k 7300 m l 2500 m
   m 4900 m n 4900 m o 6700 m
6. a 87 km/h b 93 km/h c 85 km/h
   d 78 km/h

3:03

1. A 23 mm, 2·3 cm B 42 mm, 4·2 cm C 67 mm, 6·7 cm
   D 99 mm, 9·9 cm E 128 mm, 12·8 cm
2. a 6·3 cm b 9·5 cm c 7·9 cm d 3·6 cm
   e 8·1 cm f 10·3 cm g 4·1 cm h 16·2 cm
3. a 1·35 m b 4·31 m c 5·65 m d 9·84 m
   e 7·29 m f 3·85 m g 8·95 m h 10·75 m
4. a m b cm c km d cm
   e km f mm
5.

Metres	1659	4752	9225	5076	3855	7125
Kilometres	1·659	4·752	9·225	5·076	3·855	7·125

6.

Kilometres	3·575	9·635	2·749	7·36	6·4	8
Metres	3575	9635	2749	7360	6400	8000

3:04

1. a 4·9 cm b 6·7 cm c 9·6 cm d 3·2 cm
   e 8·3 cm f 5·4 cm g 14·5 cm h 46·3 cm
2. a 1·35 m b 3·26 m c 9·51 m d 7·96 m
   e 4·78 m f 6·59 m g 12·34 m h 38·65 m
3. a 1657 m b 3425 m c 2976 m d 5125 m
   e 9816 m f 4291 m g 6950 m h 8750 m
   i 7100 m
4. a m b mm c km d cm e m f km
5.

	Object	Millimetres	Centimetres	Metres
a	Watchband Width	17 mm	1·7 cm	0·017 m
b	Pencil Length	94 mm	9·4 cm	0·094 m
c	Eraser Length	38 mm	3·8 cm	0·038 m
d	Felt Pen Length	145 mm	14·5 cm	0·145 m
e	Paper Width	210 mm	21 cm	0·21
f	Calculator Thickness	12 mm	1·2 cm	0·012 m

3:05

1. a km b cm c m d mm
   e cm (or mm) f m g m h cm
2. a 1·56 m b 3·39 m c 2·94 m d 5·67 m
   e 4·86 m f 7·24 m g 9·45 m h 8·17 m
3. a 6·27 m b 4·95 m c 2·61 m d 4·78 m
   e 8·94 m f 1·34 m g 7·16 m h 5·48 m
4. a 3·1 km b B, 6·4 km c 22·1 km

3:06

1. a 95 mm. Three of the measurements equal 95 mm and another measurement has only a 1 mm difference.
   b 19 cm 6 mm. This measurement is about 10 cm longer than the other measurements.
2. Answers will vary.
3. Answers will vary.

3:07

1. a 14 cm b 12 cm c 9 cm
2. a 20 cm b 20 cm c 19 cm d 15 cm
3.

Length	Width	Perimeter
6 cm	4 cm	20 cm
9 cm	7 cm	32 cm
8 cm	5 cm	26 cm
7·5 cm	4·5 cm	24 cm
6·5 cm	3·5 cm	20 cm

4. a 32 cm b 30 m c 28 cm d 24 km
5. a 16 cm b 26 m c 28 cm
   d 18 km e 19 m f 38 cm
6. shape **d**

3:08

1. a 24 m b 24 m c 28 m d 21 cm e 24 m f 14 m
2. a 30 cm b 48 m c 21 cm d 42 km
3.

	Shape	Length of Sides	Perimeter
a	Square	6·4 cm	25·6 cm
b	Regular Hexagon	9 cm	54 cm
c	Regular Pentagon	4·9 cm	24·5 cm
d	Regular Octagon	12 m	96 m
e	Regular Heptagon	10·2 cm	71·4 cm

f shape **d**

3:09

1. a 13 h b 8 h c 6 h 50 min d 5 h 40 min e 23 h f 5 h 25 min g 21 h 58 min h 12 h 46 min i 15 h 28 min j 24 h
2. a 5:39 pm b 8:28 am c 6:03 pm d 12:35 am
3. a 10:07 am b 12:31 pm c 10:39 am d 12:59 pm e 9:46 am f 1:48 pm g 11:49 am
4. a 32 minutes b 28 minutes c 37 minutes d 31 minutes

3:10

1. a 25 minutes b 19 minutes c 11:54 am d 12:09 pm e 11:27 am f 9 minutes
2. a 07:15 b 13:32 c 04:50 d 11:13 e 17:21 f 18:48 g 14:19 h 20:03
3. a 6:33 am b 10:19 pm c 10:04 am d 7:39 pm e 11:24 am f 7:44 am g 1:56 pm h 8:34 pm

3:11

1. a 10 cm^2 b 12 cm^2 c 9 cm^2 d 12 cm^2 e 4 cm^2
2.

	Length in cm	Width in cm	Area in cm^2
a	5	3	15 cm^2
b	4	4	16 cm^2
c	5	6	30 cm^2
d	6	5	30 cm^2
e	4	6	24 cm^2
f	7	2	14 cm^2
g	8	3	24 cm^2
h	5	4	20 cm^2

3. a 40 cm^2 b 30 cm^2 c 32 cm^2 d 21 cm^2 e 24 cm^2 f 60 cm^2

3:12

1. a 9 m^2 b 12 m^2 c 8 m^2 d 15 m^2 e 4 m^2
2. a 30 m^2 b 70 m^2 c 56 m^2 d 72 m^2
3.

	Length in m	Width in m	Area in m^2
a	1	7	7
b	2	6	12
c	3	5	15
d	4	4	16
e	6	6	36
f	7	5	35
g	8	4	32
h	9	3	27

i square

3:13

1. a 4 cm b 8 cm c 12 cm d 16 cm
2. a 2 b 3 c 4
3. a 1 cm^2 b 4 cm^2 c 9 cm^2 d 16 cm^2
4. a 4 b 9 c 16
5. a A = 21 m^2, P = 22 m b A = 18 m^2, P = 22 m c A = 30 m^2, P = 24 m d A = 18 m^2, P = 22 m
6. The volume is multiplied by 8.

3:14

1. a 6 b 3 c 18
   d Answers will vary. The box is a rectangular prism and is easy to stack and carry. It has as much of its space used as possible.
2. Answers will vary.

Number of Levels	Cans in the Length	Cans in the Width	Number of Cans
1	3	6	18
1	2	9	18
2	3	3	18
1	18	1	18

3. Answers will vary.
4. Answers will vary.

3:15

1 a cm^3 b m^3 c m^3 d cm^3 e cm^3 f cm^3
2 a 27 cm^3 b 16 cm^3 c 36 cm^3 d 36 cm^3
3 a 128 m^3 b 486 m^3 c 324 m^3 d 768 m^3

3:16

1 a 80 m^3 b 192 m^3 c 64 cm^3 d 336 cm^3
e 168 cm^3 f 160 cm^3
2 a 120 cm^3 b 48 cm^3 c 200 cm^3 d 54 cm^3
e 36 cm^3 f 50 m^3 g 160 m^3 h 180 m^3
3 a 18 blocks b 2 times

3:17

1

	Length	Width	Height	Volume
a	8 cm	3 cm	2 cm	48 cm^3
b	4 cm	3 cm	3 cm	36 cm^3
c	5 cm	5 cm	3 cm	75 cm^3
d	5 cm	2 cm	3 cm	30 cm^3
e	7 cm	2 cm	2 cm	28 cm^3
f	4 cm	3 cm	2 cm	24 cm^3
g	12 cm	2 cm	2 cm	48 cm^3

2 a 18 cm^3 b 8 cm^3 c 20 cm^3 d 16 cm^3

3:18

1 a 6 m^3 b 30 m^3 c 15 m^3 d 20 m^3 e 75 m^3
f 34 m^3 g 300 cm^3 h 76 cm^3
2 a m^3 b cm^3 c m^3 d m^3 e cm^3 f m^3
3 a cm^3 b cm^3 c m^3 d m^3 e m^3 f cm^3
g cm^3 h m^3 i m^3
4 a 36 m^3 b 48 m^3 c 40 m^3 d 120 m^3
e 63 m^3 f 48 m^3

3:19

1 a 0·439 L b 0·835 L c 0·146 L d 0·340 L
e 0·207 L f 0·067 L
2 a 641 mL b 793 mL c 145 mL d 839 mL
e 95 mL f 26 mL
3 a 2·346 L b 4·475 L c 3·149 L d 4·207 L
e 1·947 L f 6·503 L
4 a 3104 mL b 5624 mL c 2891 mL
d 6516 mL e 1487 mL f 7235 mL
g 8705 mL h 3065 mL i 9094 mL
5 600 mL (E), 955 mL (F), 375 mL (D), 0·4 L (C), 0·55 L (G), 635 mL (H), 0·25 L (A), 0·175 L (B)

3:20

1 a 1 kL b 9 kL c 4 kL
d 15 kL e 36 kL f 42 kL
2 a 3·5 kL b 13·57 kL
c 29·591 kL d 76·208 kL
e 64·25 kL f 812·75 kL
3 a 4 ML b 14 ML c 53 ML
d 50 ML e 725 ML f 491 ML
4 a 6·5 ML b 39·8 ML
c 32·629 ML d 94·25 ML
e 640·75 ML f 371·423 ML
5 a 1000 L b 3000 L c 6300 L
d 29450 L e 32513 L f 4662 L
6 a 3000 kL b 17000 kL
c 8300 kL d 28240 kL
e 87560 kL f 97465 kL
7 a 2·24 kL b 68000 ML

3:21

1 a 20 cm^3 b 95 cm^3 c 50 cm^3 d 65 cm^3
e 140 cm^3 f 650 cm^3 g 825 cm^3 h 975 cm^3
2 a 45 mL b 75 mL c 90 mL d 60 mL
e 350 mL f 590 mL g 750 mL h 980 mL
3 a 3000 cm^3 b 5000 cm^3 c 4000 cm^3
d 9000 cm^3 e 10000 cm^3 f 12000 cm^3
g 15000 cm^3 h 20000 cm^3
4 a 6 L b 8 L c 2 L d 7 L
e 9 L f 10 L g 11 L h 16 L
5 a 2 L b 450 mL c 3 L

3:22

1 a 0·457 L b 0·379 L c 0·955 L d 0·157 L
e 0·058 L f 0·734 L g 0·029 L h 0·005 mL
2 a 216 mL b 351 mL c 625 mL d 475 mL
e 750 mL f 500 mL g 800 mL h 150 mL
3 a 1·255 L b 3·005 L c 2·025 L
d 4·675 L e 0·247 L f 0·075 L
4 a 2104 mL b 1768 mL c 4500 mL d 3430 mL
e 7230 mL f 5800 mL
5 a 355 cm^3 b 985 cm^3 c 600 cm^3
d 1256 cm^3 e 1470 cm^3 f 2825 cm^3

3:23

1 a 35 mL b 40 mL c 95 mL d 65 mL
e 100 mL f 125 mL g 500 mL h 750 mL
2 a 25 cm^3 b 80 cm^3 c 45 cm^3 d 75 cm^3
e 60 cm^3 f 100 cm^3 g 30 cm^3 h 90 cm^3
i 250 cm^3 j 850 cm^3 k 425 cm^3 l 950 cm^3
3 a 1 L b 3 L c 5 L d 8 L e 4 L
f 6 L g 9 L h 7 L

4 a 1000 cm^3 b 6000 cm^3 c 9000 cm^3 d 10 000 cm^3
e 8000 cm^3 f 5000 cm^3 g 12 000 cm^3 h 15 000 cm^3

5 100 cm^3 (or 100 mL)

6 600 mL

7 500 mL

3:24

1 a 26 years b 25 years c 44 years
d 60 years e 87 years

2

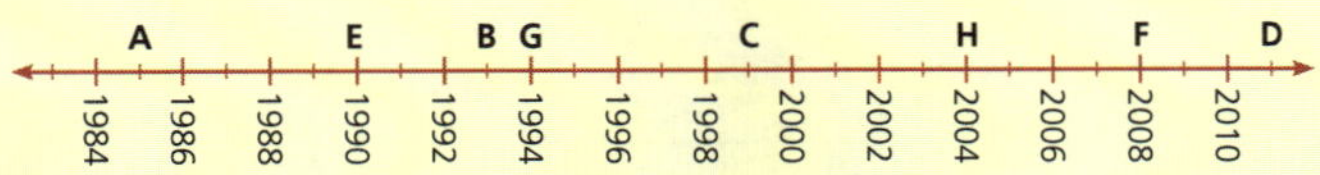

3:25

1 a 3000 kg b 6000 kg c 10 000 kg d 7000 kg
e 5000 kg f 9000 kg g 15 000 kg h 20 000 kg

2 a 4 t b 6 t c 2 t d 8 t
e 11 t f 16 t g 25 t h 40 t

3 a 2750 kg b 5375 kg c 3495 kg d 6550 kg
e 9625 kg f 4125 kg g 1500 kg h 3500 kg
i 7500 kg

4

Kilograms	4 kg	7 kg	9 kg	3·5 kg	6·5 kg	8·5 kg
Grams	4000 g	7000 g	9000 g	3500 g	6500 g	8500 g

5 a F b B or A c A d C or B e D f E

3:26

1 a 6000 kg b 3500 kg c 9500 kg d 23 000 kg
e 7250 kg f 5750 kg g 11 500 kg h 16 000 kg

2 a 3 t b 6·5 t c 8·75 t d 10·5 t
e 4·5 t f 7·25 t g 9·5 t h 5·75 t

3

a	1 t 965 kg	1·965 t
b	2 t 645 kg	2·645 t
c	1 t 375 kg	1·375 t
d	5 t 697 kg	5·697 t
e	3 t 125 kg	3·125 t
f	7 t 850 kg	7·85 t
g	1 kg 375 g	1·375 kg
h	1 kg 962 g	1·962 kg
i	2 kg 395 g	2·395 kg
j	3 kg 625 g	3·625 kg
k	4 kg 625 g	4·625 kg
l	5 kg 835 g	5·835 kg

4 a 6 L b 9 L c 5 L d 10 L
e 350 mL f 750 mL g 825 mL h 125 mL

5 a 3 kg b 7 kg c 12 kg d 8 kg
e 650 g f 375 g g 925 g h 635 g

6

	Mass	g	kg	t
a	a frog	✓		
b	a bus			✓
c	a boy		✓	
d	a ferry			✓
e	a woman		✓	
f	a plane			✓

3:27

1 a 357 g b 624 g c 506 g d 975 g e 167 g
f 719 g g 412 g h 832 g i 235 g

2 a 1·359 kg b 1·725 kg c 3·403 kg d 2·635 kg
e 5·943 kg f 7·875 kg g 4·264 kg h 9·517 kg
i 6·125 kg

3 a 2124 g b 5275 g
c 3506 g d 8407 g
e 1327 g f 6712 g
g 4851 g h 7963 g
i 9615 g

4

a	1 t 357 kg	1·357 t	g	4 t 125 kg	4·125 t
b	1 t 945 kg	1·945 t	h	6 t 456 kg	6·456 t
c	1 t 625 kg	1·625 t	i	5 t 364 kg	5·364 t
d	2 t 375 kg	2·375 t	j	7 t 056 kg	7·056 t
e	3 t 285 kg	3·285 t	k	9 t 835 kg	9·835 t
f	5 t 649 kg	5·649 t	l	8 t 224 kg	8·224 t

5 Answers will vary.

3:28

1 a mg b mg c kg d mg
e t f mg g g h g

2 a 2000 mg b 23 000 mg c 9 g
d 10 000 mg e 1·8 g f 5·6 g
g 5000 mg h 8·923 g i 3·402 g
j 4000 mg k 9 g l 97 000 mg

3 a 7000 g b 2·37 kg c 2000 kg
d 5 t e 5·419 g f 65 000 mg
g 12 500 g h 4·823 kg i 45 000 kg
j 1·446 t k 2·406 g l 10 000 mg

4 a 33 mg b 30 mg c 6 mg d 5 g

3:29

1 a 5 ha b 9 ha c 23 ha d 48 ha e 3·6 ha
f 9·5 ha

2 a m^2 b ha c cm^2 d m^2 e ha f ha g ha h ha i cm^2

3 a 1 ha b 2 ha c 6 ha d 9 ha e 7 ha f 5 ha g 11 ha h 13 ha i 18 ha

4 a 40 000 m^2 b 80 000 m^2 c 30 000 m^2 d 100 000 m^2 e 70 000 m^2 f 90 000 m^2 g 120 000 m^2 h 160 000 m^2 i 240 000 m^2

3:30

1 a 100 ha b 300 ha c 400 ha d 900 ha e 600 ha f 500 ha g 1 100 ha h 2 000 ha i 1 200 ha

2 a 2 km^2 b 6 km^2 c 10 km^2 d 12 km^2 e 8 km^2 f 17 km^2 g 30 km^2 h 25 km^2

3 a m^2 b ha c m^2 d km^2 e ha f ha g m^2 h ha i km^2

4 a Netherlands b France c 23 830 km^2 d 56 616 km^2 e 584 061 km^2 f 908 868 km^2 g France

3:31

1

	WST	CST	EST
a	06:30	08:00	08:30
b	06:45	08:15	08:45
c	11:45	13:15	13:45
d	17:20	18:50	19:20

	WST	CST	EST
e	10:25	11:55	12:25
f	08:20	09:50	10:20
g	14:40	16:10	16:40
h	15:05	16:35	17:05

	WST	CST	EST
i	23:10	00:40	01:10
j	18:13	19:43	20:13
k	09:17	10:47	11:17
l	23:55	01:25	01:55

2 a 8:07 am b 10:07 am c 9:37 am d 11:33 am e 10:03 am f 12:03 pm g 1:54 am h 1:24 am i 11:54 pm

3 a

b

c

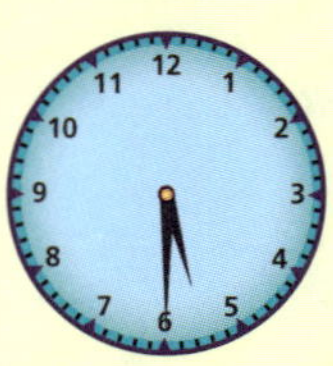

d

e

4:01

1 a B b A c C d D

2 a 0 b 1 c 0 d 5 e 8

3 a 11 b 64

4

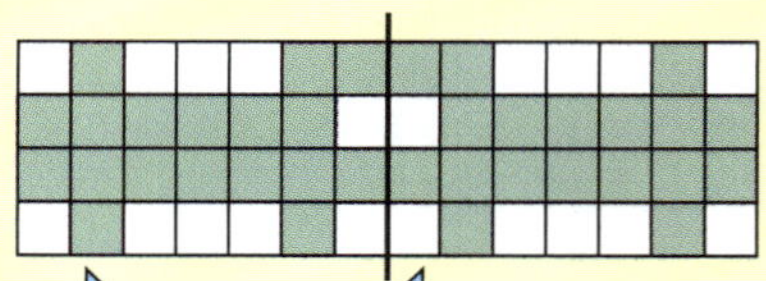

5 a

b

6

	Number of Surfaces	Number of Vertices	Number of Edges
A	5	5	8
B	5	6	9
C	2	1	1

7 a rectangular pyramid b triangular prism c cone d B (triangular prism)

4:02

2 a E b B c D d A e C f F

3 E, B, D, A, C, F

4 a 2 b 3 c 9 d 4

5 a about $1\frac{1}{2}$ b almost 5 c about $2\frac{1}{2}$

4:03

1 a 60° b 120° c 40° d 110° e 70° f 160°

4:04

1 a acute b straight c revolution d reflex e acute f reflex g right h obtuse i obtuse j reflex k reflex l straight m actute n acute o obtuse p revolution

2 a 40°, acute b 120°, obtuse c 160°, obtuse d 50°, acute e 110°, obtuse f 135°, obtuse g 95°, obtuse h 85°, acute

4:05

1

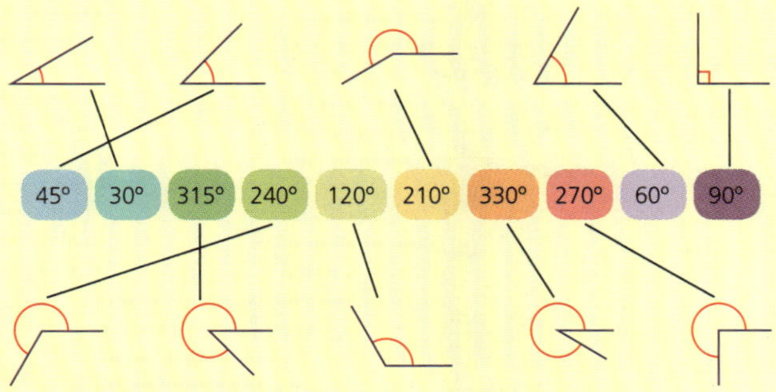

2 **a** 60°

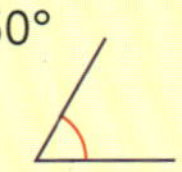

b 135°

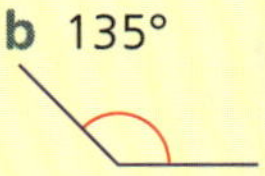

c 150°

d 210°

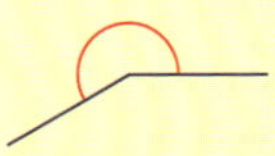

e 300°

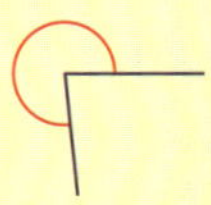

f 240°

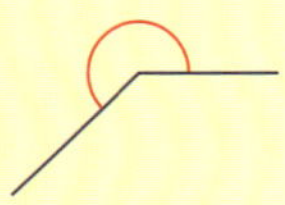

4:06

1 **a** 150°, 30°, 180° **b** 55°, 125°, 180°

2 **a** 155° **b** 35° **c** 90° **d** 25° **e** 75° **f** 20° **g** 87° **h** 96° **i** 47°

4:07

1 **a** 90°, 270°, 360° **b** 160°, 132°, 68°, 360°

2 **a** 105° **b** 170° **c** 40° **d** 155° **e** 30° **f** 120° **g** 115° **h** 116° **i** 49° **j** 65°

4:08

1 **a** 95°, 85°, 95°, 85°, 3, 4 **b** 35°, 145°, 35°, 145°, 3, 4

2 **a** 104° **b** 135° **c** 46° **d** 93°

3 **a** 122° **b** 325° **c** 107° **d** 220° **e** 142° **f** 242° **g** 158° **h** 160°

4:09

1 **a** F **b** A **c** E **d** B

2 **a** A, C, E, G **b** A, C **c** A, E

3 **a** L **b** Q (or O) **c** K (or O) **d** S
e P or R (T, H, J, K, O and Q are special trapeziums.)
f I

4 yes

5 yes

4:10

1 **a** home **b** shops **c i** D3 or E3 **ii** H2 or I2
d The shortest paths will be coloured.
e i about 160 m **ii** about 200 m

2 **a** flower **b** bird **c** turtle **d** P **e** E **f** N **g** D **h** B **i** H

3 **a** bus **b** rocks **c** bus **d** tree **e** girl, house **f** bus, car

4:11

1 **a** southerly **b** westerly **c** easterly

2 **a** north-west **b** south-east **c** south-west

3 **a** west **b** east **c** south

4 east to west

5 **a** 90° **b** 90° **c** 180°

6 North, as the sun is at its highest point and we are well into the Southern Hemisphere.

7 **a**

b

c

d

4:12

1 **a** south-east **b** north-west

2 **a** Caleb **b** Luke **c** Alana **d** Isabel **e** west **f** north **g** north-west **h** south-east **i** south-west **j** north-east

3 **a** Northern Territory (NT)
b Queensland (Qld)
c Western Australia (WA)
d South Australia (SA)
e New South Wales (and Victoria)
f Queensland (Qld)
g Western Australia (WA)

4

4:13

1 **a** Cup Island **b** Tooth Island **c** Cup Island **d** Shell Island **e** Star Island **f** Tooth Island **g** Crab Island

2 **a** Cup Island **b** Tooth Island **c** Palm Island

3 **a** 4 km **b** 2 km **c** 2 km **d** 3 km

4 **a** north **b** south **c** Palm Island

5 **a** 7 **b** 6:13 pm

 • *AUSTRALIAN SIGNPOST MATHS 6* • ISBN 978 1 4886 2193 2

4:14

1. Answers will vary.
   a about 480 km b about 390 km
   c about 320 km d about 230 km
2. 

   a E2 b I4 c E4 d B6 (or C6)
   e C5 f D3
3. a Warracknabeal, Ouyen
   b Ballarat, Bendigo or Melbourne
   c Geelong, Warrnambool, Portland
   d Warrnambool, Geelong, Ballarat
   e Bendigo, Echuca, Swan Hill
4. Answers will vary.
   a about 270 km b about 410 km
   c about 440 km d about 570 km
5. a E b NW c SE or E
   d SE e SW f NE
6. 3b, 3a, 3c, 3d and 3e

4:15

1. a Iran b Tanzania c Ethiopia
   d South Africa e Algeria f Niger
   g Madagascar h Saudi Arabia
2. a J11 or K11 b G14 c E14 or F14
   d C19 or D19
3. L13, M14, N14 (or N13)
4. a 1900 km b 1500 km c 1600 km d 800 km
5. a 1800 km b 1200 km c 1200 km d 700 km
6. Approximately:
   a 4950 km b 2400 km c 2850 km
   d 2400 km e 4200 km f 6450 km

4:16

1. a A, B b B, C, D c A, C
2. a b c
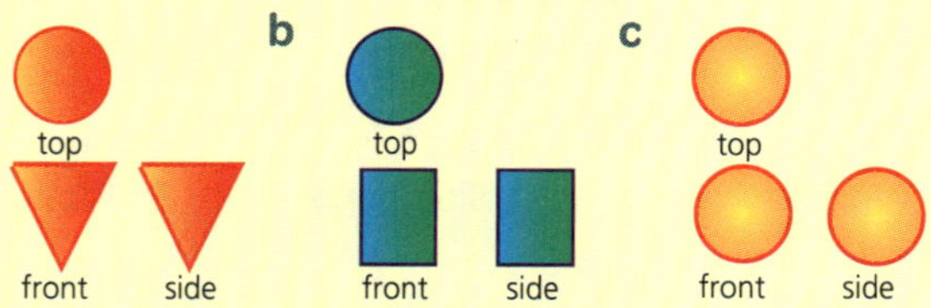

4:17

1. 

2. a A, B, C, H, J, K b D, E, F, G, I, L
   c A, D, E, K, L d B, F, I, J
   e A, B, C, E, H, J, K, L f C, K
3. Answers will vary.
   a b c
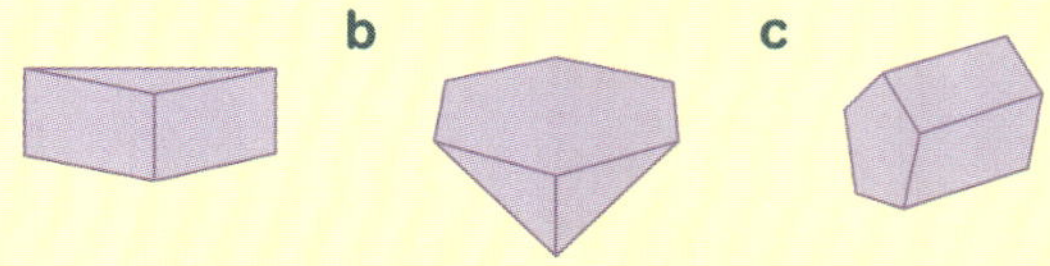

4:18

1. 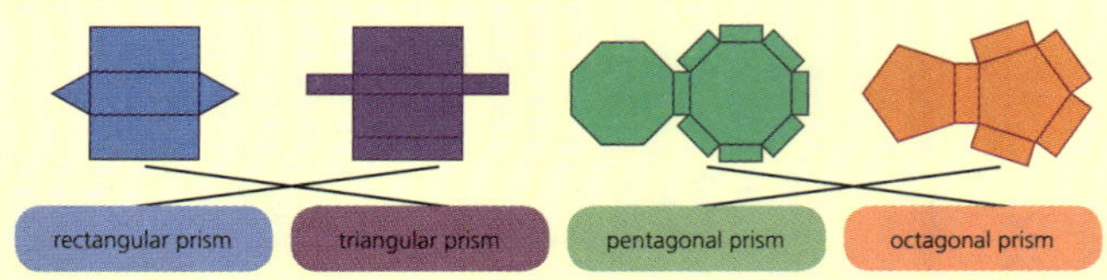

2. a 5 b 6 c 9 d triangular prism

4:19

1. 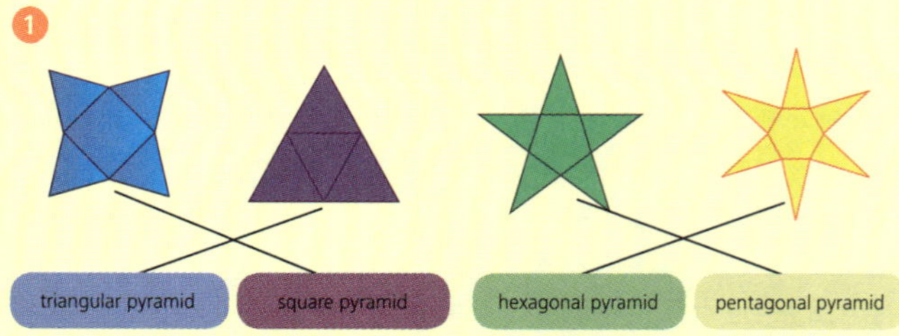

2. a 5 b 5 c 8 d rectangular pyramid

4:20

1. a N b V c H d C e D
   f B g Y h F i R j Q
   k O l K
2. a (1, 5) b (4, 4) c (7, 4)
   d (0, 3) e (3, 3) f (3, 2)
   g (2, 1) h (5, 3) i (6, 2)
   j (7, 3) k (10, 3) l (10, 1)
3. a (7, 2) b (0, 2) c (4, 0) d (5, 3)
   e (10, 4) f (3, 2)

4:21

1. **a** A **b** C **c** L **d** P **e** O **f** T **g** H **h** M **i** Q **j** S
2. **a** (0, 3) **b** (2, 0) **c** (–3, 5) **d** (–4, –6) **e** (–8, 8) **f** (–7, 4) **g** (–7, 1) **h** (–4, 0) **i** (6, –7) **j** (2, –8)
3. **a** (4, 7) **b** (–8, 5) **c** (–4, –1) **d** (–4, –4) **e** (6, –5) **f** (2, –4) **g** (–7, 0) **h** (–1, 4)
4. **a** 2 **b** 3 **c** 12 **d** 12 **e** 13 **f** 5

4:22

1. **a** Spain **b** Sweden **c** Italy **d** Morocco
2. **a** B3 **b** L2 **c** M3 or N4 **d** K5
3. K7, K8, L7, L8
4. H6, H7, I6, I7, I8
5. **a** 500 km **b** 700 km **c** 600 km
6. **a** 200 km **b** 800 km **c** 500 km
7. **a** 900 km **b** 1 100 km **c** 4000 km **d** 2 100 km

4:23

1. **a** about 2 km **b** Hadrian Av, Keyworth Dr, Blacktown Rd, Vesuvius St, Columbia Rd, Emerald Rd
   c Answers will vary. (About half an hour)
2. **a** 1 850 km **b** 3 400 km **c** 3 000 km **d** 3 hours **e** 6 hours **f** 5 hours
3. Answers will vary. Turn into Lucretia Rd, take the second road on the right (Athabaska), then first left (Lawrence St), then second left (Paroo Pl).

4:24

1. **a**

Each tile is flipped onto the tile on its right.

   b
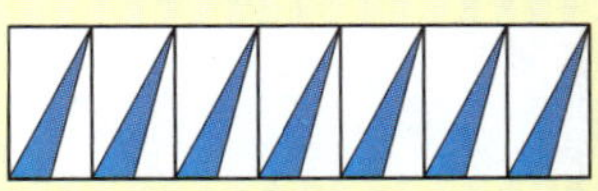
Each tile is slid onto the tile on its right.

   c
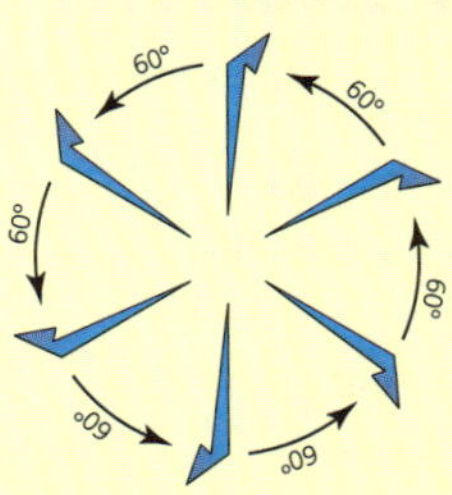

Each shape is turned 60° anticlockwise.

   Descriptions will vary.
2. **a** rotation of 120° **b** rotation of 180° **c** rotation of 240° **d** rotation of 60° **e** translation of 5 cm right **f** no change **g** translation of 6 cm right

4:25

1. Answers will vary.
   a The shape and size are the same but it is facing the opposite direction and has moved to the right.
   b The shape, size and orientation are the same but it has been moved to the right.
   c The shape and size are the same but it has been turned around the point shown. It has not been flipped.
2. **a** translation of 3·8 cm to the right; reflection about a line; reflection about a line
   b translation of 3·8 cm to the right; translation of 1·9 cm to the right; translation of 1·9 cm to the right
   c translation of 3·8 cm to the right; rotation about a point; rotation about a point
   d yes
3. The pattern will be copied.
   The transformations used are: a reflection, followed by a rotation, followed by a translation, followed by a reflection.

4:26

1. **a** flip **b** slide **c** turn

2–4 Answers will vary.

4:27

1. **a**
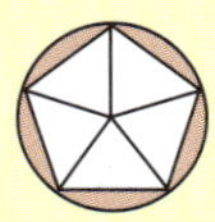
   b
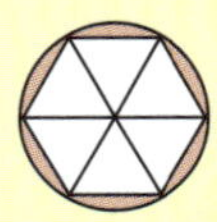
   c
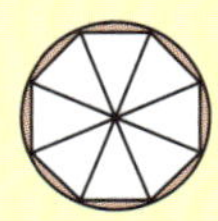
2.
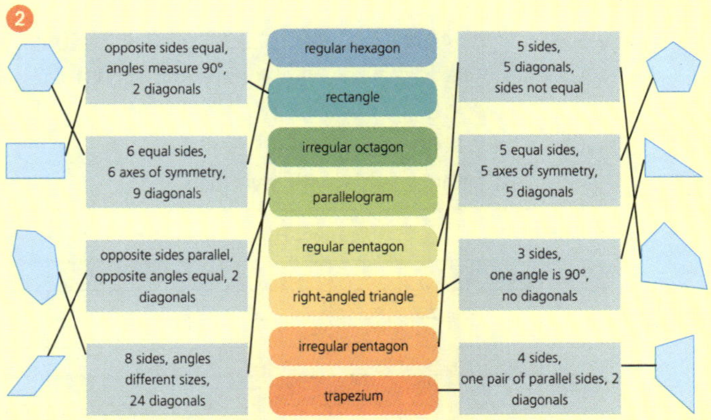

4:28

1. Answers will vary. They both have a rectangular base. A rectangular cross-section can be obtained from both solids. The pyramid's cross-section changes size. They differ in the number of faces, edges and vertices.
2. The 3D objects will be drawn.
3. **a** a cylinder and a sphere **b** a cone and a cylinder
   c part of a cone and a sphere
   d two cylinders and a cone
   e a cone, a cylinder and a hexagonal prism

4:29

1 a cylinder b sphere

2 a D b B c A d C e E
f in a circle g in a straight line

3

Shape	Number of Edges	Number of Surfaces	Number of Corners	Does It Roll?	Does It Have a Curved Surface?
cone	1	2	1	Yes	Yes
cylinder	2	3	0	Yes	Yes
sphere	0	1	0	Yes	Yes
cube	12	6	8	No	No

4 a cylinders
b

Stack	Height Order	Width Order	Value
5c	6	6	50c
10c	5	4	$1
20c	4	2	$2
50c	3	1	$5
$1	2	3	$10
$2	1	5	$20

5:01

1 a January b April c $1 200 d $350
e i April and May ii January and March

2 a Chris b 18 c 13 d 14
e David f 100

3 a The smallest total possible with two dice is 2.
b

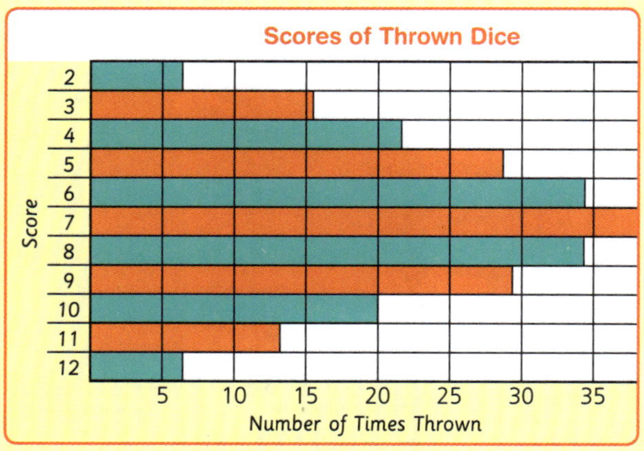

5:02

1 a 8 b 21 c weeks 2 and 3 d 12
e 142 f week 6
g The same passage was used.
h 9 or less. Week 8 had 10 errors and the number is still decreasing.
i Conveys more information than most graphs. Shows trends. Easy to read.

2 a February b about 570 c 100
d 30 e March, April, May f About 250

3 a

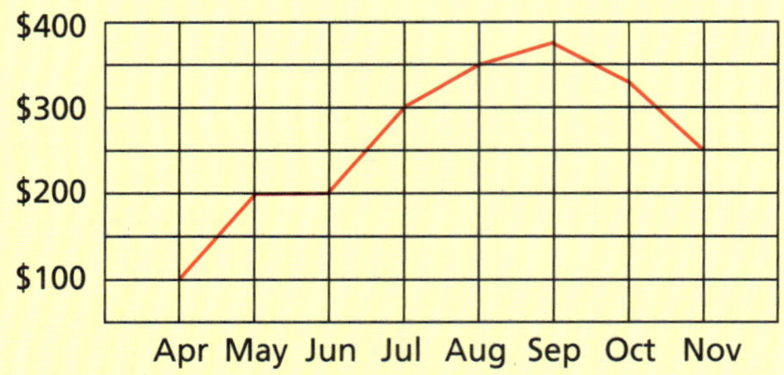

b It will decrease because of Christmas spending.

5:03

1 22%, or $\frac{22}{100}$ or $\frac{11}{50}$

2 30%, or $\frac{30}{100}$ or $\frac{3}{10}$

3 10·5%

4 a 36% b 49%

5 I. Speakalot

6 men

7 100

8 a percentage of total for I. Speakalot
b

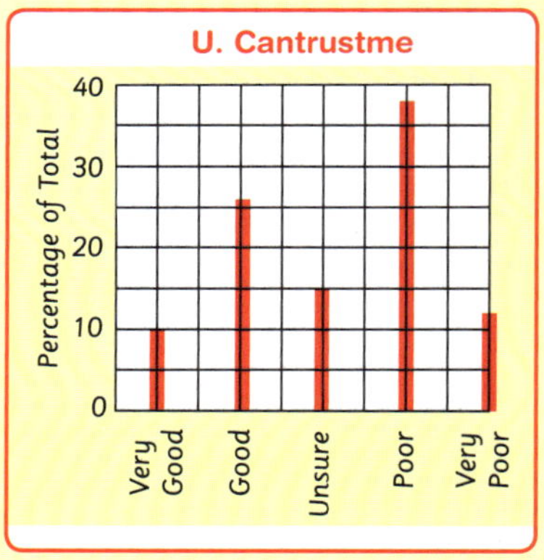

c good d very good

9 a 12 b 3 c 21

5:04

1 a i 55 000 (meals) ii 85 000 (meals)
b 220 000 (meals) c The categories are obvious.
d yes, your choice
e your choice (drought, famine, war etc.)
f North America, Australia, Antarctica

2 a kangaroo b 83 c 17
d The gate (卌) is like a picture that stands for 5.

3 a i 16 ii 12 iii 6 iv 26

b

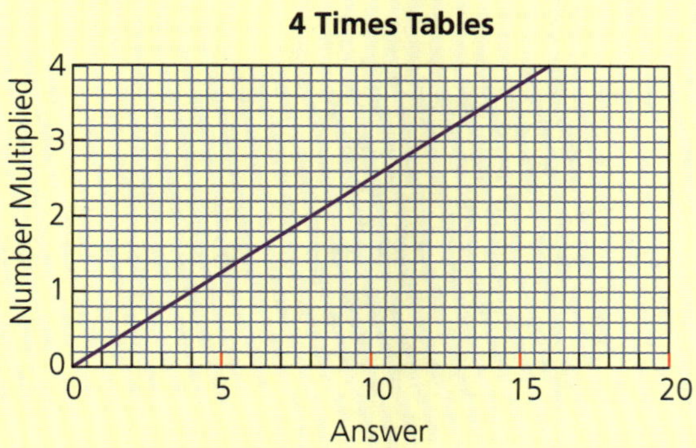

c **i** 12 **ii** 14 **iii** 6 **iv** 9

5:05

1. Answers are approximate and will vary.
   a 50% **b** 17% **c** 0% **d** 100% **e** 70–100%
2. **a** 0·5 **b** 0·6 **c** 0·1 **d** 0·3 **e** 1
3. **a** 0·4 **b** 0·2 **c** 0·5 **d** 0·1 **e** 0
4. **a** 0·1 **b** 0·2 **c** 0·4 **d** 0·6 **e** 0

5:06

1. **a** C **b** A **c** D **d** B
2. **a** D **b** B **c** A **d** C
3. **a** A **b** A **c** C **d** B and C
4. **a** 0·2 **b** 0·1 **c** 0·4 **d** 0·3 **e** 0
5. **a** 7 **b** 2 or 12 **c** 2, 12 **d** 3, 11 **e** 4, 10
   f 5, 9 **g** 7

5:07

1. **a** **i** $\frac{1}{10}$ or $\frac{100}{1000}$ **ii** $\frac{5}{10}$ or $\frac{500}{1000}$ or $\frac{1}{2}$
   iii $\frac{4}{10}$ or $\frac{400}{1000}$
   b **i** 100 mL (or 100 cm^3) **ii** 500 mL (or 500 cm^3) **iii** 400 mL (or 400 cm^3)
   c Liquid in the jug. **d** Yes. The jug is a vertical bar.
2. **a** sector (or pie) graph **b** Flavour Preferred
   c **i** $\frac{1}{2}$ **ii** $\frac{1}{4}$ **d** **i** 50% **ii** 25%
   e **i** 120 **ii** 60 **iii** 60
3. **a** encyclopedias **b** 100 mm **c** **i** 20 **ii** 50
   d No. All books do not have the same mass.
   e The graphs drawn will differ.

5:08

1. 96°, 144°, 120°

Naomi's Actions

2. 360° divided by 60 = 6.
   Each minute is represented by 6° at the centre.

Calculating the angles at the centre of the graph:

Eating Lunch	Playing Handball	Talking to Friends
Angle = 15 x 6°	Angle = 20 x 6°	Angle = 25 x 6°
= 90°	= 120°	= 150°

Luke's Lunchtime

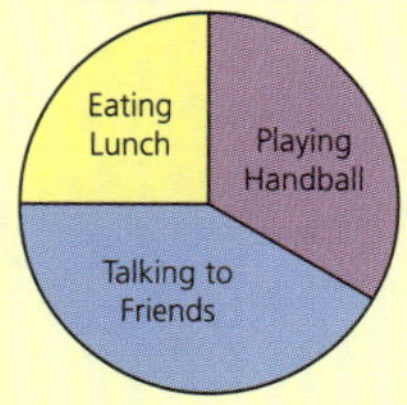

3. The angle used for horses should be 54°.
   The angle used for cows should be 205·2°.
   The angle used for goats should be 100·8°.

5:09

1. **a** 0·5 (or $\frac{1}{2}$) **b** 0 **c** 0·1 **d** 0·3
   e 0·3 **f** 300
   g

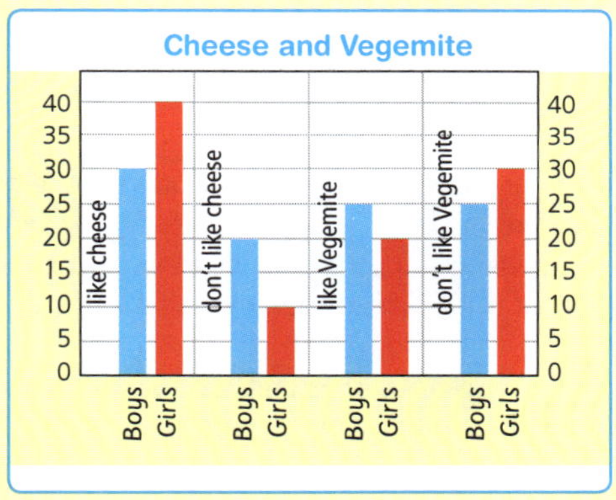

2. **a** 96 **b** 76
   c Answers will vary. Different people would be present with varied preferences for ice-cream flavours.
3. **a** 30 **b** 40 **c** 10 **d** 20
   e Answers will vary. It is unlikely to be exactly the same. From looking at the bag, the most likely result is 4 red, 3 green, 2 blue and 1 yellow. Results would tend to be like this.

5:10

1. **a** $300 **b** $600 **c** June
   d April
   e

Money Earned						
Month	Jan	Feb	Mar	Apr	May	June
Money Earned	$500	$300	$300	$200	$400	$600

2. **a** 40 **b** 30 **c** Tuesday **d** Friday

e

Customers						
Day	Mon	Tues	Wed	Thurs	Fri	Sat
Number of Customers	30	50	30	40	20	40

3 a

Money Earned				
Month	March	April	May	June
	$2906	$4392	$1896	$879
Rounded to the Nearest $1000	$3000	$4000	$2000	$1000

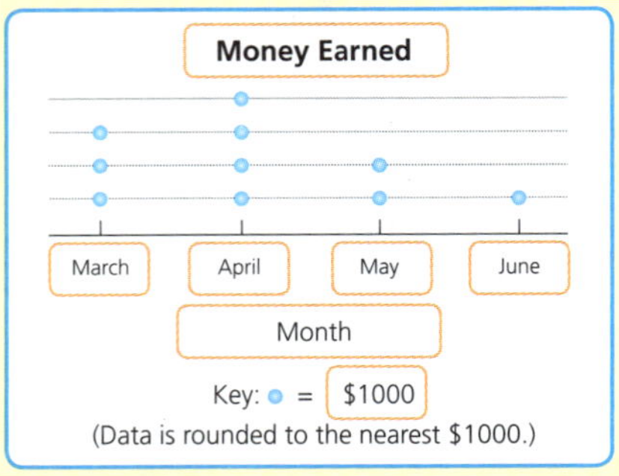

b Answers will vary.

5:11

Answers will vary.

1 People from only four countries are included in the graph. Countries where hunger is more likely are not included.

2 a The broken vertical axis makes the money spent in both years appear to be similar, even though there is a great difference.

b A non-linear scale is used so units on the vertical axis are not regular. The columns give a false impression.

c The whole pie does not represent the whole population about which the claims are made.

3 B

5:12

Answers will vary.

1 The graph does not indicate how many people were asked and so we are inclined to assume many were asked, and to assume that everyone likes maths.

2 a Only a small part of the 'profit' axis is shown. The graph gives the impression that profit is increasing rapidly but over these 3 years, profits are similar. (This is an example of a truncated axis.)

b The symbols are very different in size which gives a false impression.

c The columns should only vary in height, not thickness or depth. BM appears to be producing more cars than ACD.

5:13

1 a head, tail **b** $\frac{1}{2}$
c Answers will vary.

2 a 1, 2, 3, 4, 5, 6 **b** $\frac{1}{6}$
c Answers will vary.

3 Marcus Peter Pieris, Marcus John Pieris, Alan Peter Pieris, Alan John Pieris

4 3%, $\frac{3}{100}$

5 a $\frac{5}{10}$ or $\frac{1}{2}$ **b** $\frac{3}{10}$ **c** $\frac{2}{10}$ or $\frac{1}{5}$ **d** Tom

6 Answers will vary. Naomi is most likely to win.

5:14

1 2, 3, 4, 5, 6, 7, 8, 9, 10, 11, 12

2 a 1 **b** 2 **c** 3 **d** 4 **e** 5 **f** 6
g 5 **h** 4 **i** 3 **j** 2 **k** 1

3 36

4 a $\frac{1}{36}$ **b** $\frac{2}{36}$ **c** $\frac{3}{36}$ **d** $\frac{4}{36}$ **e** $\frac{4}{36}$
f $\frac{6}{36}$ **g** $\frac{5}{36}$ **h** $\frac{5}{36}$

5:15

1 a 18 **b** 28 **c** 12 **d** 42
e 120 **f** 420 **g** 180 **h** 280

2 a 20 **b** 20 **c** 10 **d** 50
e 10% **f** 50% **g** 20% **h** 20%
i 70% **j** 30% **k** 90%
l yes (Answers will vary.)
m Answers will vary. If the purpose was to sample the school, some from each year would be a better sample.

5:16

Answers will vary.

5:17

1 Answers will vary.

2 Trial 1 plus Trial 2. The more times it is trialled, the greater is the likelihood that the answer will reflect the real probability.

5:18

1 Calculated probability: chance of red = $\frac{3}{4}$, chance of blue = $\frac{1}{4}$. Tally will vary.
a $\frac{30}{40}$ **b** $\frac{10}{40}$
c–e Answers will vary.
f yes, no **g** We are calculating the expected outcome.
h Answers will vary.

2 Calculated probability: chance of heads = $\frac{1}{2}$, chance of tails = $\frac{1}{2}$. Tally will vary.
a 50% **b** 10 **c** Answers will vary.

 • *AUSTRALIAN SIGNPOST MATHS 6* • ISBN 978 1 4886 2193 2

5:19

1 **a** 3 **b** 4 **c** 1 **d** $\frac{3}{12}$ or $\frac{1}{4}$

e $\frac{4}{12}$ or $\frac{1}{3}$ **f** $\frac{6}{12}$ or $\frac{1}{2}$

2 **a**

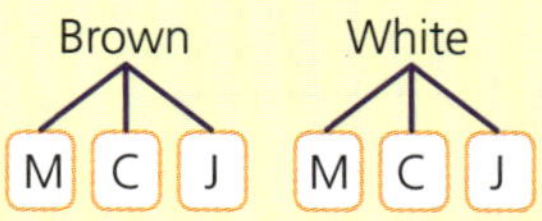

b

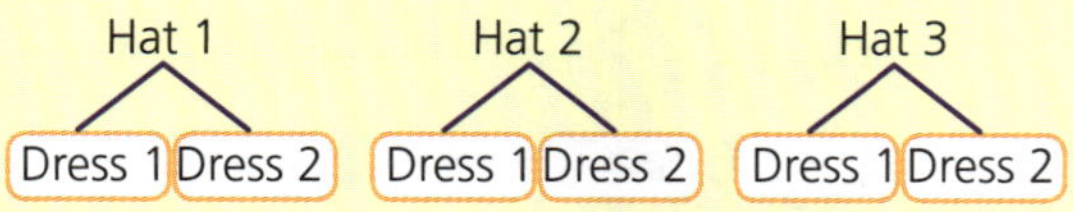

c 3 **d** 2 **e** 4 **f** $\frac{1}{2}$ **g** $\frac{2}{6}$ or $\frac{1}{3}$

h $\frac{4}{6}$ or $\frac{2}{3}$

3 **a** 8 **b** $\frac{4}{8}$ or $\frac{1}{2}$ **c** $\frac{1}{8}$ **d** $\frac{2}{8}$ or $\frac{1}{4}$

e $\frac{6}{8}$ or $\frac{3}{4}$ **f** yes

5:20

1 **a** $\frac{40}{100}$ or 40% **b** $\frac{36}{100}$ or 36%

2 **a** $\frac{3}{8}$ or $37\frac{1}{2}$% or 0·375

b $\frac{1}{8}$ or $12\frac{1}{2}$% or 0·125

3 **a** True. A coin landing on 'heads' has the same chance as a boy being born, so 2 heads from 2 coins is like 2 boys from 2 births.

b False. A 6 has the same chance of being thrown as a 3. They each have one chance out of six.

5:21

1 Answers will vary.
2 Callum
3 Answers will vary.

5:22

1 column (or bar) graph
2 **a** 249 m **b** 236 m
3 Empire State Building. The width is about 130 m.
4 4250 million
5 yes
6 1900
7 3250 million
8 between 1966 and 1983 (approximately)
9 between 1000 and 1600
10 From 1000 to about 1983, about 983 years.
11 The answer will change with time. In 2010 it was 6 900 000 000 correct to the nearest 100 000 000.

Diagnostic Test 1

Number

see page

1 Use numerals to write: 1

a thirty-seven million four hundred and fifty-two thousand one hundred and sixty-nine

b ninety-two million one hundred and five thousand six hundred and seventy-eight

2 Write the value of 5 in: 1

a 2 561 394

b 5 749 890

3 2 3

a $6 \times 6 =$

b $4 \times 4 =$

c $5 \times 5 =$

d 9 = ___ squared

e 25 = ___ squared

Operations

4 30

a 7×9

b 6×8

c 5×6

d 8×7

e 9×6

f 9×9

g 4^2

h 7^2

i 6^2

j List all factors of 16.

5 31

a $63 \div 9$

b $56 \div 8$

c $64 \div 8$

d $72 \div 8$

e $54 \div 6$

f $48 \div 8$

see page

6 32

a
```
  4095
+ 2768
```

b
```
  5361
  2975
  4063
+ 7392
```

c
```
  64357
+ 29865
```

d
```
  $31.75
  $27.60
+ $12.85
```

e
```
  $26.92
  $15.37
+ $33.49
```

f
```
  $19.67
  $13.06
+ $40.79
```

7 33

a
```
  5314
−  967
```

b
```
  9000
−  436
```

c
```
  35216
− 19407
```

8 Find: 34

a $16 - (4 \times 3)$

b $36 - 10 + 4$

c $12 \div 3 \times 5$

d $(6 - 4) \times (16 - 8)$

e $6 + 4 \times 3$

f $(18 - 6) \div 4$ 35

g $(53 + 21) \div 2$

h $9 \times (16 - 7)$

i $60 - (5 + 5)$

j $70 - 4 \times 6$

 • *AUSTRALIAN SIGNPOST MATHS 6* • ISBN 978 1 4886 2193 2

Diagnostic Test 1 (continued)

Measurement

see page

82

9 Write as centimetres:

a 69 mm ☐ **b** 105 mm ☐

Write as milliimetres:

a 78 cm ☐ **b** 11·3 cm ☐

83

10 Write as metres:

a 7 km ☐ **b** 9 km ☐

Write as kilometres:

c 5000 m ☐ **d** 8000 m ☐

Choose the most suitable unit of length (**cm, m, km**) to measure:

e the length of your arm ☐

f the width of Lake Eyre ☐

g the width of a classroom ☐

Find the speed (**km/h**) for each.

h 84 km in one hour ☐

i 360 km in four hours ☐

j 950 km in five hours ☐

Geometry

113

11 **a** How many cubes have been used to build this model? ☐

b How many cubes cannot be seen? ☐

115

12 What is the size of this angle? ☐

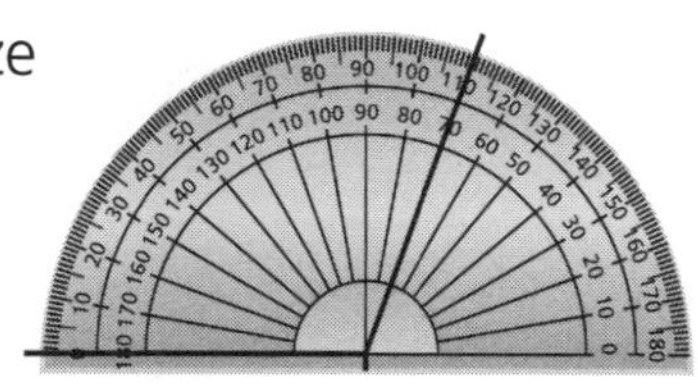

Statistics

see page

142

13 **The Circus Crowd**

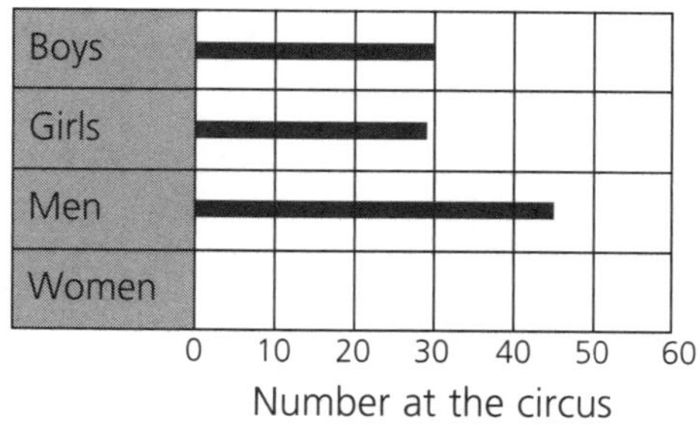

a How many girls were at the circus? ☐

b How many men were at the circus? ☐

c There were 42 women at the circus. Draw in the bar for women.

d What was the total number of people in the crowd? ☐

142

14 What is the tally?

a  ☐

b  ☐

143

15

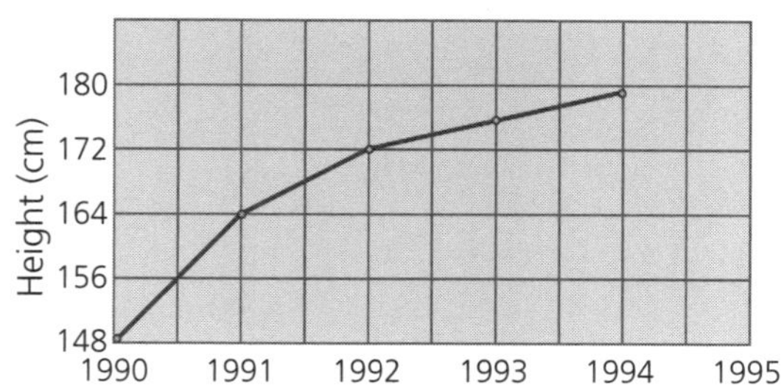

What was Andrew's height in:

a 1990? ☐ **b** 1993? ☐

c 1994? ☐

d Estimate his height in 1995. ☐

 • *AUSTRALIAN SIGNPOST MATHS 6* • ISBN 978 1 4886 2193 2

Diagnostic Test 2

Number

see page

1 What percentage is coloured? 4 5

a ☐ b ☐

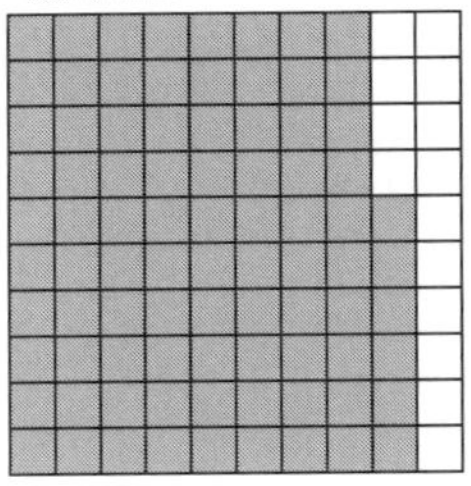
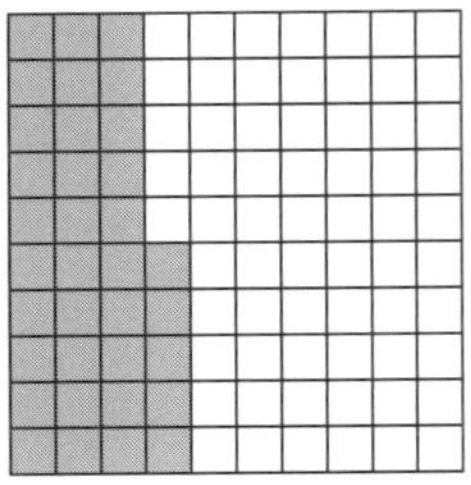

c Complete:

0·15	$\frac{\quad}{100}$	%
0·65	$\frac{\quad}{100}$	%

2 Write the percentage for: 6

a $\frac{65}{100}$ ☐ b $\frac{4}{10}$ ☐

c $\frac{9}{10}$ ☐ d 0·55 ☐

e 0·85 ☐ f 0·7 ☐

Write the decimal for:

g 35% ☐ h 60% ☐

i 95% ☐

3 For each square, colour and write the equivalent percentage. 6

a
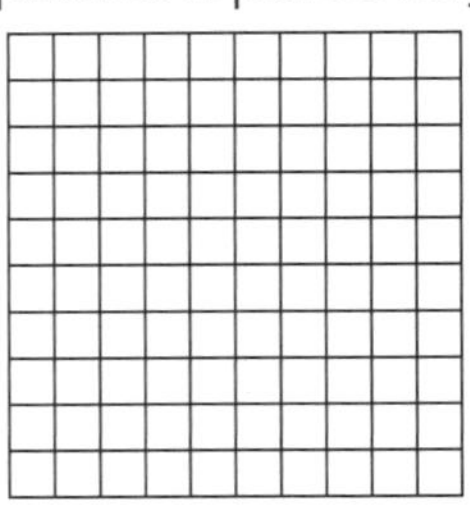
b
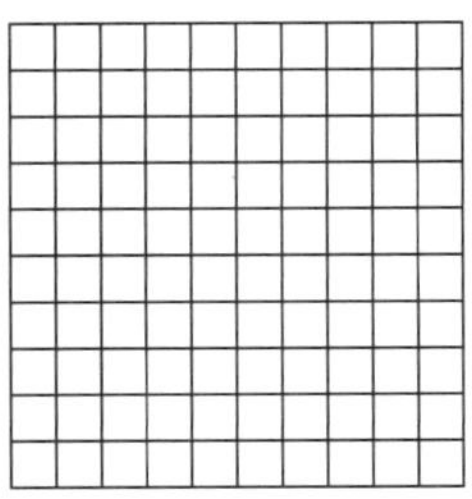

0·82 ☐ 0·05 ☐

Operations

see page

4 36

a 90×6 b 700×8 c 8000×7

5 Use counting on to find: 38

a 47 − 43 ☐ b 69 − 62 ☐

c 33 − 28 ☐ d 52 − 47 ☐

e 91 − 84 ☐ f 72 − 65 ☐

Use a suitable strategy to find:

g 41 − 36 ☐ h 63 − 42 ☐

i 72 − 25 ☐ j 95 − 46 ☐

k 84 − 53 ☐ l 51 − 22 ☐

6 Use the compensation strategy to find: 38

a $86 - 38$ b $173 - 29$ c $294 - 47$

7 37

a 128×5 b 307×9 c 659×8

d $\$6.25 \times 6$ e $\$9.73 \times 5$ f $\$7.08 \times 9$

8 39

a Eight children each had $11.35. How much money did they have altogether? ☐

b A farmer had 3 230 sheep and sold 345. How many did he have left? ☐

Measurement

see page 84

9 Complete the table:

Metres		3950	8654	
Kilometres	5·275			9·85

85

10 Complete the table:

Object	Millimetres	Centimetres	Metres
Length of a pen		13·5 cm	
Thickness of a pen	9 mm		

85

11 Write as metres:

a 3·496 km

b 7·5 km

86

c 947 cm

d 506 cm

e 3 m 42 cm

f 12 m 95 cm

Geometry

116

12

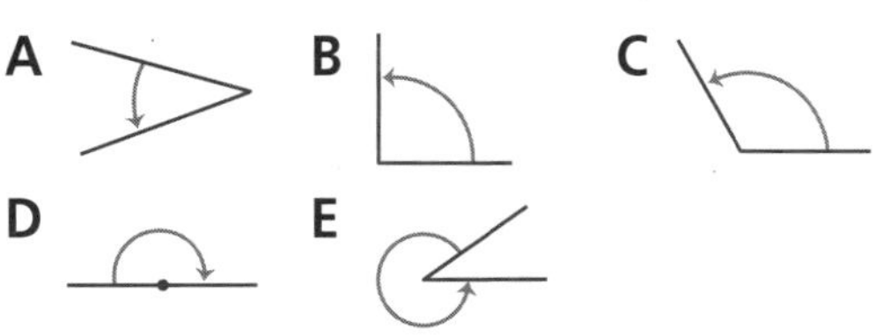

Which angle is:

a a right angle?

b less than a right angle?

c a straight angle?

d more than a straight angle?

e more than a right angle but less than a straight angle?

see page 117

13 Use a protractor to draw this angle.

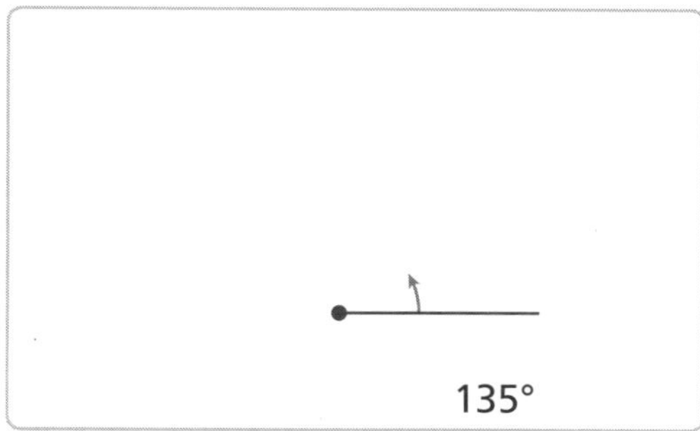

Statistics

144

14

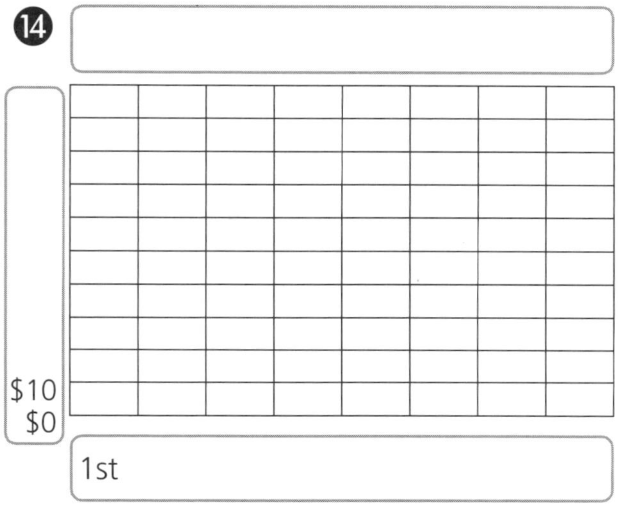

Use the grid to graph the prize money.

1st = $90, 2nd = $70, 3rd = $50,

4th = $20, 5th = $10, 6th = $10,

7th = $5, 8th = $5

145

15 a Find the totals for the tallies.

Ford	𝍸 𝍸 𝍸 𝍸 𝍸 𝍸 𝍸 |	
Holden	𝍸 𝍸 𝍸 𝍸 𝍸 𝍸 𝍸 𝍸 ||||	
Toyota	𝍸 𝍸 𝍸 𝍸 𝍸 𝍸 𝍸 𝍸 𝍸 ||	

Which car was tallied:

b least?

c more than 45 times?

 • *AUSTRALIAN SIGNPOST MATHS 6* • ISBN 978 1 4886 2193 2

Diagnostic Test 3

Number

see page

1 Write the numeral for: 7

80 000 + 5000 + 600 + 20 + 7

2 Write the numeral for: 7

a $(4 \times 10^4) + (9 \times 10^3) + (7 \times 10^2) + (5 \times 10^1) + 8$

b $(6 \times 10^4) + (7 \times 10^3) + (2 \times 10^2) + (4 \times 10^1) + 3$

c 70 000 + 9000 + 600 + 40 + 9

3 Record the temperatures shown. 8

a

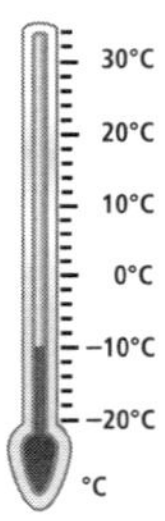

b

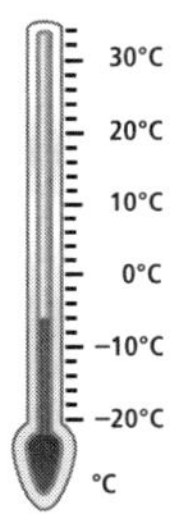

a ☐ °C   b ☐ °C

4 Record the number shown. 8

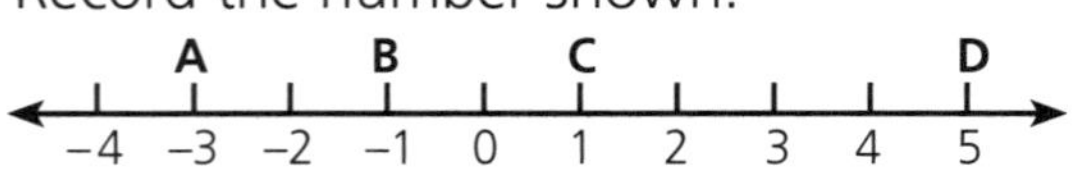

a D ☐ b A ☐ c B ☐ d C ☐

5 Fill in the missing numbers. 9

a

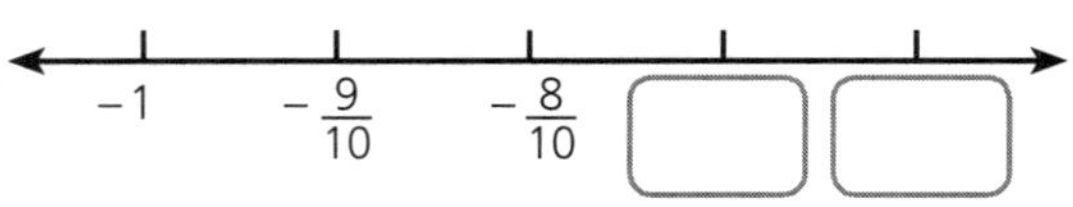

b

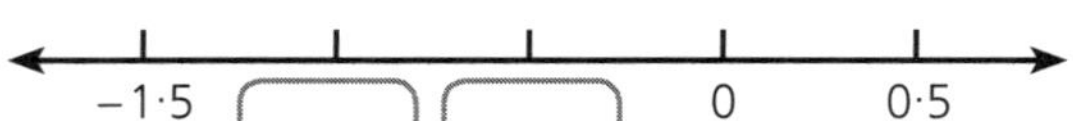

Operations

see page

6 Find the average of: 43

a 19, 16, 21, 34 and 35

b 876, 1254, 1362

c $7.45, $9.62, $6.48, $8.33

d What is my average speed if I travelled 340 km in 5 hours?

7 44

a
```
   76947
   39605
    7842
+  19351
```

b
```
  463925
   34691
  186307
+  68432
```

8 45

a
```
  97263
− 48975
```

b
```
  654219
− 197652
```

c 420 374 refugees left their country. If only 187 196 returned, how many did not return?

9 46

a
```
  60000
− 24785
```

b
```
  90000
− 63479
```

10 40 41

a $4\overline{)720}$

b $6\overline{)906}$

c $8\overline{)752}$

d $5\overline{)6785}$

Write the answer as a mixed number: 42

e $8\overline{)4075}$

f $7\overline{)9000}$

ISBN 978 1 4886 2193 2

Measurement

see page 88 89

11 Find the perimeter of each shape.

a

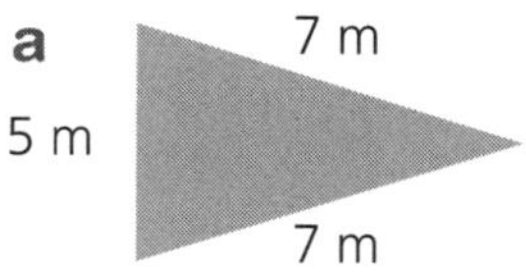

b

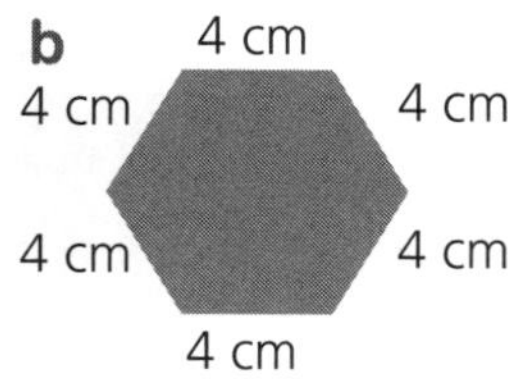

P = ☐ P = ☐

c Find the perimeter of each rectangle.

Length	Width	Perimeter
12 cm	8 cm	
15 m	6 m	
25 cm	5 cm	

90 91

12 Write as 24-hour time.

a 11:19 am ☐ **b** 3:24 pm ☐

c 9:36 pm ☐ **d** 4:12 am ☐

Use **am** or **pm** to write:

e 16:31 ☐ **f** 05:43 ☐

g 10:25 ☐ **h** 23:51 ☐

Geometry

118

13 Find the value of the unknown angle.

a

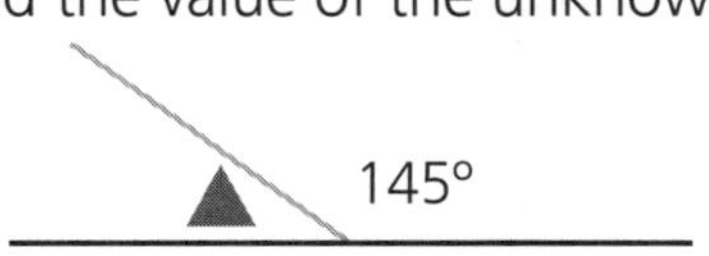

▲ = ☐

119

b

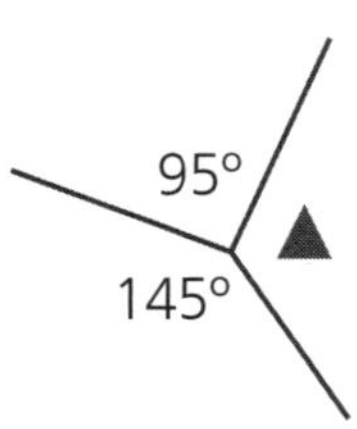

c

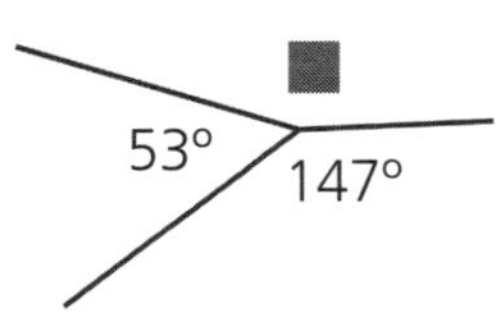

▲ = ☐ ■ = ☐

see page 120

d

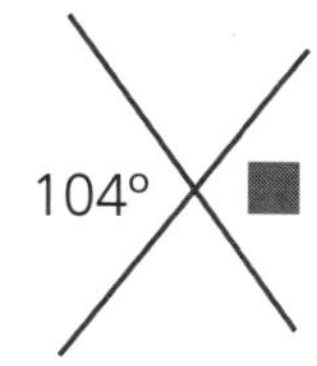

e

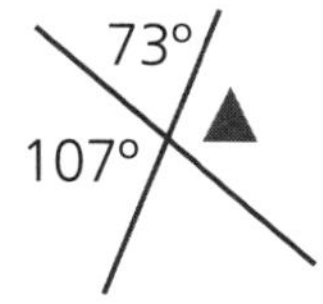

■ = ☐ ▲ = ☐

Probability

147

14 Use the scale 0 to 1 to rate the probability of the spinner landing on:

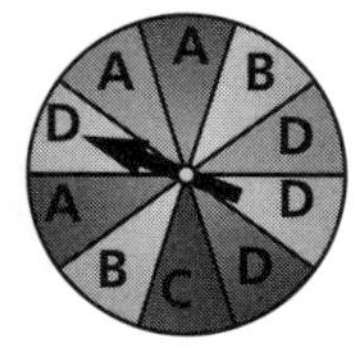

a **A** ☐ **b** **B** ☐

c **C** ☐ **d** **D** ☐

e either **B** or **C** ☐

f either **A** or **B** ☐

g either **D** or **A** ☐

h either **D**, **B** or **A** ☐

147

15

Dice One (columns) / Dice Two (rows)

	1	2	3	4	5	6
1	2	3	4	5	6	7
2	3	4	5	6	7	8
3	4	5	6	7	8	9
4	5	6	7	8	9	10
5	6	7	8	9	10	11
6	7	8	9	10	11	12

Jake listed all the possible totals for a pair of dice on this table. Which total is:

a most likely? ☐

b least likely? ☐

Diagnostic Test 4

Number

see page 12

1 Complete the pattern.

−5 [ ] −3 [ ] −1 [ ] 1

12

2 Which is the lower temperature:

a −2°C or 2°C? [ ]

b −5°C or −2°C? [ ]

10

3 Write the improper fraction for:

a $3\frac{1}{2}$ [ ] b $4\frac{3}{4}$ [ ] c $2\frac{5}{6}$ [ ]

Write the mixed number for:

d $\frac{13}{4}$ [ ] e $\frac{11}{3}$ [ ] f $\frac{17}{5}$ [ ]

11

4 Complete each pattern using decimals.

a 0·6, 0·8, 1, [ ], [ ], [ ]

b 0·1, 0·3, 0·5, [ ], [ ], [ ]

c 2·1, 2·2, 2·3, [ ], [ ], [ ]

11

5 Complete each pattern using fractions.

a $\frac{1}{5}$, $\frac{3}{5}$, $\frac{5}{5}$, [ ], [ ], [ ]

b $\frac{1}{3}$, $\frac{2}{3}$, $\frac{3}{3}$, [ ], [ ], [ ]

Operations

47

6

a
```
  $ 6.34
  $ 9.21
+ $13.40
```

b
```
  $67.49
  $19.06
+ $ 8.60
```

c
```
  $29.45
  $ 8.32
+ $75.69
```

d
```
  $93.74
− $68.95
```

see page 47

e
```
  $469.51
− $ 87.63
```

f
```
  $679.34
− $199.86
```

48

7 a If a car uses 1 L of petrol for each 10 km travelled, how much will be used in travelling 670 km? [ ]

b What is the total of 176, 197, 205 and 324? [ ]

50

8 Use mental strategies to complete these decimals.

a 0·08 + 0·02 [ ]

b 0·05 + 0·05 [ ]

c 0·09 + 0·91 [ ]

d 1·06 + 0·94 [ ]

e 1·4 + 0·60 [ ]

51

9

a
```
  6·037
+ 3·385
```

b
```
  9·508
+ 3·681
```

c
```
  7·911 L
+ 3·68  L
```

d
```
  4·699 m
+ 6·73  m
```

49

10 a 3 × (5+3) [ ]

b 24 + 10 ÷ 2 [ ]

c 25 ÷ 5 + 16 [ ]

d 42 ÷ 6 × 4 [ ]

e 14 × 4 + 20 [ ]

f 90 ÷ 3 + (6 × 2) [ ]

Measurement

11 Find the area of each rectangle. see page 93

a
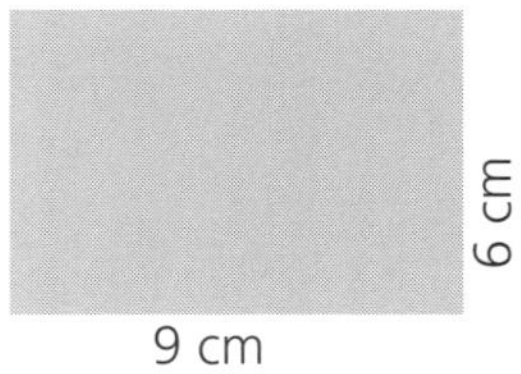

b
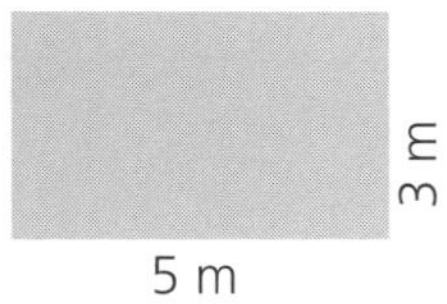

Area = ☐ Area = ☐

12 Find the area of each rectangle. see page 92

Length	Width	Area
7 cm	6 cm	
10 m	8 m	
12 cm	5 cm	

13 Find the area and perimeter of the shape if each square on the grid represents one square metre. see page 94

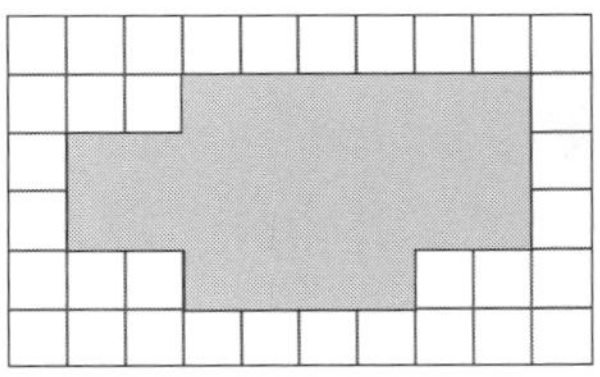

Area ☐

Perimeter ☐

Geometry

14 see page 121

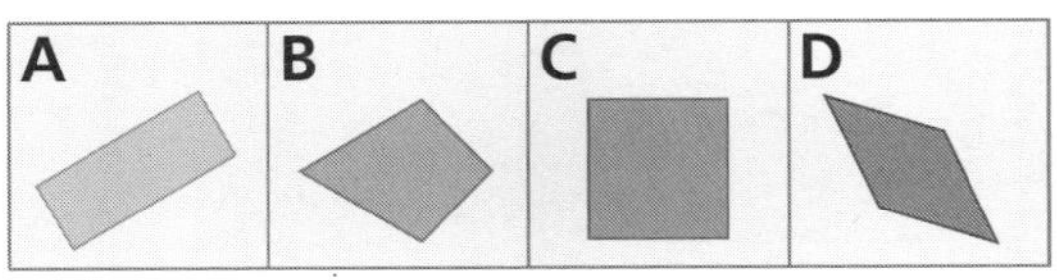

Name the above shapes.

a A ☐

b B ☐

c C ☐

d D ☐

15 see page 122

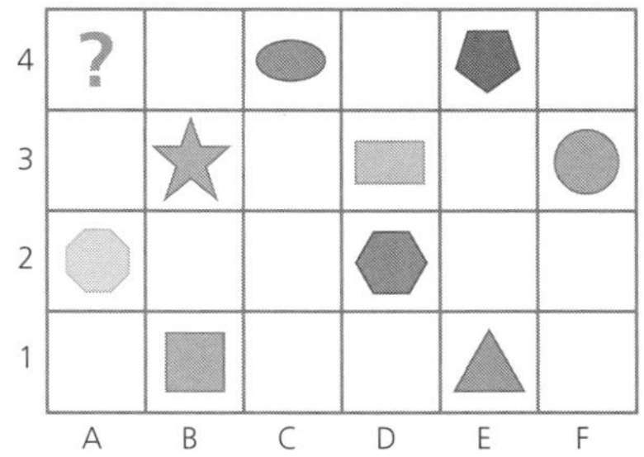

What is found in:

a E1? ☐ b B3? ☐

In which square is the:

c circle? ☐

d question mark? ☐

16 What letter is found: see page 123 124

a west of **A**? ☐

b east of **G**? ☐

c north of **E**? ☐

d south-west of **A**? ☐

e north-east of **A**? ☐

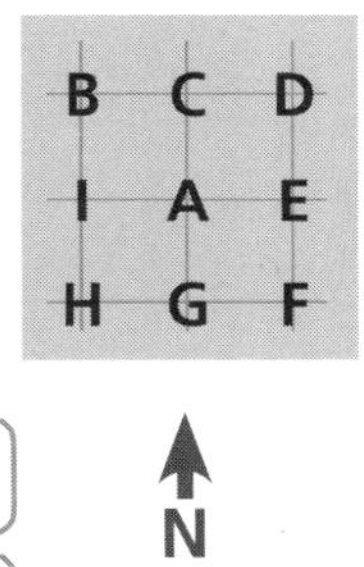

Statistics

17 Name each type of graph. see page 148

a

☐

b
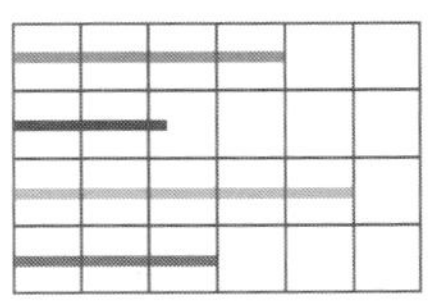
☐ 150

c
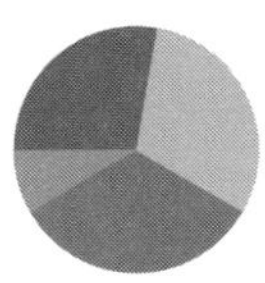
☐ 149

Diagnostic Test 5

Number

see page

1 Write an integer for each statement. 13

a a loss of 10 kg

b a rise of 30 points

c \$100 withdrawal

2 What temperature is shown on each thermometer below? 13

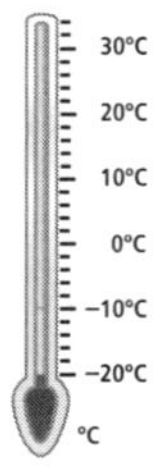

3 14

Write the negative number shown on the number line at:

a B b F c A

d D e C f E

4 a Complete the table.  … 15

Number of hexagons	1	2	3	4	5	6
Number of sides	6					

Rule:

b Complete the table.  …

Number of decagons	1	2	3	4	5	6
Number of sides	10					

Rule:

Operations

see page

5 54

a 3624×7

b 5096×6

c 49319×5

d 30283×8

6 58

a $5\overline{)7642}$

b $7\overline{)31462}$

59

c $8\overline{)6725}$

d $9\overline{)40517}$

7 52

a $63{\cdot}75 + 9{\cdot}6 + 15{\cdot}96$

b $67{\cdot}49 + 14{\cdot}7 + 2{\cdot}643$

8 53

a $102{\cdot}3 - 99{\cdot}68$

b $35{\cdot}621 - 19{\cdot}835$

9 55

a $3 \times 0{\cdot}2$

b $6 \times 0{\cdot}5$

c $4{\cdot}5 \times 4$

d $8{\cdot}3 \times 9$

10 56 57

a $34{\cdot}69 \times 7$

b $3{\cdot}406 \times 9$

c $\$231.95 \times 6$

d $167{\cdot}2\text{m} \times 8$

Measurement

see page 96 97

11 **a** Find the volume of this prism.

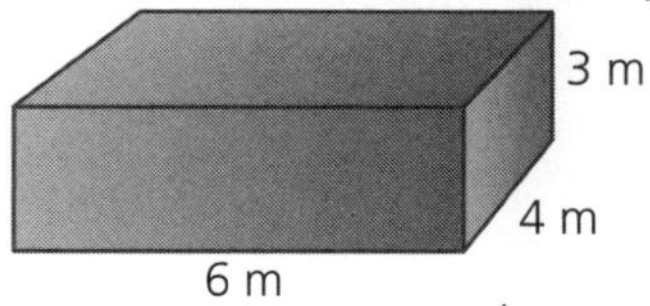

Volume =

b Find the volume of these rectangular prisms.

Length	Width	Height	Volume
9 cm	5 cm	8 cm	
9 m	7 m	4 m	

98 99

c Find the volume of a rectangular prism that is 6 cm long, 4 cm wide and 3 cm high.

d

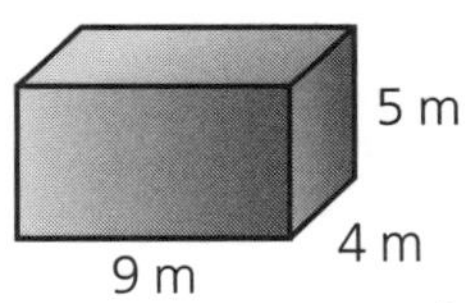

Volume =

Geometry

125

12

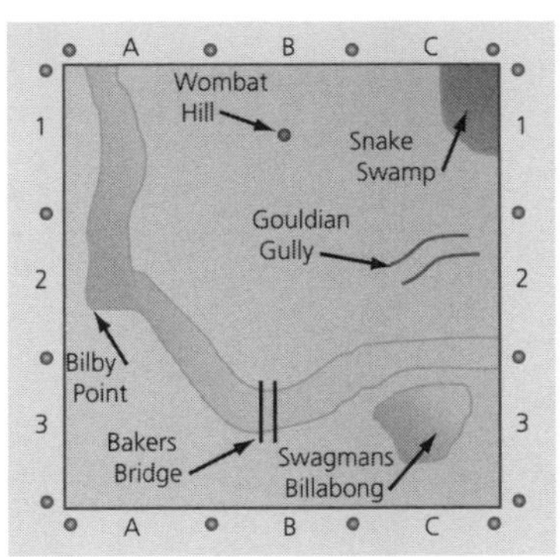

What is found at:

a C3?

b B1?

What are the coordinates of:

c Gouldian Gully?

d Bakers Bridge?

see page 126 127

13

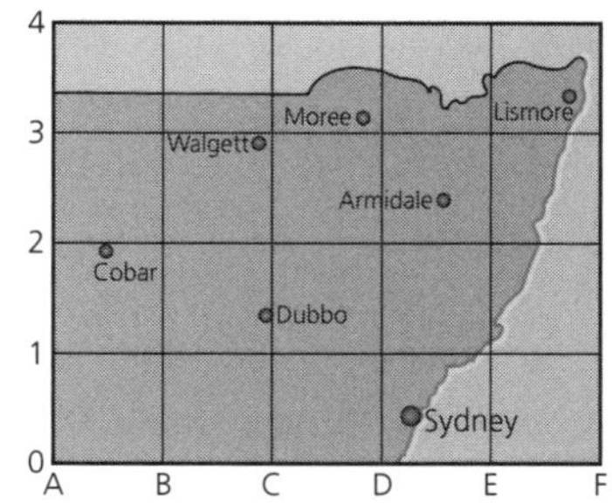

Write down the place closest to:

a D3

b E2

c D1

Find the best coordinates for:

d Cobar **e** Walgett

f Sydney **g** Moree

Statistics

151

14

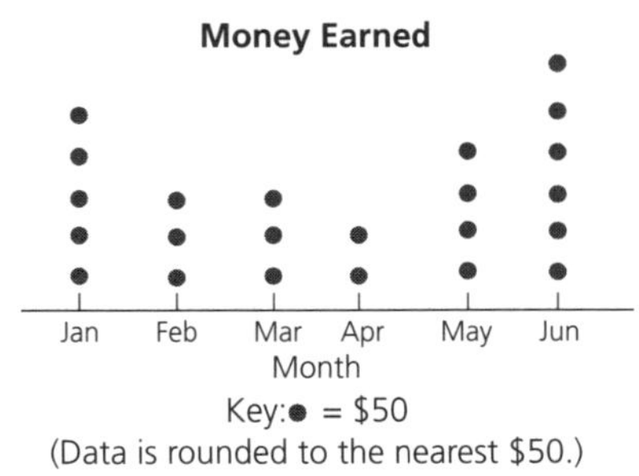

How much money was earned in April and May altogether?

152

15

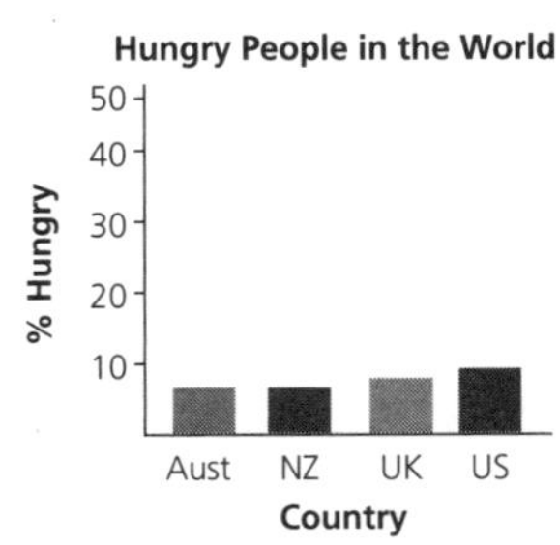

Why is this graph misleading?

Diagnostic Test 6

Number

see page

1 Use the number lines to show equivalent fractions. 20

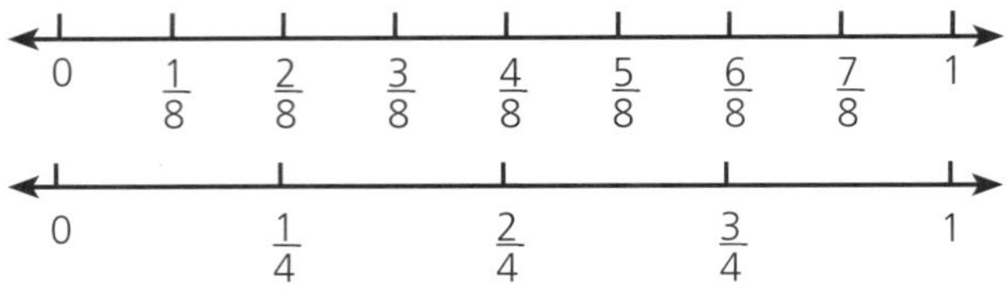

a $\frac{1}{4} = \frac{\square}{\square}$ b $\frac{6}{8} = \frac{\square}{\square}$ c $\frac{2}{4} = \frac{\square}{\square}$

2 a $\frac{3}{8} + \frac{1}{8} = \square$ b $\frac{6}{10} + \frac{3}{10} = \square$ 18 19

c $\frac{1}{4} + \frac{2}{4} = \square$ d $\frac{1}{6} + \frac{2}{6} = \square$

3 a $\frac{11}{12} - \frac{6}{12} = \square$ b $\frac{7}{8} - \frac{4}{8} = \square$ 18 19

c $\frac{8}{10} - \frac{3}{10} = \square$ d $3 - \frac{1}{5} = \square$

4 Find the fraction of each group. 16

a  b 

$\frac{1}{4}$ of 8 = $\square$ $\frac{1}{6}$ of 12 = $\square$

c $\frac{2}{3}$ of 12 = $\square$ d $\frac{5}{8}$ of 40 = $\square$ 17

5 a Sue answered $\frac{9}{10}$ of her emails. If she had 200, how many did she answer? $\square$ 17

b Con said that $\frac{4}{5}$ of 90 was 72. Was he right? $\square$

Operations

6 Complete each pattern. 66

a

6 × 4	
6 × 40	
6 × 400	

b

7 × 5	
70 × 5	
700 × 5	

see page

7 a $10\overline{)57631}$ b $10\overline{)43565}$ 60

c $30\overline{)1410}$ d $70\overline{)4970}$ 62

8 a $2\overline{)0{\cdot}806}$ b $3\overline{)0{\cdot}027}$ 64

c $9\overline{)2{\cdot}142}$ d $8\overline{)19{\cdot}84}$

e $5\overline{)5{\cdot}60}$ f $4\overline{)1{\cdot}8}$ 65

9 a 7·7931 × 100 $\square$ 61

b 94·274 × 100 $\square$

c 8·7935 × 1000 $\square$

d 40·3571 × 1000 $\square$

10 a 83·251 ÷ 10 $\square$ 61

b 836·73 ÷ 10 $\square$

c 725·63 ÷ 100 $\square$

d 5623·9 ÷ 1000 $\square$

11 Construct a number sentence to match each problem. 63

a Increase a number by 6 to make 13. $\square$

b 45 is the product of 9 and another number. $\square$

c A number minus 38 is equal to 34. $\square$

Measurement

see page

12 Complete the table. 100

mL	3 578 mL		1 795 mL	945 mL
L	3·578 L	2·575 L		

13 Write as millilitres (**mL**). 103

a 3·759 L ☐ **b** 6·35 L ☐

Write as litres (**L**).

c 5695 mL ☐ **d** 4300 mL ☐

Write as kilolitres. 101

e 3500 L ☐

f 76 208 L ☐

Write as megalitres.

g 4000 kL ☐

h 50 000 kL ☐

14 How many cubic centimetres (**cm³**) are there in: 102

a 60 mL? ☐ **b** 375 mL? ☐

c 3 L? ☐ **d** 7 L? ☐

How many millilitres (**mL**)?

e 75 cm³ ☐ **f** 960 cm³ ☐

How many litres (**L**)?

g 4000 cm³ ☐ **h** 8000 cm³ ☐

i How much water would be in ten 300 mL bottles? ☐

j How much would each receive if ten children shared 950 mL of drink? ☐

k How much milk is in six 600 mL cartons? ☐

Geometry

see page

15 Draw the front, side and top elevations of this 3D object. 128

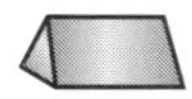

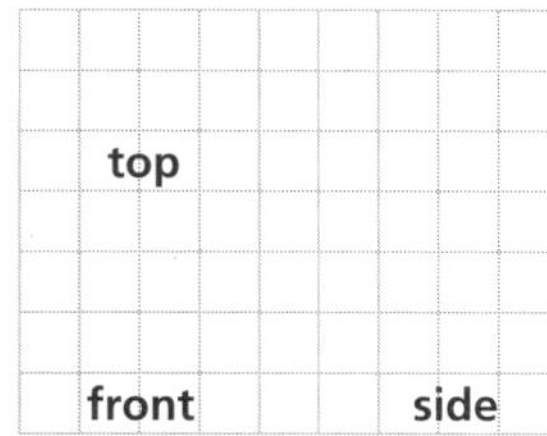

16 **a** Match each 3D object with its name. 129

hexagonal pyramid ☐

octagonal prism ☐

pentagonal prism ☐

triangular prism ☐

rectangular prism ☐

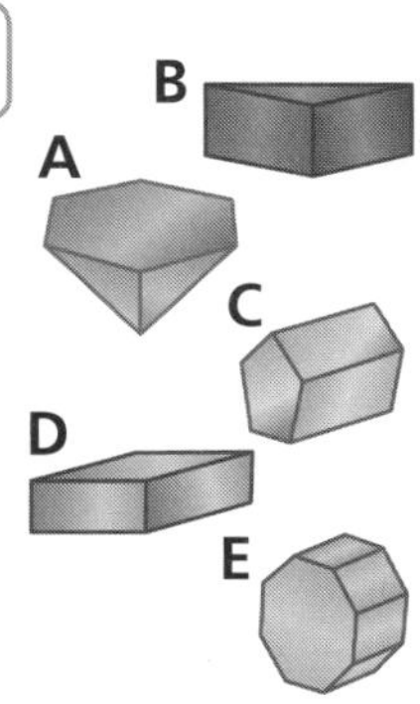

How many faces has:

b an octagonal pyramid? ☐ 130 131

c a pentagonal pyramid? ☐

d a hexagonal prism? ☐

17 Write the letter used to name each point. 132

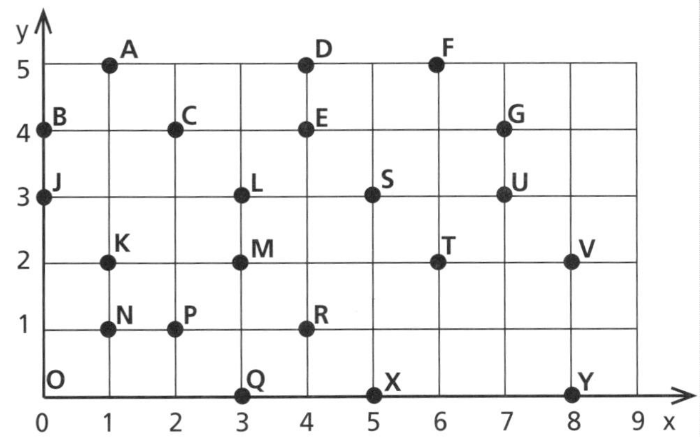

a (4, 5) ☐ **b** (0, 4) ☐

c (8, 0) ☐ **d** (6, 5) ☐

Write the coordinates of each point named.

e A ☐ **f** E ☐

Diagnostic Test 7

Number

see page

1 Complete using fractions. 21

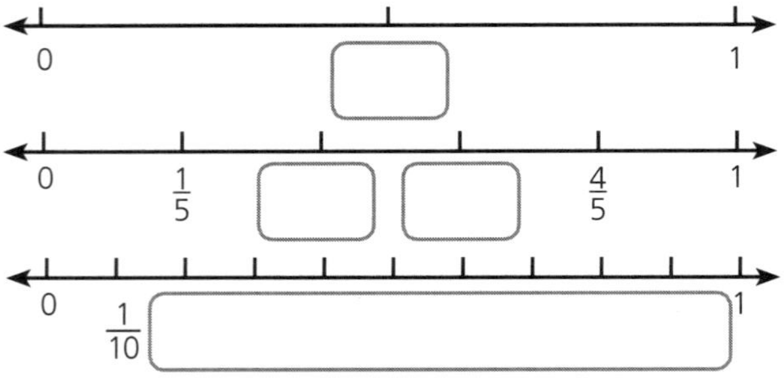

2 Use the number lines above to answer true (T) or false (F). 21

a $\frac{3}{5} = \frac{6}{10}$ ☐ **b** $\frac{2}{5} = \frac{5}{10}$ ☐

Write an equivalent fraction for:

c $\frac{8}{10} =$ ☐ **d** $\frac{2}{10} =$ ☐

3 Write an equivalent fraction for: 22

a $\frac{3\ (\times 3)}{5\ (\times 3)} =$ ☐ **b** $\frac{5\ (\times 2)}{6\ (\times 2)} =$ ☐

c $\frac{1}{2} = \frac{\square}{8}$ **d** $\frac{3}{4} = \frac{\square}{12}$

e $\frac{6\ (\div 2)}{8\ (\div 2)} = \frac{\square}{\square}$ **f** $\frac{5\ (\div 5)}{20\ (\div 5)} = \frac{\square}{\square}$

g $\frac{9}{12} = \frac{\square}{4}$ **h** $\frac{4}{10} = \frac{\square}{5}$

4 Complete the number lines. 21

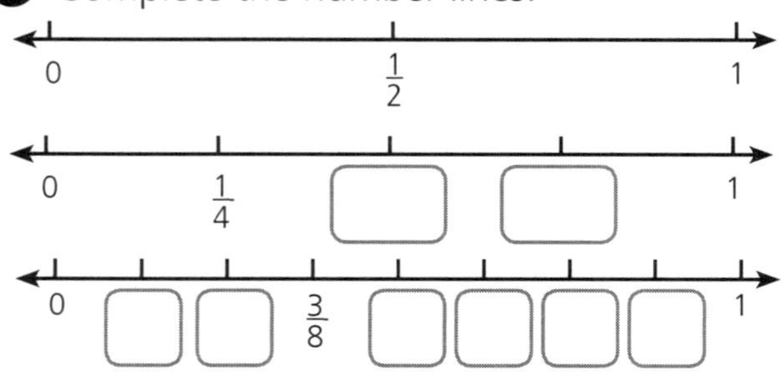

5 Complete: 23 24

a $\frac{1}{4} + \frac{1}{2} = \frac{\square}{4} + \frac{\square}{4} = \frac{\square}{\square}$

b $\frac{3}{10} + \frac{1}{5} = \frac{\square}{\square}$ **c** $\frac{3}{8} + \frac{1}{4} = \frac{\square}{\square}$

d $\frac{9}{10} - \frac{3}{5} = \frac{\square}{\square}$ **e** $\frac{3}{4} - \frac{5}{8} = \frac{\square}{\square}$

Operations

see page

6 **a** 20 × 35 ☐ **b** 52 × 30 ☐ 67

c 40 × 30 ☐ **d** 60 × 40 ☐

e Complete the table.

×	40	60	70
50			
200			
500			

7 68

a 72 × 30

b 83 × 60

c 96 × 4

8 69 70

a 36 × 57
(7 × 36)
(50 × 36)

b 45 × 64
(4 ×)
(60 ×)

c 38 × 75 71

d 960 × 87

9 72

a ■ + 10 = 14. So ■ = ☐

Does 14 − 10 = ■? ☐

b ▲ − 38 = 17. So ▲ = ☐

Does 17 + 38 = ▲? ☐

c ■ × 6 = 54. So ■ = ☐

Does 54 ÷ 6 = ■? ☐

d ⬢ ÷ 9 = 8. So ⬢ = ☐

Does 9 × 8 = ⬢? ☐

 • *AUSTRALIAN SIGNPOST MATHS 6* • ISBN 978 1 4886 2193 2

Measurement

see page 106

10 Write as kilograms.

a 7 t ☐ b 12 t ☐

c 6·5 t ☐ d 3 t 250 kg ☐

Write as tonnes.

e 5000 kg ☐ f 9000 kg ☐

see page 104

11 Write as cubic centimetres (**cm³**).

a 45 mL ☐ b 375 mL ☐

c 3 L ☐ d 6 L ☐

Write as millilitres (**mL**).

e 75 cm^3 ☐ f 490 cm^3 ☐

Write true (T) or false (F).

g 1 L occupies 1000 cm^3? ☐

h 5 L occupies 5500 cm^3? ☐

i 750 mL occupies 750 cm^3? ☐

Geometry

see page 133

12

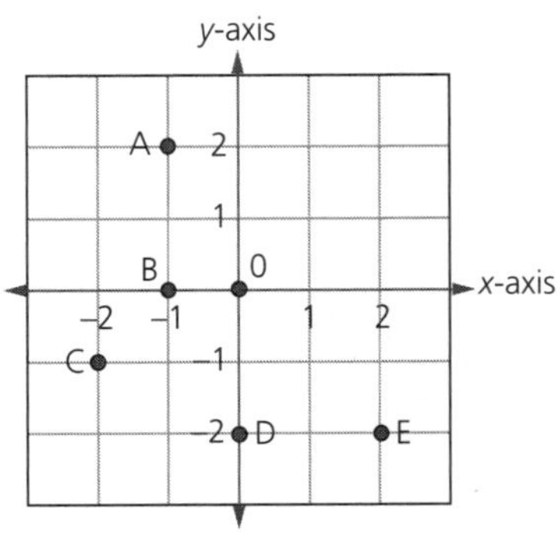

a **A** is the point ☐

b **B** is the point ☐

c **C** is the point ☐

d **D** is the point ☐

e (0, 0) is called the ☐

see page 134

13

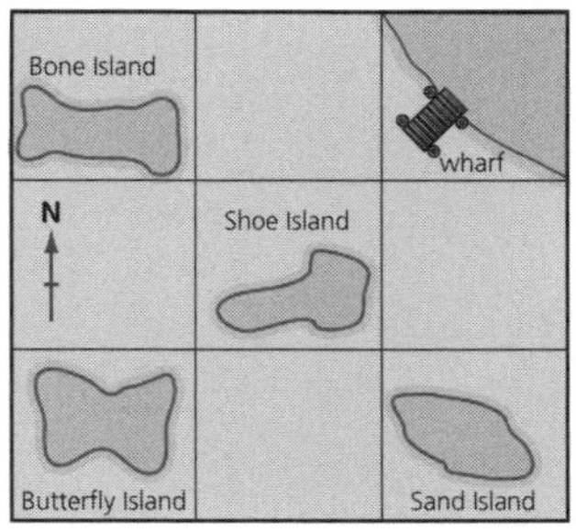

From the wharf, what is:

a south? ☐

b west? ☐

From Shoe Island, what is:

c north-west? ☐

d south-east? ☐

e south-west? ☐

Statistics

see page 156

14 A survey of 50 children was conducted to find colour preference.

Blue	Green	Orange	Purple	Red
16	8	11	2	13

Use the results to predict how many children in 300 would prefer:

a green ☐ b purple ☐

c blue ☐ d red ☐

e orange ☐

see page 158

15 Here are some results from an experiment. Complete the table.

Number	Fraction	Percentage
23	$\frac{\ }{50}$	$\frac{\ }{100}$ or %
27	$\frac{\ }{50}$	$\frac{\ }{100}$ or %

 • *AUSTRALIAN SIGNPOST MATHS 6* • ISBN 978 1 4886 2193 2

Diagnostic Test 8

Number

see page

1 26

a 10% of 30 ☐

b 25% of 80 ☐

c 20% of $40 ☐

d 50% of $90 ☐

Find the discount.

e $150, discount is 10% ☐

f $160, discount is 25% ☐

2 25

a Mum bought $\frac{1}{4}$ kg of cheese. It cost $18 per kilogram. How much did she pay? ☐

b A petrol drum holds 100 litres. If the drum is $\frac{3}{4}$ full, how many litres of petrol have been used? ☐

c Dad bought 18 cartons of milk. We used $\frac{1}{3}$ of the cartons. How many cartons are left? ☐

Operations

3 List all the factors of: 73

a 20 ☐

b 24 ☐

Are these numbers divisible by 9?

c 3463 ☐ d 7245 ☐

4 Round each decimal to the nearest whole number. 74

a 95·7 ☐ b 28·2 ☐

c 17·1 ☐ d 5·6 ☐

e 46·3 ☐ f 14·8 ☐

g 83·4 ☐ h 39·5 ☐

see page

5 Complete the written algorithms. 75

Use rounding to check your answers.

a

$$\begin{array}{r} 43\cdot653 \\ +\ 27\cdot402 \\ \hline \end{array}$$

Round: ☐ +

b

$$\begin{array}{r} 63\cdot972 \\ -\ 58\cdot601 \\ \hline \end{array}$$

Round: ☐ −

6 Multiply these decimals. 75

a 0·2 × 8 ☐ b 7 × 0·06 ☐

c 4 × 0·09 ☐ d 0·03 × 5 ☐

7 77

a 240 ÷ 60 ☐ b 8·8 ÷ 1·1 ☐

c 280 ÷ 40 ☐ d 810 ÷ 90 ☐

e 400 ÷ 80 ☐ f 2·4 ÷ 0·8 ☐

Measurement

8 Write as kilograms. 107 108

a 9 t ☐ b 6·5 t ☐

c 1·725 t ☐ d 3·956 t ☐

Write as tonnes.

e 12 365 kg ☐

f 8900 kg ☐

Write the mass equivalent of:

g 450 mL of water ☐

h 3 L of water ☐

see page 109

9 Choose the best unit to measure the mass of each item (**mg**, **g**, **kg** or **t**).

a a cat ☐

b a raindrop ☐

c a bus ☐ d a cup ☐

109

10 Convert:

a 1 446 kg = ☐ t

b 5 000 kg = ☐ t

c 12·5 kg = ☐ g

d 4 g = ☐ mg

e 7 kg = ☐ g

f 5 g = ☐ mg

Geometry

135

11

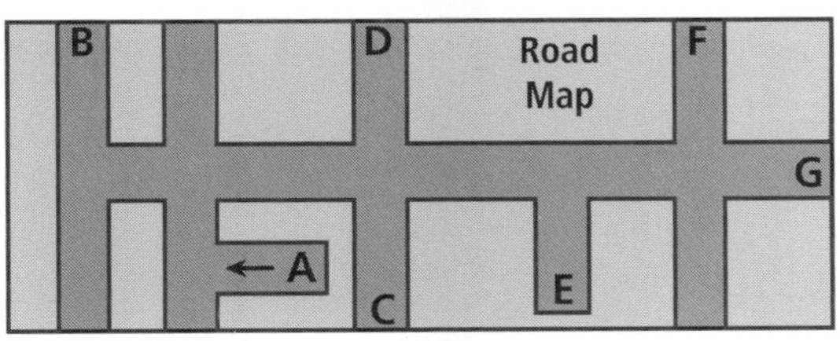

a How would I get from **A** to **F**?

☐

☐

b Which is the shorter journey, **A** to **B** or **C** to **G**? ☐

136

12 Continue this pattern. Describe how it was made.

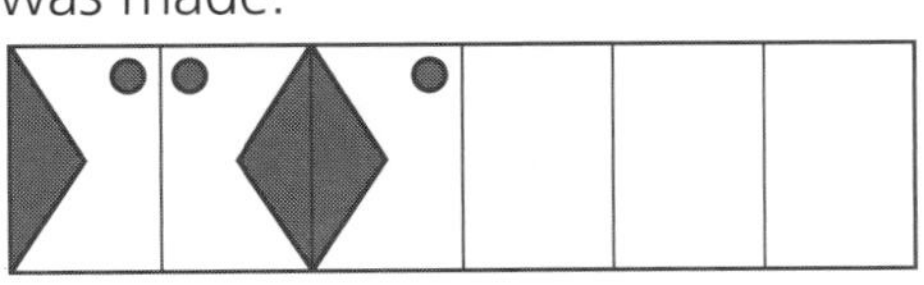

☐

☐

Probability

see page 160

13 a Use the tree diagram to list all the possible results from tossing a coin twice.

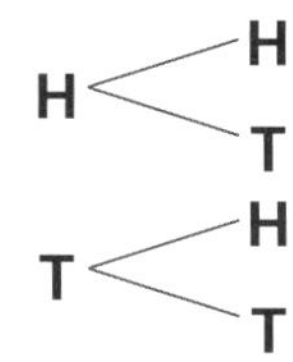

☐

b How many of the outcomes contain one head and one tail? ☐

161

14 a Use this tree diagram to list all possibilities if Sue has three children.

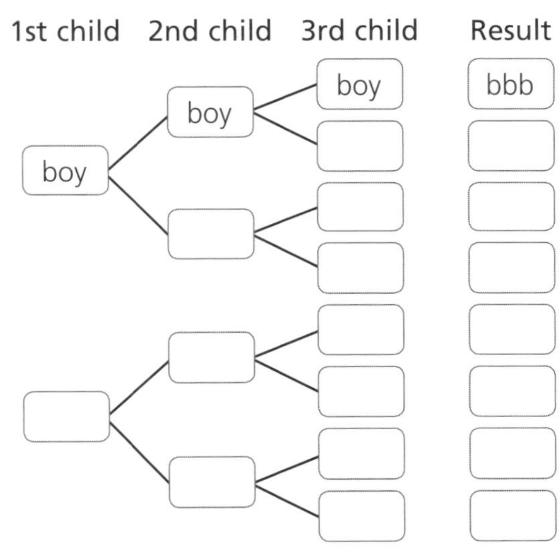

b How many possibilities have two boys and a girl? ☐

c How many possibilities are there altogether? ☐

d If we choose one possibility at random, what is the probability of choosing two boys and a girl? ☐

 • *AUSTRALIAN SIGNPOST MATHS 6* • ISBN 978 1 4886 2193 2